S0-ACB-785

Horizons in Physical Geography

Horizons in Physical Geography

Edited by

Michael J. Clark, Kenneth J. Gregory and Angela M. Gurnell

BARNES & NOBLE BOOKS
TOTOWA, New Jersey

First published 1987

First published in the USA 1987 by
BARNES & NOBLE BOOKS
81 ADAMS DRIVE
TOTOWA, NEW JERSEY, 07512

ISBN 0–389–20752–7

Typeset in Great Britain by
TecSet Ltd, Wallington Surrey

Printed in Hong Kong

Library of Congress Cataloging-in-Publication Data
Horizons in physical geography.
Includes bibliographies and index.
1. Physical geography. I. Clark, M. J. II. Gregory,
K. J. (Kenneth John) III. Gornell, Angela M.
GB54.5.H67 1987 910′.02 87–11427
ISBN 0–389–20752–7

Contents

NEW PHYSICAL FRONTIERS: INSTABILITY AND CHANGE

Preface

It is now more than twenty years since Richard Chorley and Peter Haggett compiled their path-breaking survey of *Frontiers in Geographical Teaching* (1965), which, together with *Models in Geography* published just two years later (1967), did so much to consolidate the foundations of what had come to be called the 'New Geography'. The intervening two decades have seen further (and, on occasion, dramatically different) developments in the discipline. Indeed, one of the lasting contributions of both *Frontiers* and *Models* was to recognise and welcome change as the very life-blood of intellectual inquiry. To be sure, many of these changes collided awkwardly with individuals, institutions and even governments which, in their various ways, sought to impose some sort of stability or direction – a semblance of order – on the shifting kaleidoscope of discovery and debate. But there can be no doubt that since 1965 geography has continued (and is continuing) to change, and *Horizons in Human Geography* and *Horizons in Physical Geography* are intended to introduce some of the most exciting challenges of the contemporary subject to a wider audience. Like *Frontiers*, the *Horizons* volumes are directed primarily at teachers, although we naturally hope for a wider readership. We regard geography in the schools and geography in the colleges, polytechnics and universities as parts of a corporate project. Their aims and audiences are of course different, and it would be quite wrong to think of school geography as no more than a conveyor belt into further and higher education. But each has a measure of responsibility for the other, and we hope that the contributions to these volumes will help to promote a sustained dialogue between them.

For all that they have in common, however, these two books depart from the original volumes in a number of ways which we want to signpost in advance. In the first place, we have accepted that geography is a bipolar subject – like so many others – and we have therefore divided the contributions into human geography and physical geography. Some will no doubt regard this as a betrayal of the integrity of the subject, others as a mere convenience which reflects little more than the conventions of teaching and research. We see it, rather, as a way of drawing attention to the substantial differences between a human geography modelled on the humanities and the social sciences and a physical geography modelled on the natural sciences. This is not to say that there are no contacts or connections between the two, and we have attempted to chart some of the most important in both volumes. But we believe that the differences which remain – between a human geography which, as one of its central tasks, has to make sense of a 'preinterpreted' world which is intrinsically meaningful to the people who live within it, and a physical geography which seeks to explain a shifting, changing but none the less 'object' world – are of vital significance to the future development of both human and physical geographies. That some of the most exciting developments in geography lie now, as they have for the past century, at the point where these two worlds and their different intellectual traditions intersect stengthens rather than weakens the case for a bipolar approach. The integrities of each must be respected, not erased by the casual translation of one into the other.

In the second place, and in addition to these differences between the two volumes, there are differences within them. Neither has been conceived as a manifesto for some new orthodoxy. Each essay is of its author's making, and although we have drawn contributors' attention to cross-references and cross-connections, we have deliberately made no attempt to ensure uniformity of viewpoint. Common threads do emerge, but these have not been imposed by editorial design. One of the hall-marks of geography today, and a sign of its maturity, is its diversity. For this reason we have not aimed at an encyclopaedic coverage, but have preferred instead to identify a series of key topics and themes to be approached by different authors from different perspectives. These themes, and the structures that they create, are of course editorial artifacts, and should not be taken to represent a fixed or self-evident map of geography's intellectual landscape. It is, after all, a characteristic of a kaleidoscope that its component parts can build an infinite number of different but often equally satisfying patterns. The result, we believe, is thought-provoking testimony to the continuing power of that deep-rooted 'concern' for geography which Haggett and Chorley accentuated in the 1960s, and of their own reminder that 'to stand still is to retreat'.

Cambridge and Southampton

Michael Clark
Kenneth Gregory
Derek Gregory
Angela Gurnell
Rex Walford

Acknowledgements

The author and publishers wish to thank the following who have kindly given permission for the use of copyright material: Australian Geographical Studies for Fig. 1 from 'Any Milleniums Today, Lady? The Geographic Bandwaggon Parade' by J. N. Jennings 1973 Vol. 11; Department of Geography, University of Guelph for Fig. 1 from 'Research in polar and alpine geomorphology' by M. M. Miller, 3rd Guelph Symposium, 1973, Geo Books (Geo Abstracts Ltd); Gebruder Borntraeger for Figs. 2, 12 from 'Modelling cliff development in South Wales' by M. J. Kirby, *Zeitschrift für Geomorphologie*, 28, 4; D. K. Jones for Fig. 9 from *TransInst. Brit. Geogr.* NS 8, 1983, Institute of British Geographers; J. Haans and G. Westerveldt for Fig. 12 from 'The application of soil survey in the Netherlands', *Geoderma*, 1970, Elsevier Science Publishers B. V.; Hessischen Landesamt fur Bodenforschung for maps I–IV from *Die Standortkartierung der Hessischen Weinbaugebiete* by H. Zakosek, W. Kreutz, W. Bauer, H. Becker and E. Schroder (1967); J. D. Ives for Fig. 7 and Table 2 from 'The Nature of Mountain Geomorphology' by D. Barsch and N. Caine, *Mountain Research and Development*, 4(4) 1984; Ministry of Agriculture, Fisheries and Food for extracts from *Agriculturual Land Classification Maps*, Sheets 157–8, © British Crown Copyright; Quaternary Research Association for Fig 1. from 'A classification of till' by G. S. Boulton, *Quaternary Newsletter* No. 31, May 1980; Royal Meteorological Society for Figs. 1, 3 from 'Visual presentation of weather forecasting for personal comfort' by A. Auliciems and F. K. Hare, *Weather*, 28, 1973; W. R. D. Sewell for figure 'Adjustment to the flood hazard'; Soil Survey of England and Wales for extracts from *Soil Survey Map*, Sheet 253, 1: 63 360; J. R. G. Townshend for Figs. 2.1, 2.6 from *Terrain Analysis and Remote Sensing*, George Allen & Unwin; M. Toshino for Fig. 2.1 from *Climate in a Small Scale*, University of Tokyo Press (1961). Every effort has been made to trace all the copyright-holders, but if any have been inadvertently overlooked the publishers will be pleased to make the necessary arrangement at the first opportunity.

Notes on the Contributers

Atkinson, B. W. Professor of Geography, Queen Mary College, University of London

Barry, Roger G. Professor of Geography, University of Colorado, USA

Chorley, Richard J. Professor of Geography, University of Cambridge

Clark, Michael J. Senior lecturer in Geography, University of Southampton

Clayton, Keith M. Professor of Environmental Sciences, University of East Anglia

Cooke, Ronald U. Professor of Geography, University College, University of London

Embleton, Clifford Professor of Geography, King's College, University of London

Gardiner, Vincent Lecturer in Geography, University of Leicester

Goudie, Andrew S. Professor of Geography, University of Oxford

Gregory, Kenneth J. Professor of Geography, University of Southampton

Gurnell, Angela M. Senior lecturer in Geography, University of Southampton

Ives, Jack D. Professor of Geography, University of Colorado, USA

Kirkby, Michael J. Professor of Physical Geography, University of Leeds

Lewin, John Professor of Geography, University College of Wales, Aberystwyth

Newson, Malcolm D. Professor of Physical Geography, University of Newcastle upon Tyne

Oldfield, Frank John Rankin Professor of Geography, University of Liverpool

Penning-Rowsell, Edmund C. Professor of Geography and Planning, Middlesex Polytechnic

Ritchie, William Professor of Geography, University of Aberdeen

Simmons, Ian G. Professor of Geography, University of Durham

Sugden, David E. Professor of Geography, University of Edinburgh

Thornes, John B. Professor of Geography, University of Bristol

Townshend, John R. G. Reader in Geography, University of Reading

Trudgill, Stephen T. Senior lecturer in Geography, University of Sheffield

Walling, Desmond Professor of Physical Geography, University of Exeter

Warren, Andrew Reader in Geography, University College, University of London

Whalley, W. Brian Reader in Geography, Queen's University of Belfast

Whittow, John Senior lecturer in Geography, University of Reading

Introduction: Change and Continuity in Physical Geography

Michael J. Clark, Kenneth J. Gregory and Angela M. Gurnell
University of Southampton

AN UNEASY ALLIANCE?

Over the past twenty years it has been realised that, during recent geological time at least, periods of real stability in landform systems have been relatively rare. More often than was formerly assumed, the power and dynamism responsible for change have been concentrated into the relatively brief periods during which adjustment takes place from one system steady state to another, optimising energy expenditure in responding to new controlling conditions.

This idea, developed with reference to the natural environment, may also be appropriate as a commentary on the study of that environment. The vigour of physical geography in the 1960s heralded a period during which change in the targets, methods and aims of study came to be the norm. New ideas became fashionable and conservatism was increasingly seen as counterproductive. The 'steady state' of the first half of the century, which had been characterised by the cumulative elaboration and refinement of a research and teaching curriculum which was accorded near-universal acceptance, suddenly yielded to new pressures.

Since the 1960s, geographers have looked in vain for a new equilibrium that could mark the end of the turmoil triggered by this so-called 'new geography' – a period exciting and bewildering in equal measure. However, it may seem that no end is in sight and that consensus has been replaced by conflict. Familiar aims and approaches have been examined and found wanting, though their replacements often lack reassuring solidity. Embroiled as always in the broader struggle to achieve credibility for their subject, geographers have felt themselves to be diverted and perhaps weakened by apparent internal strife: human against physical, positivist against humanist, pure against applied, process against form. Perhaps most worrying to those whose geographical upbringing took place in more ordered times, the subject appears to have lost the desire for convergence and stability. Nevertheless, whilst such concerns appear to offer a ready focus for debate, there are signs of a great coherence in the 1980s following the multiparadigmatic approaches of the last two decades (although the paradox of unity in diversity had been avowed even in the 1960s).

Explanation of this apparently aberrant behaviour is the central purpose of *Horizons in Physical Geography*. It is explanation motivated by the belief that, despite all fears to the contrary, physical geography has gained strength from its diversity, and inherent stability from its acceptance of constant change. In more senses than one, chaos seems to be here to stay. At the very least, it can be noted that flexible response and defence in depth have been strategies of success in many contexts, and they appear to offer the same potential for success to physical geography. Nevertheless, a faith in the value of diversity (even to the point of conflict) does not relieve us of the obligation to make serious efforts to maximise coherence and to stress supportive rather than destructive activity. To this end, attention needs to be paid briefly to the problem which has come to symbolise the concern and confusion of the 1970s and 1980s, and which remains the focus of a somewhat uneasy alliance, namely the relationship between human and physical geography.

Physical geography has contrasted strongly with human geography throughout the last twenty years, and whilst there continue to be substantial overlaps of aim there is no indication that the two branches of the subject will (or should) converge completely. Of all the roots of distinction, that which is deepest and thus most influential is the methodological contrast between the humanist and structuralist focus of much modern human geography and the scientific (positivist-type) approach to which physical geography has maintained a strong adherence. The undiminished acceptance of the scientific mode by physical geographers does not signify that the alternatives have been overlooked, but rather that they have been found to be less than ideal for many of the purposes of the physical geographer. Although this is undoubtedly the single most important diagnostic characteristic of the difference between human and physical geography (far more fundamental than the fact that one focuses on people and the other on nature), it does not in any way diminish the importance of non-scientific aspects of physical geography, nor does it deny the value of the merging of philosophy and methodology at the interface between human and physical geography.

Nevertheless, both in theory and in practice human and physical geography display clearly different leanings, and this creative distinction is fully reflected in the differing content and approach of the two volumes in the *Horizons* series. However, it should not be overemphasised, for whilst Section 1 of this volume demonstrates the impressive advances achieved by 'positivist' physical geography, the remainder of the book is strongly coloured by a necessary recognition that scientific techniques are to be employed in solving a menu of problems determined by humanist ideals and understood through explanations which have clear structuralist overtones. The physical basis of physical geography remains strong but it is certainly not used as an excuse for an introspective philosophy.

EXPLORATION IN A TECHNOLOGICAL AGE: NEW SCALES – NEW DIMENSIONS

Given the retention of a scientific core, it is not surprising that physical geography has placed considerable emphasis on the importance of technique and methodology, both as constraints on progress and as triggers for innovation – aspects which are fully explored in Section 1 of this book. Of particular importance have been developments in the scales at which effective investigation could take place, ranging from global patterns to the microscopic realist world of material and process. Across this whole spectrum it is possible to see new perspectives yielding new awareness, as is also the case with the very important advances that have been made in the conceptualisation and observation of time sequences. In the 1960s, the development of techniques in physical geography took place largely in order to permit implementation of the potential of important philosophical shifts that had already taken place in the subject. By the 1980s, however, there were signs that the aims of the subject were themselves being moulded by the potential of new techniques and technologies. This reversal, albeit temporary, of the relationship between philosophy and technique has far-reaching implications for physical geography as a whole, and goes some way to explaining the continuing focus of interest on the nature of technique

EVOLVING PRIORITIES: NEW WORLDS – NEW ENVIRONMENTS

Clearly, the application of techniques within a scientific investigative mode could be directed towards any aspect of the physical environment, but if one target typified the concerns of the last twenty years it would be process – both natural and man-dominated. No element has triggered greater advance in explanatory power, and none has better served the needs of prediction and management. Since process is inherently subject-specific, it is understandable that the broad generality which characterises physical methodology is replaced in Section 2 of the book by a series of substantive divisions in terms of process detail. Thus it has been deemed necessary to cover a number of branches of the subject under the process heading – atmosphere, the hydrological system, glacial and cryogenic processes, the coast and the soil system, At the same time it is clear that these process case-studies incorporate both the detail of an individual subfield (e.g. fluvial process) and a broad expression of the general methodological issues discussed in Section 1. This concern with technical and methodological matters is characteristic of recent physical geography.

It would, nevertheless, be extremely misleading to characterise the work of the last two decades as being either divisive or narrowly focused, since, paradoxically, the diversity of interest in process has been widely linked with a direct or indirect adoption of a systems framework which has provided greater unity and a general move towards the type of spatially and thematically integrated studies which form the subject of Section 3. Such rediscovery of synthesis has been achieved at a variety of scales – global, regional and local. Some of the resulting units have familiar titles (the temperate zone, the glacial lands) whilst others reflect newly identified foci related to shifts in world social, economic or strategic priorities (the arid lands, the mountain zone, the northlands). In all cases the approach is dynamic – change is the overriding characteristic, and impact the repeated keyword. Seen in juxtaposition, Sections 2 and 3 present a fascinating contrast between the introvert and extrovert, or divisive and integrative, facets of modern physical geography. More important, the relationship between the two demonstrates that the strength and confidence of the macroscale and integrative approaches that are now seen as commonplace in physical geography are actually the hard-won product of two decades of detailed physical process study. It is the symbiosis of the two that might well dominate the physical geography of the next decade.

AN EVOLVING PHILOSOPHY

Whilst an integrated approach can be seen to grow out of methodological developments, it also relates directly to the fundamental shift from a dominance of 'pure' study towards a greater priority accorded to management applications. For much of the 1970s, this produced a very human-centred applied physical geography – apparently paralleling the human geographers' abandonment of positivism. Particularly in the schools, this resulted in a progressive withdrawal from the study of the physical basis in its own right, with the worrying implication that a new generation of physical geographers was being trained to recognise problems, but was not being equipped with the basic skills necessary to solve the physical components of those problems. More recently, applied physical geography has shown signs of identifying human targets and selecting strategies which include a human element, but has based both on a rigorous scientific physical study. In this sense the management approach of the 1980s may be seen to differ from that of the 1970s, and Section 4 is thus as varied and as full of a blend between problems and potential as any of the previous sections. Professional application of physical geography has reinforced the integrative approach of the subject and has demonstrated little support for the earlier fears that application was somehow intellectually inferior to pure study. At the same time, it is in the management focus that we find the clearest demonstration of the link with human geography – not in the sense of physical geography being a 'useful' but passive input, but rather in the recognition that a fusion of technically rigorous human and physical components offers the best foundation for several aspects of environmental management and land planning.

Given all these trends, physical geographers would regard their subject as being extremely healthy and as having an assured professional and academic future well suited to exploiting the potential offered by the new techniques and new technologies. Nevertheless, it is imperative to attempt an overall perspective, at least for the major branches if not of the discipline as a whole – a task which is addressed in Section 5 in a series of brief contributions of greater breadth and generality than is the case in the rest of the book. Inevitably, in addition to the highlights of individual subdisciplines, the major outstanding issues revolve around the extent to which physical geographers will relate to an evolving geography over the next two decades. Is the main challenge that of resolving apparent differences between human and physical geography, or is the greater gulf the one between the scientific and non-scientific aspects of either one of the two branches? Must academic physical geography now be adjusted sharply to come into line with the emerging needs of professional physical geography? Do the objectives of school and university level physical geography converge or diverge?

This scenario of an evolving physical geography is deliberately built into the structure of the volume. Part I concerns the acquisition and ordering of new types and qualities of information, and focuses on the key advances in process understanding. Part II assesses the extent to which these newly emphasised approaches and information sources have fed new forms of integration – the system, the environmental unit, the management problem or the professional project. Finally, an overview of the present status and future prospect of physical geography is attempted. The volume does not set out to be comprehensive since that could be achieved only at the price of superficiality. It does, however, aim to be representative of the major trends and major branches of the subject. If this aim appears ambitious, then it is at least partly justified by the varied experience and expertise of the contributors. Their message reflects two decades of progress and points the way towards a horizon two decades hence. It is not a path that we are compelled to follow, but linking past, present and future it offers us a rare opportunity to locate ourselves upon the changing face of physical geography.

PART I
THE SEARCH FOR ORDER

Section 1
Frameworks for Investigation

1.1
The Future of the Past – A Perspective on Palaeoenvironmental Study

Frank Oldfield
University of Liverpool

No geographer trying to understand the physical landscape of the present can escape a deep concern for its past. The nature and breadth of this concern has changed over the last few decades as methods, perspectives, problems and opportunities have changed. Thirty years ago, great interest focused on the early Pleistocene and pre-Pleistocene elements of the landscape – the accordant summits and upland surfaces which form such conspicuous, though often still enigmatic features in many areas. During the 1960s and 1970s the main focus shifted to the Quaternary period, and physical geographers made major and varied contributions to the multidisciplinary task of unravelling the stratigraphic, chronological, ecological and palaeoclimatic record of the last 2 million years. By now, the range of concerns is even wider. The acceptance of plate-tectonics, together with its corollary of sea-floor spreading and ocean volume changes as a coherent framework for studying continental evolution, has provided a new stimulus and a new basis for elucidating the origin of the major upland surfaces which so beguiled our predecessors. At the opposite extreme in terms of spatial and temporal scales, physical geographers have become increasingly concerned with process monitoring and with small-scale local historical studies of the immediate antecedents of the present day landscape. Not only have perspectives widened, but the tools available now are very much more powerful, and we can look forward to major advances over the next decade. The full range of historical studies by physical geographers at the present day is just too diverse to present coherently in a single chapter, so this account concentrates on the recent past and on relatively small-scale studies. These have led physical geographers into exciting new areas of environmental science, and in so doing provide answers to any geographer who queries the current and future roles of the study of the past.

To understand the rationale for this new area of study we must follow several strands of reasoning. Traditionally, landform study *per se* tended to be either purely descriptive or necessarily historical; the former rather barren and the latter frequently conjectural and inconclusive. The concern with form tended, as it declined, to diverge into studies of present day processes, and the stratigraphic and sedimentologically based reconstruction of Quaternary history. For a long time the widening gulf between contemporary process study and historical reconstruction was bridged, if at all, in only one direction, by applying the Huttonian principle of uniformitarianism and thereby using the insights derived from present day process studies to illuminate problematical aspects of historical reconstruction.

The divergence between process and reconstruction reflected different training and tradition within specialised areas of research, as well as genuinely intractable problems, both conceptual and methodological, but it left a crucial gap in our knowledge. The incentives to bridge this gap have been several, and they have reflected important shifts in both academic and practical perspectives within physical geography.

John Passmore identifies two complementary areas of problem recognition for the scientist when he makes the useful distinction between 'ecological problems' and 'problems in ecology'. As physical geographers we can readily broaden, paraphrase and share this distinction, and recognise that 'environmental problems' and 'problems in environmental science, or physical geography' may not be one and the same thing. Environmental problems have effects perceived as significant in the world at large, and they invariably have cultural, political, economic and technological dimensions, since they call for responses which affect and are dependent on all aspects of society. Problems in environmental science are those internal concerns of the professionals which often make little sense to the outside world but which may (though not inevitably) be of great importance in the development of the discipline. A growing concern for recent times and small-scale studies is a response to developments in both areas of problem recognition. The rising tide of environmental concern during the 1960s and early 1970s has led to a much greater public awareness of problems such as accelerated soil erosion, desertification, atmospheric pollution and cultural eutrophication. As all these problematical developments have come to light, they have brought into focus our ignorance on their antecedents, as well as our inability to answer questions about the extent to which particular human activities are implicated in their causation. Our retrospective ignorance does nothing to enhance prospects for effective remedial action in the future.

There is a strong and essential historical component in all these issues, and this has helped to point some physical geographers in the direction of the unfamiliar contexts and new techniques required to answer the questions they raise. At the same time, out of the rather more purely academic research of many physical geographers has grown a realisation that the operation of many contemporary processes is constrained by historical endowments in the present day landscape, and also that many important processes operate on time-scales longer than those accessible to direct methods of observation and experiment. Both realisations recognise the need to link process and reconstruction in new ways, not simply to explain the past but to strengthen our understanding of the present and our ability to predict future developments. The convergence of practical and academic incentives has led to new fields of study within which the two areas of problem recognition and scientific endeavour distinguished by Passmore become increasingly complementary and interactive.

In order to meet the challenges posed by these recognised problems both practical and academic, the following are required:

1. *Contexts* in the environment where the evidence of recent ecological hydrological and biogeochemical changes are preserved.
2. *Dating techniques* which allow stages and rates of change in the recent past to be calculated for periods predating and over time-scales longer than existing monitoring programmes.
3. *Methods of reconstruction* which give detailed and quantitative information about the way environmental systems have functioned and biotic communities have responded in the recent past.
4. *Conceptual models* which provide effective frameworks both for developing research strategies and for organising and interpreting results.

LAKES AND BOGS – A CONTEXT FOR PALAEOENVIRONMENTAL EVIDENCE

Processes such as soil degradation and accelerated erosion involve loss of productivity and sustainable yield to man. They are often triggered by changes in water availability and retention, and by reductions in nutrient supply, which in turn lead to loss of vegetation cover and soil stability. On the affected areas the record of these processes will usually be lost, the evidence being destroyed along with the soil itself. Thus, in many cases, the only way to reconstruct the sequence of changes leading up to the present day is to study the record of output from the system as it is preserved in ready-made sediment traps such as lakes, alluvial sequences, reservoirs and even near-shore marine embayments. Lake sediments also preserve a record of the way changes in air and water quality have affected aquatic communities. The sediment record integrates the history of change in both the lake and its catchment. From this record alone we cannot often reconstruct directly the history of atmospheric deposition, though under special circumstances (discussed below) the sediments can record its effects on aquatic communities. In order to learn about the history of atmospheric deposition it is necessary to turn to environments where chemical inputs from ground or surface water have been excluded. Major ice and snowfields fulfil this requirement but so do many peat bogs once peat accumulation has reached the point where the growing surface is nourished entirely by the atmosphere – the so called ombrotrophic bogs.

Lake sediments and growing peat bogs can therefore provide complementary evidence of recent environmental history. Each context provides special opportunities and presents special problems though they share some important advantages. Lake sediment accumulation and peat growth continue in many environments right up to the present day, and the rates of accumulation can often be determined with some degree of precision for the last few decades and centuries. Moreover, peats and sediments preserve all kinds of evidence for changes in the environment surrounding them, as well as recording the response of aquatic and peat-forming biological communities to these changes.

DATING THE RECENT PAST

Three types of new development in peat and sediment dating have been especially important, and their application to problems of interest to physical geographers dates from the early 1970s or later. Radiometric decay has provided the geologist with chronometers on a variety of time-scales. For dating recent changes two radioisotopes are especially important. Lead-210 is a naturally occurring radioisotope formed through the radioactive decay of Radium-226 in rocks (Fig. 1.1.1a). Its half-life of 22.26 years means that it can provide time-scales of sediment accumulation for the last 100–150 years. It is especially suitable for use in lake, estuarine and rapidly accumulating marine sediments (Figs. 1.1.1b and c), though it has also been used with varying degrees of success in peat bogs, salt marshes and snow fields. By contrast, Caesium-137 is a fall-out product of atomic bomb testing and it can help to date events from 1954 onwards. Not only can it be used in favourable situations to date sedimentation, but since it becomes strongly attached to fine soil particles, it can also be used as a measure of erosive input and a guide to erosion patterns and processes on hillslopes.

Recent research has shown that seasonal changes in surface processes, chemical effects and biological activity can combine to give seasonally laminated (varved) sediments in widely scattered areas, in addition to those ice marginal sites which first attracted Swedish geologists many decades ago. Not only deep water temperate 'continental' lakes in regions of strong climatic seasonality but also many British and some tropical and subtropical lakes are now known to have varved sediments. These can provide us with a beautifully resolved chronology of sedimentation with which to date evidence for changes in the lake and its catchment. In a comparable way, though over much more narrowly defined areas, it is possible to use the annual growth increments of certain bog mosses to date each year's growth of peat for the last few decades. This provides a valuable dating tool in studies of the history of atmospheric pollution.

Palaeomagnetism can also contribute to recent sediment (not peat) chronology, since variations in the position of the earth's magnetic field can be

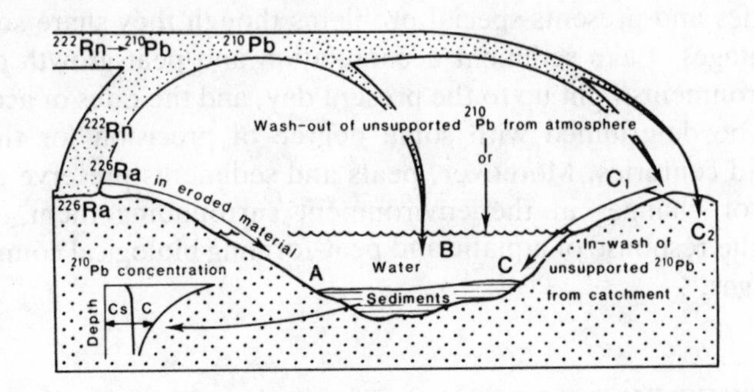

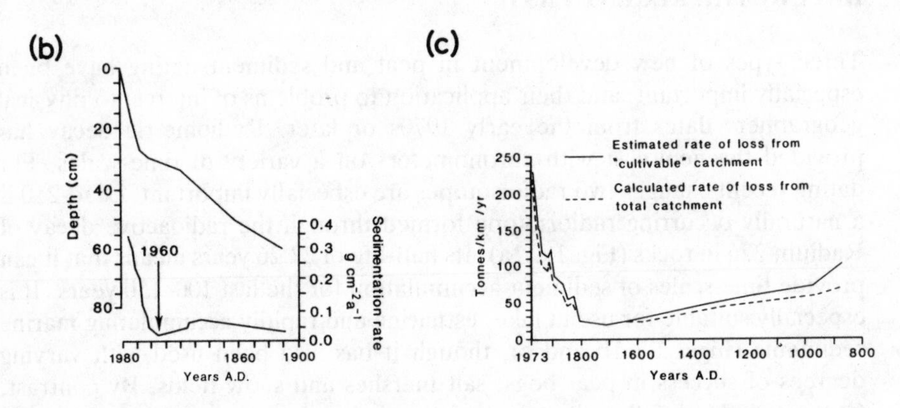

FIGURE 1.1.1 *The theory and application of lead-210 dating*

(a) Lead-210 is delivered to lake sediments. It is a daughter isotope of Radium-226 which, on decay, releases radon which escapes into the atmosphere and in turn decays to produce lead-210. Atmospheric lead-210 reaches the lake sediments either directly (B) or via the catchment (C). In order to measure this atmospherically derived 'unsupported' lead-210 activity, which provides the dating parameter, the 'supported' activity arising from Radium-226 in-wash has also to be measured and subtracted from the total lead-210 activity. The inset graph shows how the lead-210 concentration in the sediment declines with depth and how, in older sediments the unsupported component (C) falls effectively to zero whilst the supported activity (C_s) continues. The decline in unsupported lead-210 activity with age follows the radioactive decay law and allows each layer of sediment to be dated.

(b) Graphs of age versus depth and of sedimentation rate for a core from Loch Grannoch in Galloway. This acidified lake was afforested in 1960, hence the steep rise in the sedimentation rate at that point, as land drainage led to accelerated erosion. Acidification, however, began several decades earlier (see text).

(c) A reconstruction of changing erosion rates from the catchment of a small lake in the Southern Highlands of Papua New Guinea. Sediment cores from the lake bed have been correlated using magnetic measurements (cf. Fig. 1.1.4b) and dated using lead-210, carbon-14 and palaeomagnetic measurements.

reconstructed from the sediments themselves. For the last 350 years or so, the record of these variations can be dated by direct comparison with observatory records and for earlier periods less direct methods of calibration are possible.

These techniques, used in combination, have during the last fifteen years revolutionised our ability to resolve the record of environmental change and ecological response over the last few decades and centuries. As a result, they have made it possible to respond with increasing confidence to the challenge of linking process and reconstruction, and of establishing the sequence of events leading up to present day problems. Each of the methods requires sensitive use and critical evaluation though, and improving them as chronological tools is a new and vital area of research.

RECONSTRUCTING RECENT ECOSYSTEM CHANGES

Attempts to reconstruct ecosystem change have focused on one or more of the following aspects:

Biological communities Many groups of organisms leave behind in peats and sediments persistent remains such as pollen, seeds and fruit from which their former presence somewhere near the site of accumulation can be inferred. By putting together all the available evidence using techniques such as pollen analysis, it becomes possible to reconstruct aspects of past plant and animal communities. Some techniques have been used largely to trace changes in terrestrial ecosystems whilst others have been developed to study the history of aquatic ecosystems. Pollen analysis falls largely into the former category and is by now a well known and widely used technique. In the present account more attention will be devoted to diatom analysis (see below) since reconstructions based on this method have become increasingly important over the last decade.

Biological productivity Community changes often imply changes in biological productivity. Sometimes especially in the case of aquatic communities, it is possible to make the reconstruction of these changes more quantitative and realistic by devoting additional work to establishing past changes in the absolute abundance of particular species and groups in a lake.

Biogeochemical cycles These are really just the biologically important aspects of what we can broadly term 'material flux' within ecosystems. Reconstructing biogeochemical cycles for the past is a very complex business, since the chemical elements important to life must be available in soluble form, and the minerals in which they occur are often capable of undergoing rapid transformations as they pass from the soil or the atmosphere through the living systems which they help to support. An important area of study is emerging designed to use the geochemistry of lake sediments as a basis for reconstructing past patterns of biogeochemical flux within the lake and its catchment. In addition, since 1975, a completely new methodology has been developed for looking at changing erosion rates, sediment sources and soil development at the present day and for reconstructing them for the past. It is based on the magnetic properties (see below) of natural materials. Not only does it increase our insight into these processes, it also helps us to reconstruct another increasingly important aspect of material flux – particulate pollution resulting from fossil fuel combustion.

NEW METHODS OF PALAEOENVIRONMENTAL RECONSTRUCTION

Diatom analysis

Diatoms are siliceous algae which live in a very wide range of environments including lakes and rivers. The total abundance and the relative frequency of the different types of diatom present in a lake are sensitive indicators of the lake water chemistry and the prevailing level of productivity. The silica 'skeletons' or frustules of diatoms are well preserved in most sediments and they can be identified often to the species (sometimes even subspecies or variety) using a compound microscope at high magnification. Thus for any depth in a core of sediment, the diatom community present in the lake at the time when that layer of sediment was laid down can be reconstructed in great detail. Diatom analysis at different depths in the sediment allows reconstruction of changing diatom communities through time. Given a sufficiently good chronology and a method for estimating the concentration of a diatom in each slice, it also becomes possible to calculate for each level the absolute abundance of each type of diatom and to estimate the size of the total diatom crop. This type of information can be of enormous significance in reconstructing the history of cultural eutrophication and lake acidification and two types of case studies are set out in the next section.

Mineral magnetism

Measuring the magnetic properties of rocks, soils and sediments has, until recently, tended to be a very specialised affair dependent on costly and sophisticated laboratory based equipment. Now, thanks to the recent and familiar revolutionary changes in electronics and computing, magnetic measurements can be carried out virtually anywhere using portable equipment requiring a minimum of technical expertise. The results in effect open an entirely new window on order in natural systems. This remarkable development promises major contributions to a wide range of environmental disciplines but in the present context we shall concentrate on reconstructing histories of erosion and particulate pollution. Both are possible as a result of the special nature of iron compounds. Not only are they virtually ubiquitous but they are extremely sensitive to chemical and thermal changes, and the end products of these changes – whether they be involved in soil development or solid-fuel-based power generation – include oxides the magnetic properties of which are not only remarkably persistent in many depositional environments, but highly diagnostic of their past history.

CASE STUDIES

Eutrophication

During the 1950s and 1960s, aquatic scientists and others involved in the exploitation and management of lakes and rivers began to see evidence for major changes in the plant and animal communities present in lakes strongly affected by human activity. In many cases, levels of gross productivity increased, but the organisms responsible for the increase were less use to man than those they replaced; indeed, they often became of significant nuisance value, for example as 'blooms' of blue-green algae. The changes observed were consistent with a major increase in the supply of limiting nutrients such as phosphorus to the lakes, giving rise to higher productivity and decreased oxygen availability as organisms decomposed. In many lakes little or nothing was known of past aquatic communities and productivity trends, and so it was impossible to ascribe the development of the problem to any particular set of circumstances; hence equally impossible to establish which aspects of land or water management needed changing and what the prospects were for recovery given an investment in a particular treatment. Clearly historical reconstruction is an important part of the evaluation of this type of problem.

Figure 1.1.2 shows a summary of the diatom record in the sediments of Lough Erne in Northern Ireland. Both Lough Erne and nearby Lough Neagh have been affected by recent eutrophication. In both lakes, the diatom record shows that the trend in eutrophication begins at around the turn of the present century and accelerates after 1950. By relating the results to experimental work within the lakes and comparing the diatom changes with historical records, the first change can be ascribed to the development of integrated sewerage systems in the towns draining into the lakes, and the second to the great post-war increase in the use of phosphate-rich detergents for domestic and industrial processes. The results identify tertiary treatment of the sewage (including removal of phosphate from the effluent) as the most useful response to the problem.

Acidification

At the time of writing the problem of 'acid rain' is one of the most emotive environmental issues. Among the wide range of damage ascribed to acid deposition is a serious loss of productivity in some soft water lakes. This trend has been documented in southern Scandinavia, in parts of the north-eastern

FIGURE 1.1.2

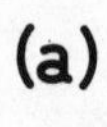
(a)

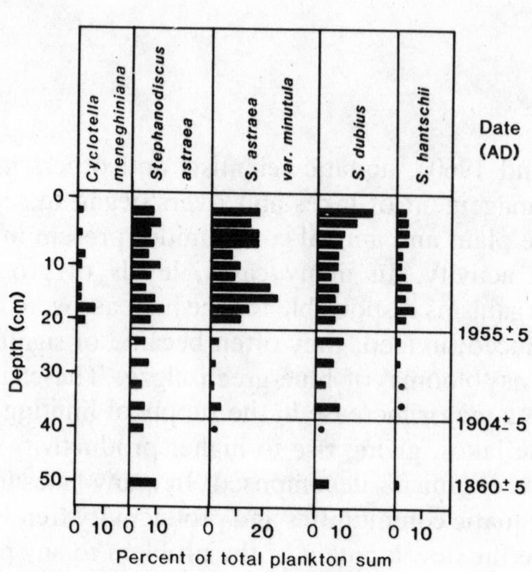
Cyclotella meneghiniana
Stephanodiscus astraea
S. astraea var. minutula
S. dubius
S. hantschii
Date (AD)
Depth (cm)
1955±5
1904±5
1860±5
Percent of total plankton sum

(b)

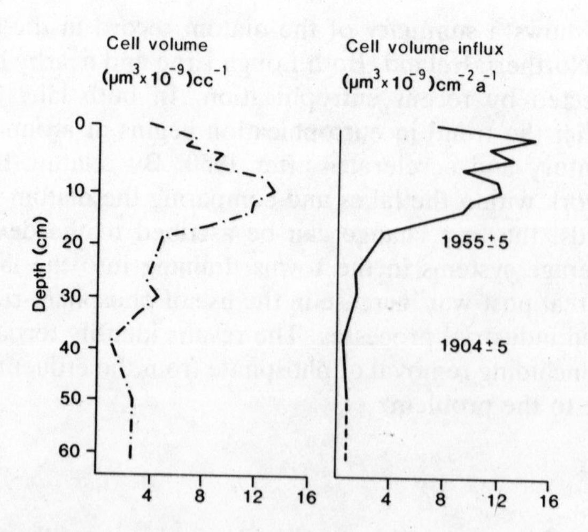
Cell volume
(μm³x10⁻⁹)cc⁻¹
Cell volume influx
(μm³x10⁻⁹)cm⁻²a⁻¹
Depth (cm)
1955±5
1904±5

FIGURE 1.1.2 *Part of the record of fossil diatom variations in a sediment core from Lough Erne, N. Ireland*
(a) Changes in the relative frequency of some of the diatom 'frustules' recognised at each depth in the sediment column. Note how types such as *Cyclotella meneghiniana, Stephanodiscus astraea* var. *minutula* and *Stephanodiscus hantschii* increase in relative abundance in the upper part of the core, especially above the depth dated by lead-210 measurements to *c.* AD. 1955.
(b) Diatom results can be converted first to an estimate of the total volume of diatom cells in each cubic centimetre of sediment, and finally, with the help of the lead-210 chronology, to an estimate of the changing volume of diatom cells deposited in each year over each square centimetre of sediment surface. This graph is one of the best estimates available of past changes in diatom *productivity*. Note how 'influx' increases after the turn of the century then more steeply after *c.*1955 in response to 'cultural' eutrophication.

United States, and eastern Canada. Lake waters have, over the last two decades, become measurably more acidic and fish populations have seriously declined alongside changes in the food chains upon which the fish depend. Such changes in the aquatic system can occur as a result of either variations in the atmospheric input or variations in the nature of the material contributed from the lake catchment. The question is complicated by the way in which atmospheric inputs and catchment processes interact. For example, changes in the chemistry of precipitation will tend to affect catchment soils and vegetation and these changes in turn can affect the nature of the water draining into the lake. Conversely, vegetation can change without any kind of 'climatic' causation (as, for example, when conifer forests are planted for commercial purposes) and the vegetation change may then affect the pattern, type and volume of atmospheric deposition.

One response to the puzzle set by such a complex and interactive situation is to try to date the acidification trends in lakes with and without afforestation in their catchment. This approach in Galloway (Fig. 1.1.3), using the diatom record in lake sediments to provide a basis for reconstructing past changes in lake pH, shows that acidification occurs irrespective of whether afforestation takes place. Moreover, in catchments affected by acidification, the trend can begin well before the afforestation programme. At present the diatom record clearly implicates atmospheric deposition and suggests that acidification of the lakes is an unwelcome by-product of the discharge of sulphur and nitrogen compounds during fossil fuel combustion and power generation.

Erosion

Eroded landscapes are increasingly familiar in many parts of the world. The prospects for their recovery will be to some extent dependent on our understanding of their origins. Different erosive processes such as gullying or surface soil wash have different implications in terms of loss of soil structure,

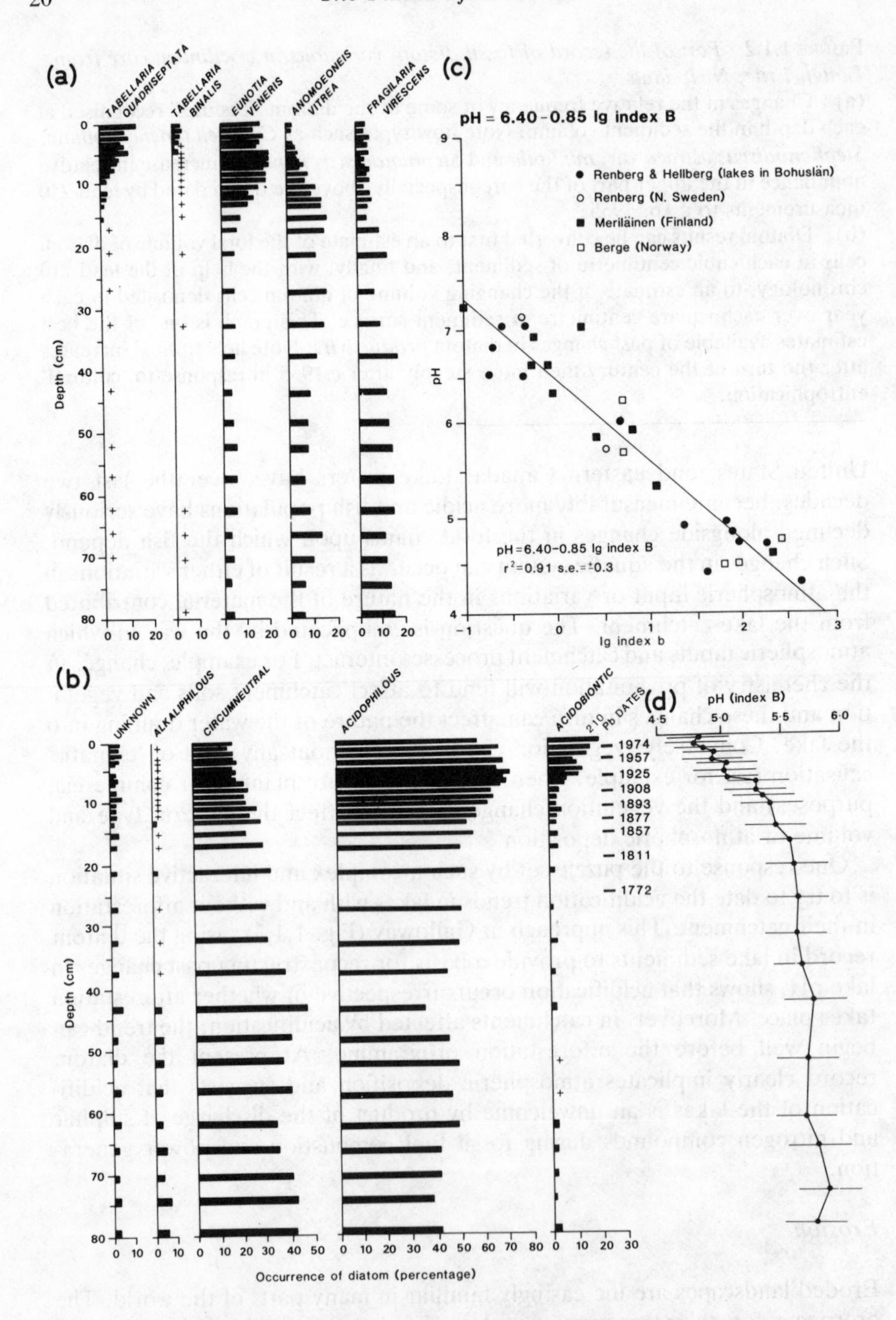

FIGURE 1.1.3 *The diatom record for a sediment core from the Round Loch of Glenhead, Galloway, Scotland, converted to estimates of changing lake water pH for the last few centuries*

(a) Changes in the relative frequency of diatom frustules. Some increase and others decline in abundance in the top 15cm of the sediment column.
(b) The diatom record for the core, grouping all the types recognised into categories with known pH preferences. The acid loving and demanding groups increase in importance in the top 15cm.
(c) The contemporary diatom records from present lakes in Scandinavia are related to the pH of the waters in which they live. The relationship can be expressed through the 'log index B' formula, through which it is possible to convert the diatom records from the Galloway Loch to changes in lake water pH through time. These are plotted against a time-scale.
(d) The pH has declined since the late nineteenth century. This catchment, unlike that of Loch Grannoch (Fig. 1.1.1b) has not been afforested.

---

moisture retention and nutrient status, since they move materials from different parts of the regolith. In consequence, our goals in reconstructing erosion patterns, processes and rates for the past must include not only quantitative estimates of total eroded material but qualitative insights into the nature and source of the eroded material, as well as of the relationships between erosive processes and the vegetation and land use changes accompanying them.

Figure 1.1.4 illustrates lake and estuarine sediment based studies of erosion. Magnetic measurements have been used in two main ways. In order to establish the changing rates of total sediment yield from the catchment, they have been used first to link together the record of sedimentation in different cores by identifying correlating horizons where changes in sediment input have given rise to changes in magnetic mineral types and concentrations common to the whole area of sediment accumulation. Secondly, they have been used to identify the source of material coming into the lake or estuary, whether surface soil, or material eroded from channel sides or cliff-sections. Chemical analyses, where presented, begin to show the implications of erosion in terms of loss of basic elements and nutrients. Pollen analyses portray the link between erosive processes and land use change. Trends towards long term soil depletion are associated with agricultural intensification in each case.

Particulate pollution

Ombrotrophic peat bogs preserve a record of atmospheric deposition during the period of their growth. The record can be affected by particle movement, solution and migration of minerals in the water contained in the peat, and a whole variety of complex chemical changes related to plant growth and decomposition. The evidence available to date suggests that magnetic minerals are less susceptible to these distortions of the record than are many others. They may therefore provide a valuable record of the history of particle deposition from the atmosphere resulting from fossil fuel combustion, iron and steel manufacture and metal smelting. Figure 1.1.5 shows

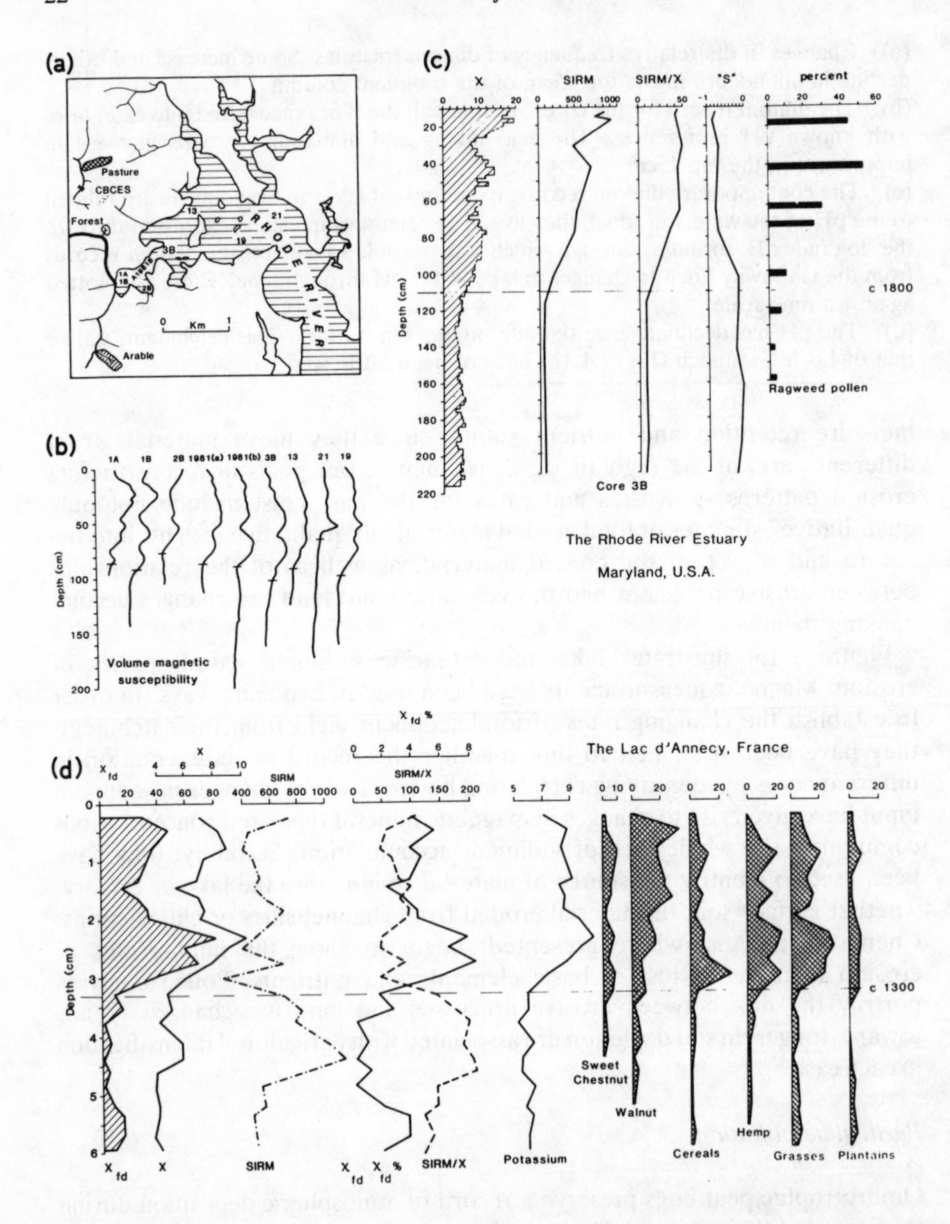

FIGURE 1.1.4 *(b) and (c) show some of the magnetic results obtained from the Rhode river estuary (a), a small tidal west bank tributary of Chesapeake Bay, in Maryland, USA*

Magnetic susceptibility scans for the cores located in (a) are shown in (b). In every core, magnetic concentrations increase irregularly from between 60 and 100 cm below the sediment surface. (c) A more detailed record of magnetic changes at and above this horizon. From the changes plotted it can be shown that the magnetic minerals present in the sediment above 110 cm are derived from the erosion of topsoil. The simultaneous increase in ragweed pollen reflects more extensive land clearance and farming beginning around AD 1800. (d) (based on unpublished work by Dr S. R. Higgitt) shows varied evidence for a core from Petit Lac d'Annecy in the French Jura. The pollen changes record a period of maximum human pressure in the large upland drainage basin of the lake, beginning with the founding of monastic houses in early medieval times. All the magnetic parameters change dramatically just below 3m, and the assemblage of magnetic minerals recorded above this identifies their source as surface soils on the limestone slopes around the lake. In addition, the sediment chemistry changes and the concentration of elements such as potassium increases testifying to some loss into the lake of mineral nutrients in the eroded soil.

---

that where a good time-scale of peat accumulation can be established from moss increment counting (see above), the magnetic record portrays well the increase in industrially derived deposition over the last one to two centuries. Where direct comparison of the magnetic and heavy metal deposition records are available, they are often closely related, so that rapid non-destructive magnetic measurements may become a useful new tool in historical pollution monitoring. Where the magnetic record comes from areas of long and complex industrial history, changes in the quality of magnetic mineral deposition probably reflect changes in industrial technology and may provide additionally a valuable new indirect dating tool.

CONCLUSIONS

Over the last fifteen years, physical geographers have made important advances in all the fields of study outlined above. These developments have begun to define a significant new area of concern in which the results of research into past processes, and rates and patterns of environmental change are of direct relevance to the understanding and amelioration of present day problems. Lakes and peat bogs are no longer afterthoughts in textbooks on geomorphology or soils, they are environmental archives which may, now that we have appropriate new techniques of dating and reconstructing, hold the key to understanding many of the environmental problems afflicting the world at large.

These new studies also have a significant role to play in the development of theory. Balanced, well-founded progress grows out of the interaction between empirical and theoretical concerns. The concepts of the theorist must ultimately relate in some way to the observations of the empiricist just as the findings should help to develop, test and refine theoretical models. Without some guiding conceptual framework based on emerging theory, the work of the empiricist all too often lacks clear focus and coherence of purpose. Equally, without the constraints of direct observation and empirical testing, the constructs of the theorist may all too readily lean away from reality towards a shallow elegance. The problem of linking process studies and

FIGURE 1.1.5

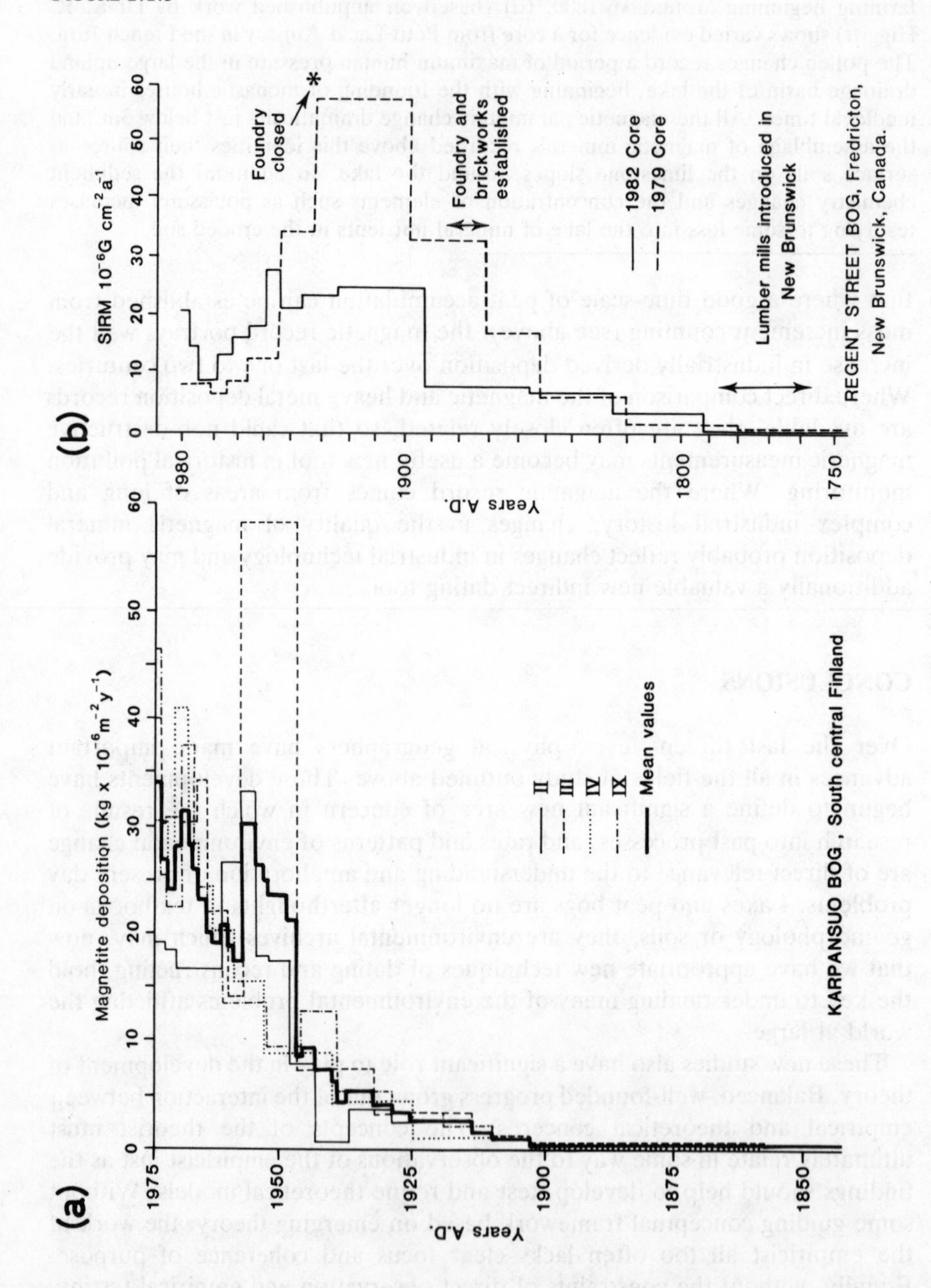

(a)
Magnetite deposition (kg x 10^{6} m^{-2} y^{-1})
0
10
20
30
40
50
60
Years A.D
1975
1950
1925
1900
1875
1850
II
III
IV
IX
Mean values
KARPANSUO BOG South central Finland
(b)
SIRM 10^{-6}G cm^{-2}a^{-1}
Years A.D.
1980
1900
1800
1750
Foundry closed
Foundry and brickworks established
1982 Core
1979 Core
Lumber mills introduced in New Brunswick
REGENT STREET BOG, Fredricton, New Brunswick, Canada.

FIGURE 1.1.5 *The record of magnetic deposition onto recent ombrotrophic peat bogs*
It has been possible to date each layer of peat at both sites by counting annual moss increments.
(a) The record of magnetic deposition in four cores of peat taken from a bog in south central Finland. There is good correspondence between the deposition records in each core. The mean values point to a slight rise in deposition from about AD 1860, steepening during the first half of the twentieth century and reaching maximum levels after the end of the Second World War. This parallels the history of industrialisation in the region.
(b) Constructed in the same way though the magnetic values have not been converted to estimated magnetite deposition. The two profiles come from different sites on the same bog close to the town of Fredricton in new Brunswick, E. Canada. Both records appear to reflect the main relevant events in the local industrial history.

reconstruction has been identified as one of the major challenges addressed by the new approaches described above. The challenge of forging the best possible links between empirical observation and theory is in one sense perennial – the task will never be completed. Nevertheless, each newly developed area of empirical endeavour and each new step in conceptual terms redefines the task by enhancing the opportunities. Grasping these is the most crucial task for the next generation of physical geographers.

FURTHER READING

Several of the author's publications can be used to provide a more detailed coverage of the themes introduced in this chapter. For a balanced and accessible introduction to the rationale and logic of palaeoenvironmental reconstruction, consult:

Oldfield F. (1983) 'Man's impact on the environment', *Geography*, vol. 68, pp. 245–56.

Oldfield F., Battarbee R. W. and Dearing J. A. (1983) 'New approaches to recent environmental change', *Geographical Journal*, vol. 149, pp. 167–81.

An indication of the detailed reasoning behind a specific technique, set in the context of a book which provides a comprehensive account of an important branch of palaeoenvironmental study, is given by:

Oldfield F. (1983) 'The role of palaeomagnetic studies in palaeohydrology', in Gregory K. J. (ed.) *Background to Palaeohydrology* (Chichester: John Wiley) pp. 141–65.

More general background reading on the new techniques and applications mentioned in the chapter can be found in the following:

Oldfield F. (1977) 'Lakes and their drainage basins as units of sediment-based ecological study', *Progress in Physical Geography*, vol. 1, pp. 460–504.

Oldfield F. (1977) 'Peats and lake sediments: formation, stratigraphy, description and nomenclature', in Goudie A. S. (ed.) *Geomorphological Techniques*, section 5.7, pp. 306–36.

Other authors have contributed to the field, and their work can be located through the reference lists of the papers quoted above, though in many cases the specialist publications involved may not be easy to find in a non-specialist library.

1.2 Environmental Systems – Patterns, Processes and Evolution

John B. Thornes
University of Bristol

It is easy to assume that we all *know* what systems are and how they work, as first-year students continually point out. In practice there are usually three ideas which new university students bring with them from the sixth form:

1. Things of interest are usually interconnected, and the key to understanding is to simplify this web of relationships (a system), and to distinguish it from things which are not of direct interest or importance.
2. A system can be understood by looking at the parts and the relationships between them, either intuitively (as a set of boxes and arrows) or statistically (as a set of correlations between the attributes of the parts of the system) or as an input–throughput–output cascade of energy, mass, money, information, etc.
3. A knowledge of the structure and operation (in a budgetary sense) provides some basis for improved understanding of past and present environmental conditions and future environmental management.

Viewed in the perspective of twenty years, this is in a very real sense a great achievement, indicating that students are able to deal with geographical problems at a level of generality and abstraction which was not even dreamed of by their teachers in the mid-sixties. Coupled with the realisation of the need to attach reliable numerical quantities to rates, masses, correlations and other measures of systems under investigation, this period represents a real scientific revolution in geography, both human and physical. The progress achieved so far must not be lost.

At the same time it seems ironic that the richer and yet more profitable side of the adoption of systems thinking, the understanding of dynamical behaviour, has been almost entirely neglected, even in university teaching and research. To some extent the failure must lie in the difficulty of communicating the more complex ideas and of absorbing them amidst the quantitative revolution. These developments have occurred not only in physical geography and in science as a whole, but also in systems analysis itself. This chapter attempts to lower the threshold for acceptance of new ideas and attract readers over the divide from a static, boxes-and-arrows, equilibrium approach towards a set of interconnected basins brimming with concepts and techniques for the study (and teaching) of real world systems.

Before going on to explore these concepts and techniques, the following section traces, as an example, the progress of the investigation of river channel morphology, with the intention of indicating the benefits of adopting a dynamical systems approach and thereby providing an incentive for the reader to read on. The mental effort required will be well justified, in that dynamical systems offer a refreshingly challenging insight to the nature of change, and a glimpse of the underlying relationship between physical geography and other sciences.

AN EXAMPLE: RIVER CHANNEL SYSTEMS

Alluvial river channels (those flowing in movable bed and bank materials) are archetypal systems. The parts are clearly interrelated – if flow increases the bed scours, the section changes, the types of bed and bank materials adjust to the flow and the dissolved, suspended and bed loads probably change. We can isolate the parts of the system to simplify our analysis, and it is not difficult to build up an intuitive box-and-arrows type of diagram to show the relationship between the parts (as in Fig. 1.2.1A). In addition, we might survey the inputs and outputs of water and sediments over a given period and attempt to account for variations in this budget spatially and through this fixed time period. These relationships may also be established as a set of correlations between, say, width, depth, velocity and sediment load against discharge and discharge against climate parameters in the basin (Fig. 1.2.1B and C). The latter approach culminated in the hydraulic geometry relations firmly established by L. B. Leopold, which continue to provide an underpinning of the understanding of the management of river systems.

By the mid-1960s, the analysis of river channel systems seemed to be firmly established and governed by two important principles. First, channel characteristics adjust immediately to bring the channel into line with new flow conditions, and small departures are brought back to the 'normal' conditions by negative feedback. Second, in both statistical (i.e. regression) and mechanical (i.e. hydraulic) models of channel behaviour, change was assumed to be continuously and proportionally related to changes in the controlling variables. So, for example, a given increase in discharge would produce a predictable change in the dependent variables. This is the principle of continuous (and usually linear) response.

These beliefs represented an equilibrium approach to channel morphology which largely ignored the ways in which the equilibrium had been reached through time. Similar views prevailed in slope and coastal studies, and

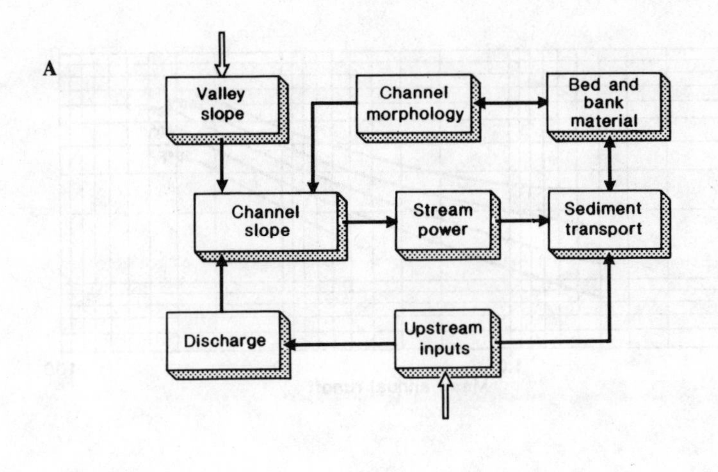
A
Valley slope
Channel morphology
Bed and bank material
Channel slope
Stream power
Sediment transport
Discharge
Upstream inputs

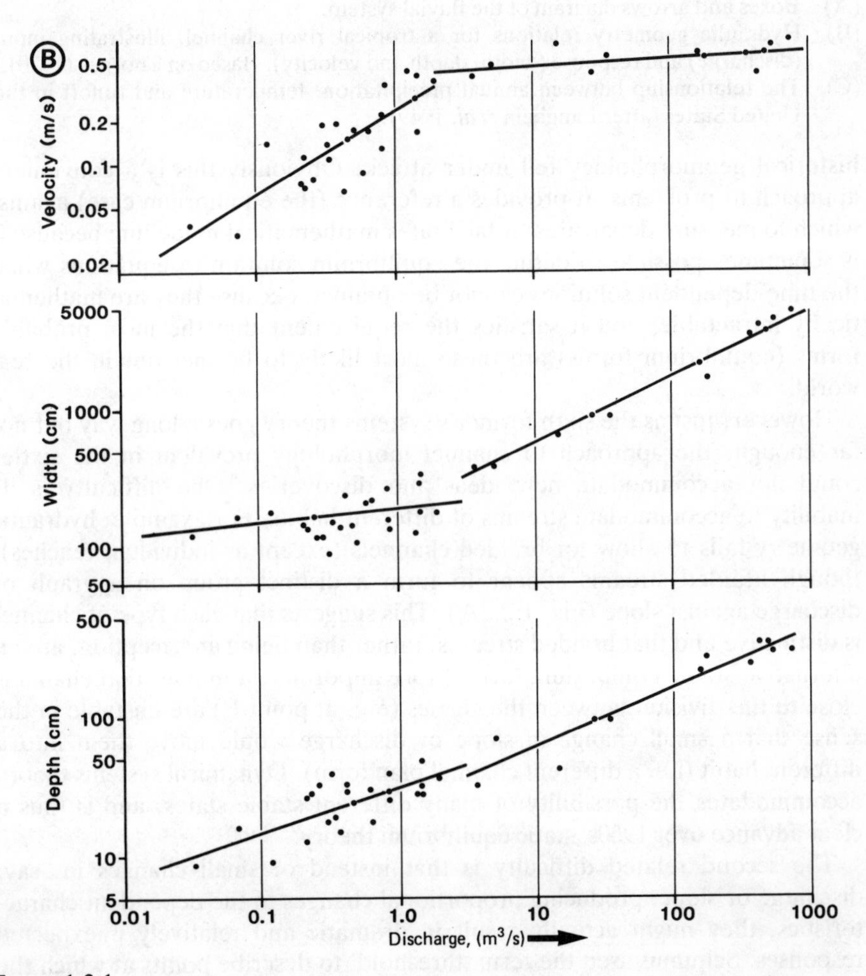
B
Velocity (m/s)
0.5
0.2
0.1
0.05
0.02
Width (cm)
5000
1000
500
100
50
Depth (cm)
500
100
50
10
5
0.01
0.1
1.0
10
100
1000
Discharge, (m³/s)

FIGURE 1.2.1

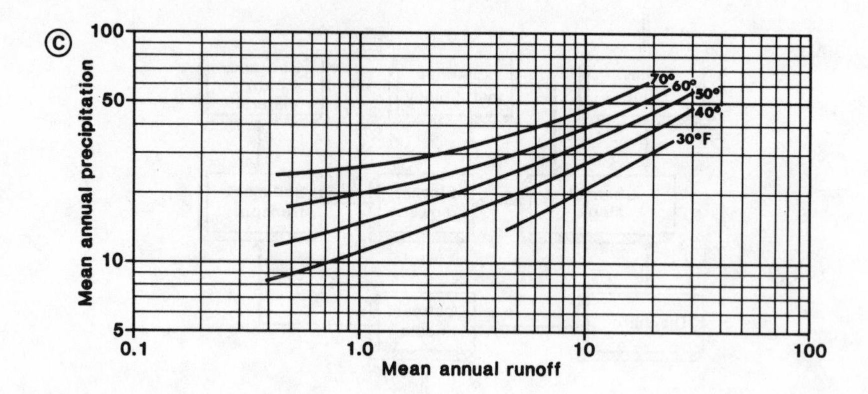

FIGURE 1.2.1
(A) Boxes and arrows diagram of the fluvial system.
(B) Hydraulic geometry relations for a tropical river channel, illustrating input (discharge) and response (width, depth and velocity). Based on Thornes (1970).
(C) The relationship between annual precipitation, temperature and runoff in the United States (after Langbein *et al.* 1949).

historical geomorphology fell under attack. Obviously this is a convenient approach to problems. It provides a reference (the equilibrium case) against which to measure departures; it facilitates mathematical modelling because it is sometimes possible to define the equilibrium solution to equations when the time-dependent solutions cannot be obtained because they are mathematically intractable; and it satisfies the requirement that the most probable forms (equilibrium forms) are those most likely to be met out in the real world.

However, just as the sixth former's systems theory goes a long way but not far enough, the approach to channel morphology prevalent in the sixties could not accommodate new ideas and discoveries. One difficulty is its inability to accommodate streams of different habits. For example, hydraulic geometry fails to allow for braided channels (except as individual reaches), though braided streams appear to form a distinct group on a graph of discharge against slope (Fig. 1.2.2A). This suggests that each type of channel is distinctive and that braided streams, rather than being an exception, are an alternative stable equilibrium form. More important, it implies that channels close to the division between the classes (e.g. at point E) are unstable in the sense that a small change in slope or discharge would move them into a different habit (i.e. a different channel plan form). Dynamical systems theory accommodates the possibility of many different stable states, and is thus a clear advance over 1960s' static equilibrium theory.

The second related difficulty is that instead of small changes in, say, discharge or slope, producing proportional changes in the dependent characteristics, they might actually result in dramatic and relatively unexpected responses. Schumm used the term 'threshold' to describe points at which the

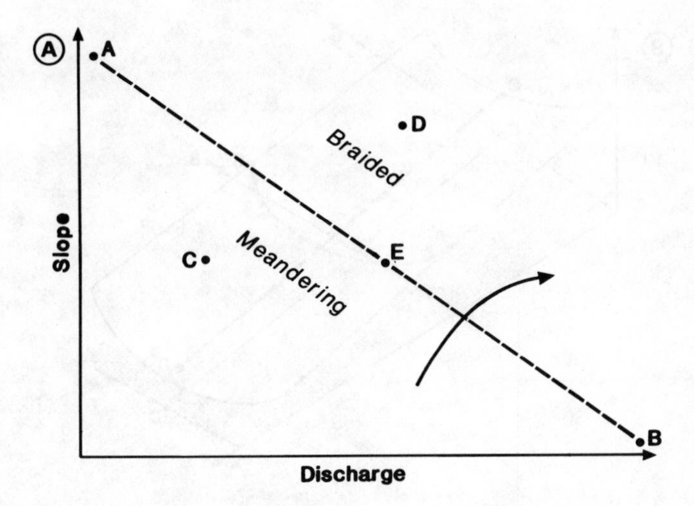

FIGURE 1.2.2

system becomes violently non-linear (i.e. departs from a simple linear relationship between input and output).

Obviously identification of such thresholds is important in management, but the non-linearity is far more important because it implies positive feedback and behaviour far from equilibrium. Such behaviour tends to magnify small deviations from the normal, allowing 'random' or incidental effects to produce significant environmental changes. In an historic interpretation, therefore, it is the dynamical properties of the channel systems which are important rather than the purely equilibrium behaviour. The incidental trigger which may vary from place to place and time to time is simply a minor fluctuation which may or may not be amplified. At a continental scale, David Sugden's consideration of ice sheet behaviour in Chapter 3.2 is an application of the same principle.

The third problem, as Frank Oldfield shows in the previous chapter, is that the trajectory of change itself is important both to the interpretation of past environments and the management of contemporary and future environments. Historical studies help us to identify actual time-trajectories of behaviour and provide calibration of the relationships which are implicit in the process laws of contemporary channel behaviour. Modern dynamical systems theory emphasises the structural relationships between variables, which in simple cases can be represented by a surface in three dimensions (Fig. 1.2.2B). The effects of changes in these variables through time can then be related to this three-dimensional response surface.

Finally, primitive systems analysis tells us little about how the system comes into existence. In the last twenty years there have been significant developments in the understanding of the stream-head problem: how does uniform steady flow across a slope change into one in which flow is

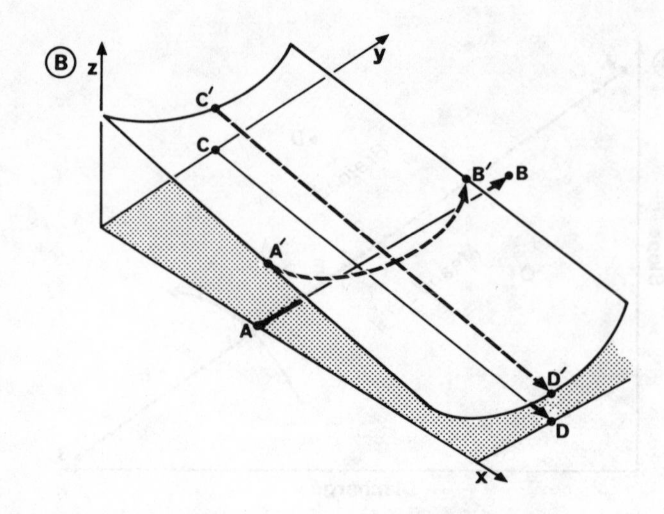

FIGURE 1.2.2

(A) Plot of discharge against slope showing the discriminant function separating braided and meandering. *A*, *E* and *B* are in the zone in which small changes will push the system to one or other habit, *C* and *D* are stable meandering and braided streams. The arrow represents a space or time trajectory moving the system from a meandering to a braided habit.

(B) The relationship between three variables *x*, *y* and *z* represented by the response surface *A′B′C′D′*. Changes in *x* and *y*, such as the paths *AB* and *CD* are mirrored by paths on the response surface *A′B′* and *C′D′* and result in corresponding variations in *z*. The curved surface is the structure to which the text refers.

concentrated into rills and gullies, and thus initiate a channel? This is precisely the type of problem which modern systems theory seeks to explore, the transition from one dominant equilibrium state (sheet flow) to a different one (channel flow) and the relative stability of these different states at different places on the hillslope.

In the rest of the chapter we explore the basis of modern dynamical systems theory in an essentially non-mathematical manner, and then outline its implications in selected areas of physical geography.

DYNAMICAL SYSTEMS

The shift of emphasis from thinking about equilibrium to exploring behaviour through time is the dynamical approach, and systems viewed in this way are called 'dynamical systems' to differentiate them from the 'dynamic systems' used in the sixties, which differentiated a functional from a static approach. In reality all systems are dynamical systems and unchanging systems can best be regarded as special cases of the more general theory. The conditions of the system through time can be monitored by the observation of certain variables, just as the state of a car is monitored through the dashboard by its speed, temperature, petrol consumption and so on. These are state variables; for example, discharge and sediment yield are state variables for the catchment system.

Time-plots

When state variables are plotted through time, certain types of behaviour are fairly common:

1. Damped behaviour, when state variables approach a fixed level and either stay at that level or fluctuate in an unsystematic fashion around it. Examples include sediment production after forest clearance or urbanisation (Fig. 1.2.3A).
2. Explosive behaviour, when state variables move progressively away from an initial value (which may have been steady for some time). The progressive increase in depths of cirques and glacial troughs, and the onset of glaciation itself are examples (Fig. 1.2.3B).
3. Periodic behaviour, when state variables move between extremes through time (Fig. 1.2.3C). Sometimes the movement takes place very sharply, as in the shift from winter to summer in continental areas. In other cases the oscillation is fairly smooth, as in the case of the ten-year cycle in the number of lynx in Arctic Canada. As with all such periodic behaviour, there is an immediate and understandable temptation to search for an external force which is itself oscillating (usually climatic). However, there are other causes of oscillation which are intrinsic to the system. This is almost certainly the case, for example, in the lynx population, and is probably important in scour and fill regimes in rivers and on beaches.
4. Unsystematic behaviour, in which there is no perceptible pattern of change through time. This is sometimes called 'random' or 'chaotic' behaviour, but these have special meanings in dynamical systems theory. Examples include long-term rainfall and river flow, short-term fluctuations in bedload transport, and volcanic activity (Fig. 1.2.3D).

Mixtures of behaviour through time are also well established. One example is the shift from complex, unsystematic (in this case truly chaotic) behaviour of rivers during the last glaciation, through a phase of large meanders in the Late Glacial to fairly straight or steady meandering rivers (Fig. 1.2.3E). Another is the punctuated equilibrium or intermittency type of behaviour, when systems exhibiting steady behaviour suddenly start behaving chaotically and then resume normal behaviour again. Rivers in apparent equilibrium may suddenly start to scour, and then settle down again into the original equilibrium state, as defined by channel state variables such as width, depth and sediment size. Schumm has coined the term 'episodic' for this type of activity; and another term is 'punctuated equilibrium'.

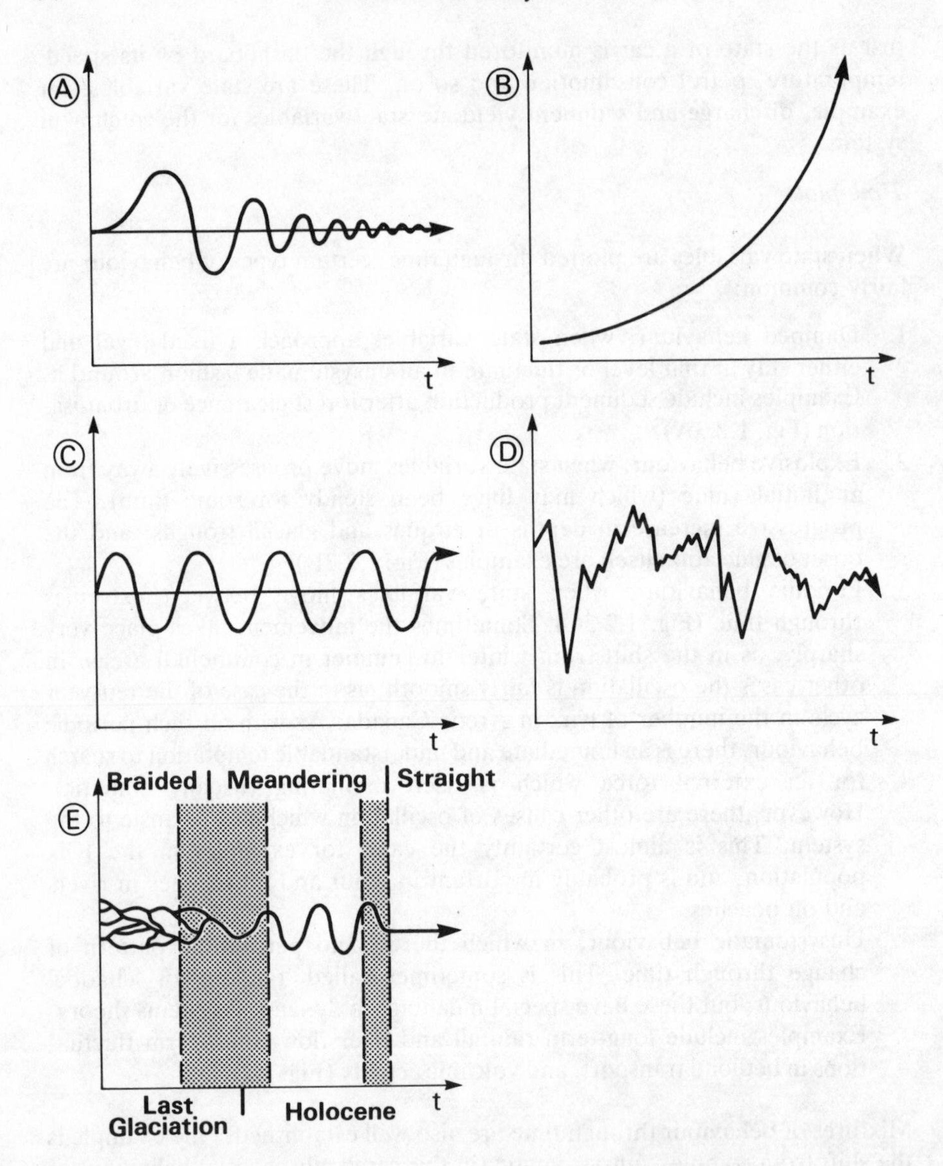

FIGURE 1.2.3
(A) Damped variables through time.
(B) Explosive behaviour.
(C) Periodic behaviour.
(D) Non-systematic behaviour.
(E) Post-glacial behaviour of temperate rivers. The shaded zones are transitions between chaotic and meandering and meandering and straight domains of behaviour.

Place-plots and variable plots

Just as we can observe behaviour through time, it can also be observed in space. A walk down a river shows riffles and pools, degrading and aggrading reaches, oscillations in width, meandering and braiding reaches and so on. We are moving through the same system, but as the values of the controlling variables changes, so do the values of the state variables (configuration, response, form).

This walk is a *trajectory*, a time-space path, just as a route across a map of Europe is a trajectory in two-dimensional space and the route taken by a climber marked on a contour map is a trajectory in three dimensional space. This could also be drawn as a block diagram; as the walker climbs up it becomes cooler, and as he heads for the valley the temperature rises. This would require a four dimensional representation to keep all the information, and is impossible to visualise. One way of handling this would be just to plot his walk in terms of elevation and temperature (Fig. 1.2.4A). This is called a variable-space or state-space (as opposed to a geographical-space) diagram. The climber's wandering (shown in a solid line) can carry him through different elevation and temperature conditions. Suppose, however, that the climber prefers to be warm and low, then wherever he started on the mountain, his route would be as shown by the dotted lines – down the topographic gradient and up the thermal gradient. We could then think of the warm, low spot as an *attractor* and the high cold place as a *repellor*. At the top, any movement would take him down to the warm, low spot and he would stay there unless forced from it, in which case he would return as soon as possible by the shortest possible route.

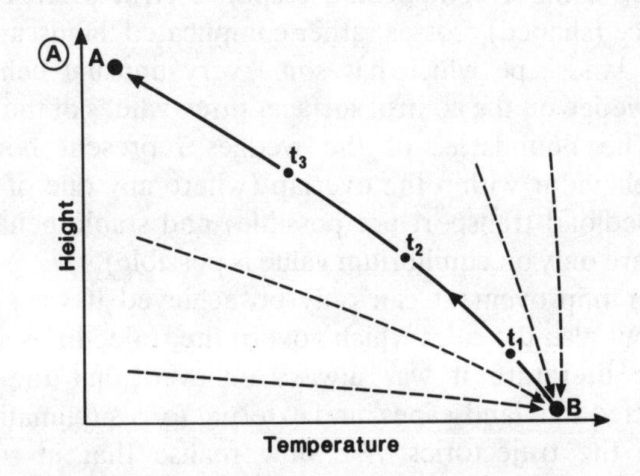

FIGURE 1.2.4

We can easily imagine some slightly more complicated landscapes in which, for example, standing on an arête he might move one way or the other to get down to warmth, ending up in quite different locations. Alternatively he might find locally warm spots before moving on to the lower, deeper, warmer, point. We see the walker operating like a stream, drawn on by the gradient to the point when any further change will again take him uphill. The inquisitive reader will note that there might be an inversion in the valley floor which will have him walking round in circles around the edge of the basin.

Notice that once he has reached the attractor, his state variables remain constant and the equilibrium is a point on the diagram.

When we observe an historical set of changes in environmental state variables (lake levels, vegetation cover density, sediment yield at a point), it is rather like observing the time plot of a walker on our landscape. We might watch him for only a few minutes, so his destination is not clear. It would be necessary to watch a lot of walkers for a long time to understand (not to mention predict) the behaviour of this system.

Domains

An improvement on this situation would be at least to know the 'landscape' i.e. the variable-space configuration. The simplest case would be a variable-space representation of geomorphic processes, such as that provided by Peltier in 1950 (Fig. 1.2.4B). The areas occupied by distinctive processes are called domains, analogous to niches in ecology. If, at a place, we are given a temporal variation of temperature and rainfall, then by plotting its trajectory on the diagram we could envisage the succession of geomorphic processes occurring there. A second case is the variable space map of fluvial processes (Fig. 1.2.4C). This indicates a point made earlier with reference to the meander – braid dichotomy, that the variable-space representation gives us some idea of stability. A condition near the boundary of a doman (Point *E* in Fig. 1.2.2A) is much more likely to change (is more unstable) than a position far from a domain boundary (as in point *C*). Small changes about *E* will produce very different behaviour. The third example (Fig. 1.2.4C) shows that the variable space represented in two dimensions may be only a simple reflection of a more complicated response structure. A trajectory in the control space (shaded) crosses rather complicated 'landscape' in the response variable, a landscape which has some very unusual behaviour. Evidently within the wedge on the control surface, three values of the sediment load are possible. The boundaries of the wedges represent boundaries between unstable behaviour within the overlap (where any one of three equilibrium values of bedload transport are possible) and stable behaviour outside the wedge (where only on equilibrium value is possible).

A further improvement can only be achieved if we know not only the landscape but also the rules which govern the trajectories across it. In earlier geomorphic literature it was always believed that the process–response models formed the 'landscapes' and external forces (climatic and diastrophic) determined the trajectories. We now realise that in some parts of the 'landscape', once certain divides are crossed, the system response must follow certain fixed trajectories.

Mathematical landscapes

The state of the system at any particular time is defined by its state variables. Where there are two of these, the state can be defined by $x - y$ co-ordinates:

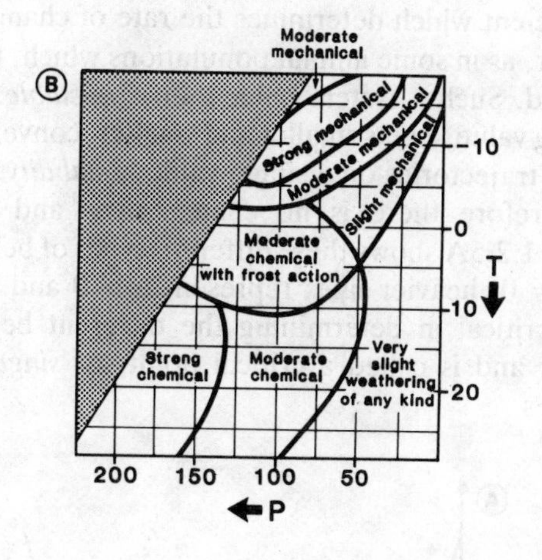

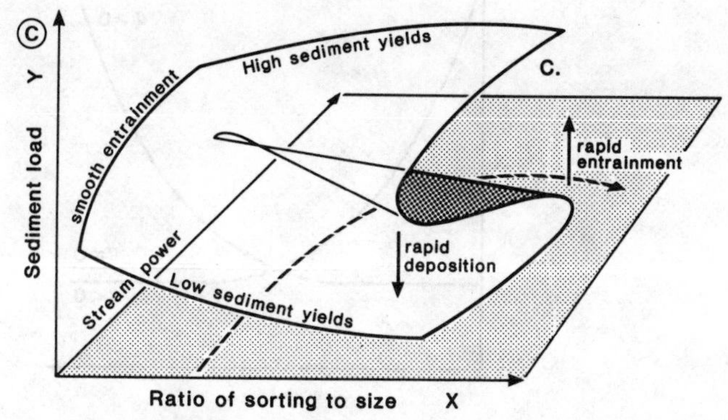

FIGURE 1.2.4
(A) Variable plot of temperature and height. The solid trajectory BA is the climber shown at different times. The dashed lines are trajectories of movement towards increased safety and warmth.
(B) Peltier's diagram of weathering domains according to mean annual temperature (T) and mean annual precipitation (P).
(C) Cusp catastrophe manifold for bedload transport. The dashed line shows a trajectory in the behaviour of stream power and sorting which controls the sediment load behaviour in the manner indicated by the 'folded' surface above, which represents possible equilibrium values for Z for given values of x and y.

in Figure 1.2.2, for example, these are slope and discharge. Producing x and y values for a future time amounts to predicting the behaviour of the system. To do this we need to write down equations describing the change in x and y over time. Such equations are called differential equations (or difference equations if the changes occur at discrete time intervals).

If the *change* in x over time (called x') depends on the magnitude of x itself, this can be written as:

$$x' = Ax$$

where A is a coefficient which determines the rate of change. This describes explosive behaviour, as in some animal populations which, for a time at least, grow without bound. Such a system behaviour is *unstable*. If A is negative, then for any starting value, x gets smaller and smaller, converging on zero. The convergence of the trajectories on a single value is *stability*. If A is zero, x' is also zero and therefore there is no change in x and the system is in *equilibrium*. Figure 1.2.5A shows these different types of behaviour. The light line represents $A = 0$, heavier lines represent $a > 0$ and $a < 0$. The value of $A = 0$ is very critical in determining the different between stable and unstable behaviour and is called a critical point, or *singularity*, where the behaviour bifurcates.

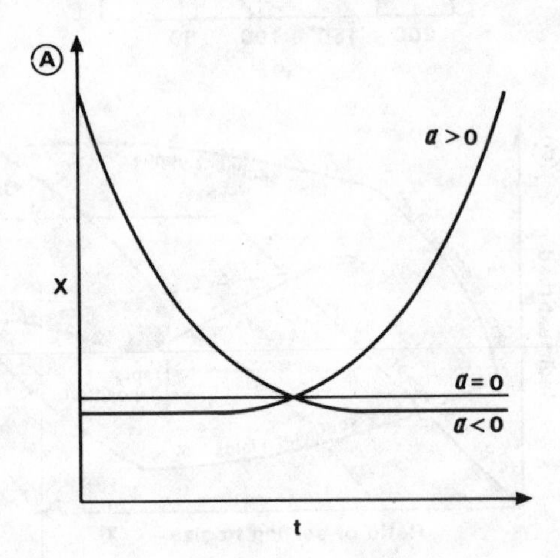

FIGURE 1.2.5

To get used to the idea of trajectories for dynamical systems, consider trajectories with state variables x and y, where the rate of change (x' and y') are:

$$x' = 2x$$

$$y' = -0.5y$$

Clearly, x' increases the bigger is x, and y works in reverse (as shown in Figure 1.2.5B where trajectories are drawn as solid lines and the dotted lines are paths dictated by the 'rules'). Notice that if we have the equations, we don't need the 'landscape' to determine the trajectories.

In most systems the differential equations are 'coupled'; the rates of change of x depend on y, and those of y on x. For example, Figure 1.2.5C shows the behaviour of a predator–prey system (for example foxes and rabbits). When the predator and prey are in a *neutrally stable* condition, the numbers oscillate, returning along the trajectory to the same starting point. Two complete oscillations are shown. If the situation *starts* with perfect equilibrium, then it will not move away from it. This is shown by the point.

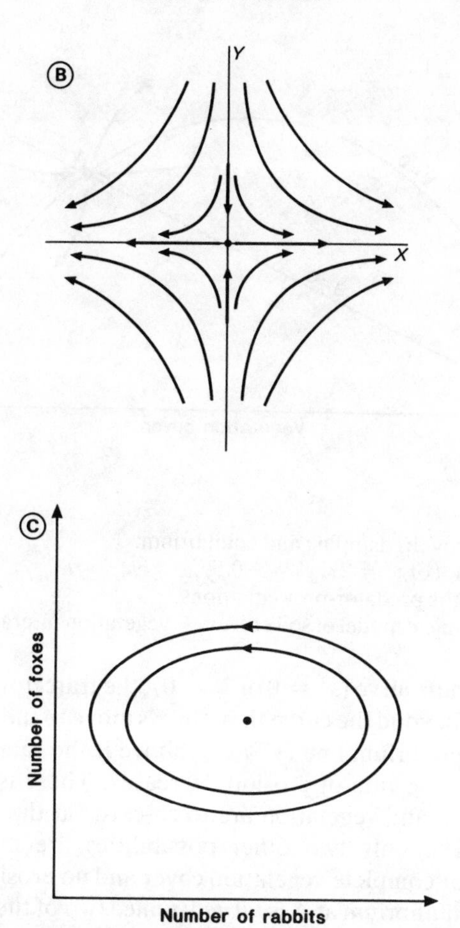

FIGURE 1.2.5 *contd.*

Asymptotic stability occurs when change brings the system to a single value whatever the initial value of the variables. That starting point is called a *stable attractor*.

Consider now a familiar problem in applied geomorphology. The rate of change of vegetation (x') is:

$$x' = (a - bx)\,x - cy$$

where y is the amount of soil erosion (considered as sediment yield). This equation expresses vegetation change as increasing up to some capacity a/b but being decreased by any soil erosion occurring ($-cy$). The rate of change of soil erosion (y') is a function of the amount of soil erosion without a capacity limit (the more there is, the more there will be) and the inhibiting effect of vegetation cover ($-kx$):

$$y' = dy - kx$$

These dynamics are shown in Figure 1.2.5D. Consider first the vegetation

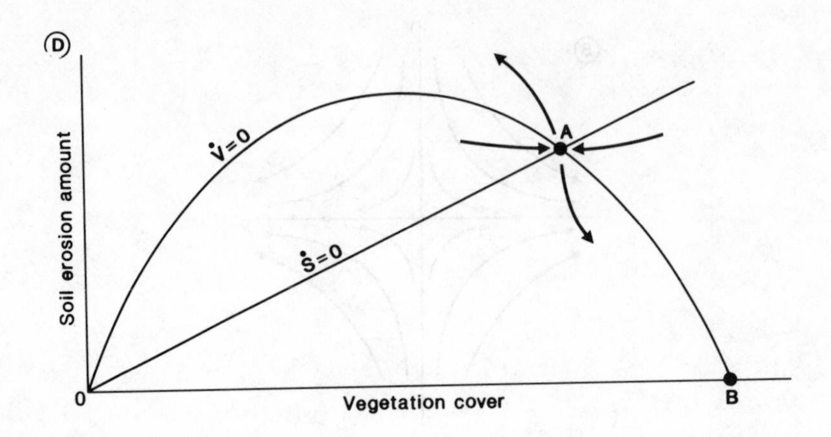

FIGURE 1.2.5
(A) Exponential growth, damping and equilibrium.
(B) Phase–space plot of $x' = 2x, y' = -0.5y$.
(C) Trajectory for the predator-prey equations.
(D) A simple dynamical model of soil erosion – vegetation interaction.

under the equilibrium curve ($x' = 0$ or $V = 0$): the trajectories always move to the right, whereas beyond the curve they always move to the left. Now consider the soil erosion equilibrium line ($y' = 0$), above it the rate of erosion always increases, below it the rate of erosion decreases. There is only *one* equilibrium if both erosion and vegetation are to co-exist (at the intersection of the two lines, (A). The only two other possibilities are no erosion and no vegetation cover, or complete vegetation cover and no erosion (B). According to this model the equilibrium at A is *unstable* since two of the trajectories move away from it and any slight shift from the point would carry us to either complete vegetation cover (with a slight increase in plant cover or decrease in erosion) or no vegetation cover (and higher and higher erosion). From a management point of view, any change of the cover which puts us into the shaded area could be expected to lead eventually to point *B* and a complete cover with no erosion. A shift to a point outside the shaded area would lead eventually to a dominance of soil erosion.

These examples illustrate an important aspect of dynamical systems theory, the capacity to gain an insight into the workings of the system without necessarily fully solving the differential equations. From a teaching point of view they illustrate qualitatively the behaviour of the system in a very graphic manner. For example, this simple model of soil erosion suggests that even with a very sparse vegetation cover it is still possible, depending on the precise starting value, for the vegetation to win the 'competition' against erosion.

Our last mathematical sketch deals with the complicated behaviour of some simple equations which could represent physical sequences, e.g.:

$$x_{t+1} = 4 \lambda x (1 - x)$$

This involves feedback because the value of the state variable at a given time (x) is used to generate the values at the next time (x at $t + 1$). If x is restricted to between 0 and 1 we can produce some values. By keeping $\lambda = 0.2$, then the succession of the first twenty values is shown in Table 1.2.1 and in the graph Figure 1.2.6, line *a*. This exercise has been repeated for $\lambda = 0.7, 0.785, 0.87$ and 0.9 and the results given as lines *b*, *c*, *d* and *e* in Figure 1.2.6. When $\lambda < 0.75$, our 'system' converges through time to a stable value no matter where we start. Between this value and 0.862 it has two stable states (0.538 and 0.780), it then has four stable states and the number of states continues to double at selected values until beyond 0.87 there is no (or infinitely many) stable states, and it is impossible to choose between them. The graph appears chaotic and this condition is called *chaos*. The point is that from a single simple equation we have the possibility of:

Table 1.2.1 For x = 0.2

	$\lambda = 0.5$	$\lambda = 0.7$	$\lambda = 0.785$	$\lambda = 0.87$	$\lambda = 0.9$
1	0.32	0.448	0.502	0.556	0.576
2	0.435	0.692	0.785	0.858	0.879
3	0.4915	0.596	9.529	0.442	0.382
4	0.4998	0.674	0.782	0.848	0.850
5	0.500	0.615	0.534	0.466	0.458
6	0.500	0.662	0.781	0.860	0.893
7	0.500	0.625	0.536	0.419	0.341
8	0.500	0.655	0.780	0.847	0.809
9	0.500	0.632	0.537	0.450	0.554
10	0.500	0.651	0.780	0.861	0.899
11	0.500	0.636	0.537	0.415	0.354
12	0.500	0.648	0.780	0.845	0.823
13	0.500	0.638	0.538	0.456	0.522
14	0.500	0.646	0.780	0.863	0.898
15	0.500	0.640	0.538	0.411	0.329
16	0.500	0.645	0.780	0.842	0.795
17	0.500	0.641	0.538	0.462	0.586
18	0.500	0.644	0.780	0.865	0.872
19	0.500	0.641	0.538	0.406	0.399
20	0.500	0.643	0.780	0.839	0.863
Stable Values	0.5	0.641	0.538 0.780	0.831 0.487 0.869 0.395	NONE

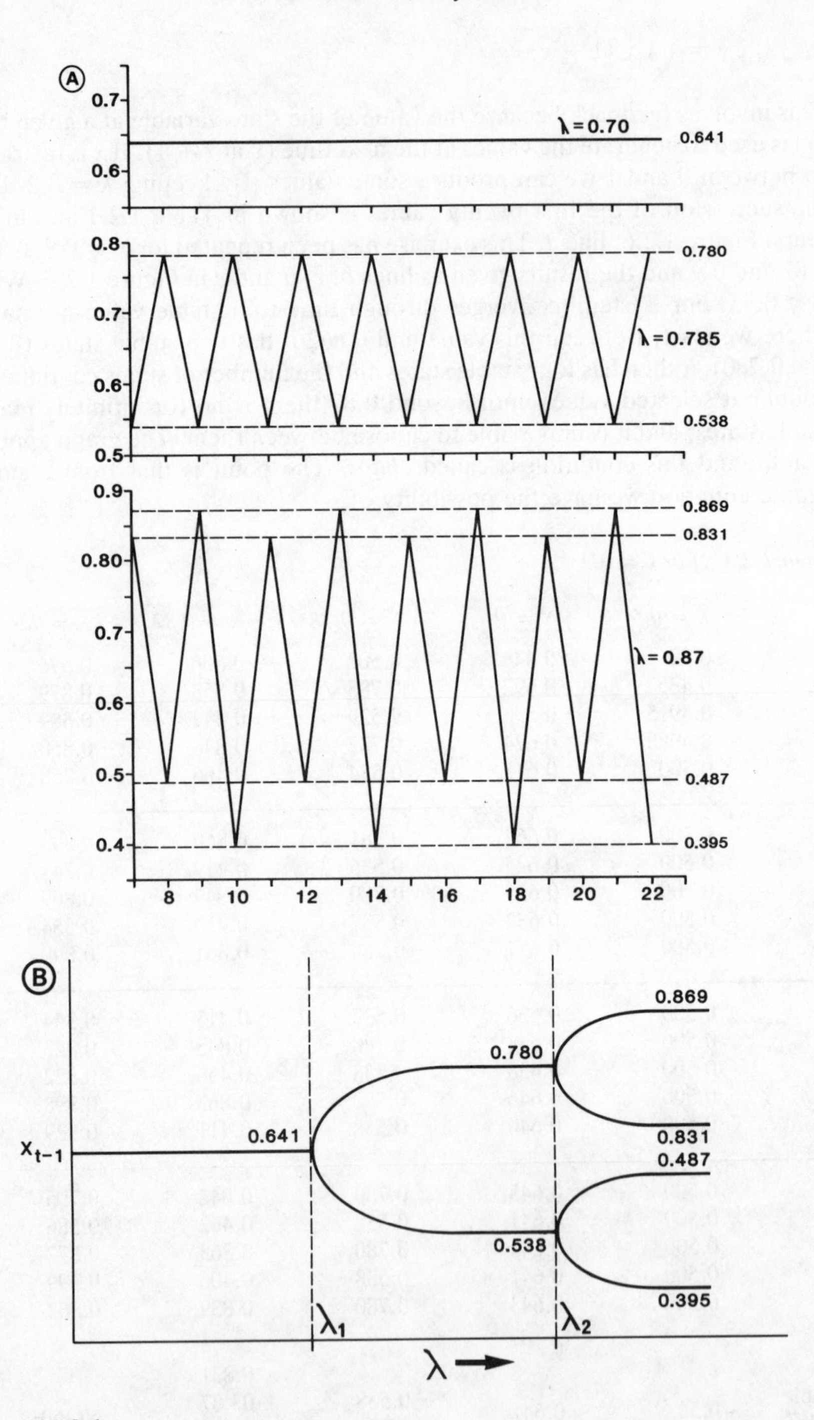

FIGURE 1.2.6
(a) Time plots of $x_{t+1} = \lambda (1 - x)$.
(b) Bifurcation diagram.

1. Stable behaviour with a single equilibrium value
2. Oscillations between 2, 4, 8, 16 . . . stable states
3. Chaotic or incoherent behaviour.

The first case is comparable to convergence on a single attractor, the second switching between several different attractors and the third switching between a very large number of alternative stable states.

To demonstrate the implications of this concept, consider a historical record, such as a cut-and-fill sequence or the temperature during the last 2 million years, both marked by switches between states (aggradation and degradation, and hot and cold respectively). The temptation is to seek an oscillation 'cause' such as climatic change in the first case and wobbling of the earth's axis in the second. What the simple case above shows is that oscillations between two states *may* be produced by some very simple equations when only a single parameter (λ) is changed. Notice too that series which might appear to be 'random' are virtually indistinguishable from those which are chaotic. Whereas random behaviour is now seen to be non-systematic behaviour for which there is no apparent cause, chaotic behaviour is reproducible irregularity generated by certain types of simple mathematical models and their physical counterparts.

Although this discovery comes as a shock to us, there are associations which help us to appreciate its importance. For example, geomorphologists will know that with flowing water, as the Reynold's number of water is increased, we pass from laminar to turbulent flow. Instead of chaos at one extreme or uniformity at the other, the earth's surface is structured and organised. In the final section of this chapter we attempt to indicate some of the implications of the dynamical approach to systems.

IMPLICATIONS AND PROSPECTS

Disorder, order and geography

The prevailing philosophy in nineteenth-century physical sciences was that in a closed system the entropy or randomness of the Universe tends to increase. Early geomorphology, and particularly the Davisian model, adopted this approach. In the absence of external inputs of matter or energy, the surface of the earth would approach an equilibrium condition. It was I. Prigogine, a chemist, who insisted in the 1940s that a new approach was required incorporating the tendency towards greater (rather than lesser) order (or differentiation) in physical systems – either because a change is in its early stages or because the system is 'open', hence subject to a flux of energy and matter from outside.

In climatology, Edward Lorenz indicated that the types of equation described in the previous section could lead to multiple stable states. In geomorphology, A. N. Strahler, W. E. H. Culling and later R. J. Chorley all showed the importance of adopting an open systems approach. In the 1960s and 1970s these ideas were used to sustain the equilibrium approach to landform processes. The existence of alternative states was demonstrated empirically, and the idea of process domains and the corresponding landform associations was firmly implanted.

Origins

Some contrasts in space arise from 'given' attributes such as bedrock type, but many arise from processes of spatial organisation initiated on otherwise undifferentiated surfaces. Modern dynamical systems theory is concerned with the way in which such dissipative (energy using) structures are initiated and evolve. As we showed in the last section, quite simple equations can give rise to very different types of behaviour at singularities or bifurcation points. It is the discovery of this *intrinsic* behaviour which opens up new horizons. The geometrical representations of the system dynamics with repellor peaks, attractor basins, arêtes and saddles need to be sought both empirically and theoretically if we are to explain the temporal behaviour of environmental systems in any coherent fashion.

Change

Dynamical systems theory takes us beyond the *origin* of structures, to the study of change. If external variables (particularly climate or human activity) change the state of the system, carrying it away from equilibrium, we assumed in the past that it would return to the same equilibrium at a rate proportional to the magnitude of the departure. We now see that the direction of change (perhaps to some alternative stable state) and its rate are determined by the 'landscape' on which the system is evolving.

Chance or choice?

For systems in stable equilibrium, small fluctuations have little effect. They simply carry the system a little way up the side of the mathematical basin and they are then carried back to equilibrium. For systems near to stable equilibria a great input of energy (a catastrophic change) such as a flood, earthquake or solar flare may be needed to initiate evolution towards a new stable state. Conversely, near a bifurcation point the smallest fluctuation may carry the system irretrievably towards a new and quite different equilibrium. From the historical geomorphology point of view, the key lies in understanding the general instability as the cause, rather than the local minutiae of the particular time or place at which change was initiated. In the past there was too much emphasis on specific cause and rates of change, and too little on the overall system dynamics. The lessons to be carried over to the management of future environmental systems from studies of the historic past must incorporate this philosophy.

CONCLUSIONS

It has been possible here to touch on only a few of the implications of dynamical systems theory. A geometrical perspective to the systems of differential equations introduced the basic concepts and indicates the potential of the approach. It provides a general framework well beyond simple boxes-and-arrows, and offers the physical geographer a complete reconceptualisation of change and differentiation – not just a reformulation of traditional ideas within a new jargon.

Setting up realistic models of dynamical systems is difficult and providing them with field-based parameters is even more daunting. Moreover, there are further intrinsic complexities in the mathematical landscape still to be explored. Simple attractors are sometimes replaced by 'strange attractors' in which the trajectories wander about the equilibrium exhibiting a 'fractal dance' but never actually get there. Notwithstanding these difficulties, whether one thinks of dynamical systems geometrically or in direct mathematical form, they provide a fresh and integrating insight into the evolutionary behaviour of the natural environment.

FURTHER READING

Perhaps the first priority is to venture beyond the geography book list to find specialist but understandable references. A delightful and easy-to-read introduction to dynamical systems and catastrophe theory, and in particular to the idea of mathematical landscapes, is provided by:

Waddington C. H. (1977) *Tools for Thought* (London: Paladin).

For a coherent and easily followed treatment of mathematical landscapes applied to ecological problems (not dissimilar to many geographical problems), the following is suitable, with chapters 1, 2 and 5 being particuarly useful:

Maynard Smith J. (1974) *Models in Ecology* (Cambridge: Cambridge University Press).

Two less conventional sources which describe the important ideas relating to chaos derived from simple models are:

Anon. (1974) 'The mathematics of mahem', *The Economist*, 8 September, pp. 83–5.

Hofstadter D. R. (1981) 'Strange attractors: mathematical patterns delicately poised between order and chaos', *Scientific American*, November, pp. 16–29.

There are a few other rather more overtly geographical sources available, the first of which provides the best introduction to a range of geographical applications based on a mathematical treatment:

Wilson A. (1981) *Catastrophe Theory and Bifurcation* (London: Croom Helm).

For a non-mathematical discussion of some ideas from dynamical systems theory to change in geomorphology, particularly ideas of relaxation after change, see the following references – the last of which discusses in detail a case study of erosion and vegetation cover:

Brunsden D. and Thornes J. B. (1980) 'Landscape sensitivity and change', *Transactions of the Institute of British Geographers*, New Series 4, pp. 463–84.
Thornes J. B. (1983) 'Evolutionary geomorphology', *Geography*, vol. 68, pp. 225–35.
Thornes J. B. (1985) 'The ecology of erosion', *Geography*, vol. 70, pp. 222–35.

1.3
Models in Physical Geography

M. J. Kirkby
University of Leeds

WHAT IS A MODEL?

Twenty years after the pioneering attempts to popularise the use of models in physical geography, it is helpful to review the progress that has been made and to consider the ever-widening scope and value of the technique. A model is considered here as any abstraction of reality. Davis's geographical cycle provides an obvious example of a model which has been highly influential, though not itself quantitative in content or in the insights it has provided. Even if a model is quantitative in its content, the insights from it may be essentially qualitative, and vice-versa. Nevertheless, many of the models used by physical geographers are expressed in quantitative terms and may be used to give numerical forecasts, and we are mainly concerned with such quantitative models here. They are usually expressed in mathematical or logical terms and often implemented through computer simulations. On this view of models, they may lie at any point on a continuum which stretches from newly formed and untested hypotheses to established physical theories or laws at the other.

THE RELATIONSHIP BETWEEN MODELS AND CASE STUDIES

Field studies and models are often seen as being in opposition to one another. It would be more correct to say that they are strongly dependent on each other, and should be still more so. Any properly designed case study is seeking to test one or a number of hypotheses which rely on explicitly stated or implicit models. By making the reliance more explicit we can only improve the choice of hypotheses and the effectiveness of the tests we make. If the result of a case study is only to describe the field site and its behaviour in greater detail, then the work has largely been wasted. We must always seek results which can be extended over some range of sites or conditions. This too implies the use of a model at some level, even if it is no more complex than the assumption that study period and site are in some way representative. By raising the generality of the model used, the extrapolation of results should be both more rational and applicable over a greater range.

An example: abandoned cliff profiles in South Wales

Savigear's work on coastal slopes near Laugharne, South Wales in 1956, and a subsequent computer model of these forms by Kirkby in 1984 illustrate some of the issues and difficulties. Figure 1.3.1a shows a sequence of surveyed profiles of an abandoned cliff which has been progressively protected from marine undercutting by an advancing coastal spit. The sequence of profiles was supposed by Savigear to have been produced by a period of marine undercutting to form a cliff, followed by subaerial degradation of the slope. From west to east the duration of undercutting is thought to have increased consistently and the period of subsequent degradation thus decreased. To model this sequence of profiles a model for slope evolution was used, which was given parameters determined from the dominant slope processes operating, their rates controlled by climate and lithology, and the initial forms of the slopes.

Figure 1.3.1b shows a set of model realisations which mimic the main features and assumptions of the surveyed profiles. This at once raises many questions. How good is the match between model and prototype, and is it good enough? How well do the modelled processes (landslides and creep/solifluction) match evidence of processes in the field, and are the rates used consistent with current and past rates for these processes? Does the field site offer additional evidence to test the model assumptions, and does the model raise additional questions for field study? Is the model applicable to other areas with suitable changes in rate parameters? These questions are raised here to show that attempting to marry case study and model usually raises more questions than it solves, while at the same time shedding a great deal of light on both.

Complex reality versus model simplicity

One of the reasons why there are few well-established theories in physical geography and why some people doubt the efficacy of model building is the complex nature of geographical reality. The natural environment contains a complex interwoven net of mutual influences, many of which we understand very poorly and therefore cannot effectively model. It might be argued that any model of the real environment is no stronger than its weakest link, and is therefore likely to be ineffective. To argue in this way is, however, to ignore the history of science which has continually shown the power of generalisa-

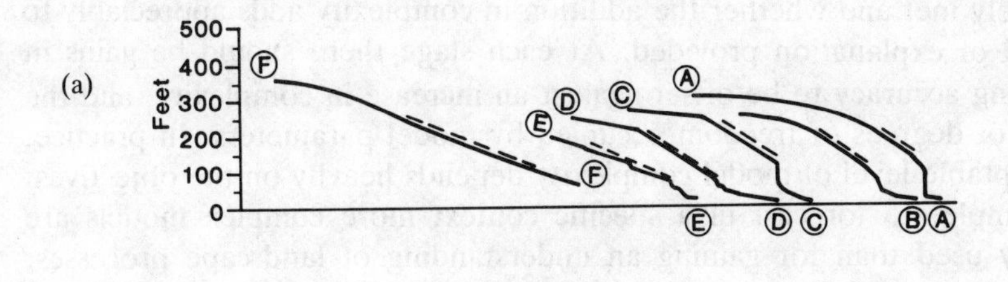

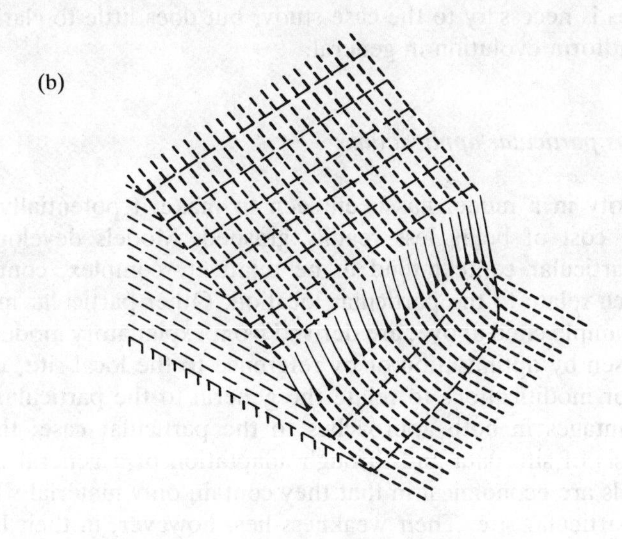

FIGURE 1.3.1
(a) A sequence of surveyed profiles of an abandoned cliff which has been progressively protected from marine undercutting by an advancing coastal spit.
(b) A set of model realisations which mimic the main features and assumptions of the surveyed profiles shown in (a).

tions which ignore a great deal of circumstantial detail. Thus Newtonian gravitation remains a powerful tool for forecasting planetary orbits, even though we are only beginning to learn about many aspects of planetary processes. In other words, drastic simplification of relationships to develop an effective model has an overwhelmingly strong pedigree.

The route taken to a suitably simplified model is largely a matter of individual preference, and may proceed by successively eliminating the least important influence from an initially complex web, or by successively adding important influences to an initially oversimplified view. It is clearer to follow the latter route for present purposes. At each stage of added complexity, a model might in principle be constructed to meet a particular stated objective. For each of this sequence of models we can ask whether the objectives are adequately met and whether the addition in complexity adds appreciably to the level of explanation provided. At each stage there should be gains in forecasting accuracy to be offset against an increase in complexity, and the number of degrees of freedom occupied by model parameters. In practice, the acceptable level of model complexity depends heavily on the objectives. For example, to forecast in a specific context more complex models are generally used than for gaining an understanding of landscape processes. Thus in the case of the South Wales cliffs the consideration of sea level and climatic histories is necessary to the case study, but does little to clarify the processes of landform evolution in general.

Generality versus particular applications

Added complexity in a model has been seen to make it potentially more accurate at the cost of being less readily grasped. Models developed to forecast in a particular context tend to be relatively complex, containing parameters which relate to the particular location. Other particular models, including the example cited above, are derived from explanatory models with parameters chosen by optimisation or by reference to the local site, usually with additions or modifications to relate the general to the particular case. There are advantages in both approaches to the particular case; through statistical analysis of site data and through adaptation of a general model. Statistical models are economical in that they contain only material which is related to the particular site. Their weakness lies, however, in their limited transferability to another location, or even outside the range of their data. The application of a general model to the particular site has the advantage that its validity has hopefully been tested over a much greater range of conditions, but it may omit significant processes operating at the site and may be unrealistic in its data demands. More seriously it may not really be applicable to the case study at all, having been constructed to meet different objectives; for example to operate at substantially different time or space scales.

The 'power' of a model is an important concept which is related to its potential for testing by refutation. A model may be refuted by any critical test of its predictions. In a spectrum from particular to general models, the more particular are able to make more exact predictions but only over their limited range of validity. More general models rarely make such accurate predictions in the particular context, but may be applied to a wider range of contexts. General models, therefore, have the greater power, but this can only be demonstrated by developing particular models from them, to test them over their full range. General models which can be tested in this way add the most to understanding of environmental processes, relationships and forms. They also come closest to providing distinctive theories or laws within physical geography.

THE ROLES OF MODELLING

Models should be constructed with clear objectives of what, if anything, they should forecast. In the same way that field studies should explicitly test particular models/hypotheses, models should be defined to forecast for particular sites, and/or particular processes at specified time and space scales. Untestable models are as unacceptable as unstructured data collection. The commonest objectives for modelling are the communication of concepts, and short-term forecasting, either to allow response to the forecasts or to compare forecasts of alternatives as a planning tool. Other objectives include the comparison of short-term process measurements with forms which develop over long time-spans, or more generally the compression of time or space, and the development of explanation towards theory.

Communication

Perhaps the most influential model in geomorphology has been W. M. Davis's Geographical Cycle of Erosion, an explanatory scheme for teaching which achieved its great success through its relative simplicity and flexibility, and through Davis' personal effectiveness as its prophet. Much work has been built upon the geographical cycle as an explanatory model for understanding the landscape, and this research potential has added greatly to the durability and value of the model, but its power as a means of communicating geomorphology has been immense. Plainly models which are designed only to communicate are rather sterile, but for some models the simplications involved in producing them allow their subject matter to be expounded to and grasped by important new audiences.

Viewed from the opposite direction, it is the duty of every scientist to communicate his ideas through outlets at all appropriate levels, ranging from professional journals to popular articles or broadcasts. Models must therefore have some role as communication, even if this is not their primary purpose.

Short-term forecasting

In many cases, models are built to give specific forecasts as a basis for immediate decision-making. For example, models are used to forecast river flows in response to current rainfall, as a basis for the management of reservoir releases and to provide flood warnings. Models similarly provide maximum warning for possible tidal surges, landslides and irrigation needs. In research, models are often used to provide a forecast which is then compared with its realisation at another site or another time. This provides an important independent check on the model before its adoption.

Forecasting models are often built from regression analysis, but may also be derived from more general models. They usually contain locally relevant detail which is essential to their particular use, but which limits their transferability to other sites and their use for communicating concepts. A widely used forecasting model based on regression is contained in the Flood Studies Report for England and Wales. The design flood of any specified return period is derived from a regression equation containing basin characteristics obtained for geographical regions related to Water Authority areas. As with other regression models, completely new parameters must be obtained from local data before the model can be applied to fresh areas. The Flood Studies model, or some equivalent, is needed to design bridges and other structures in basins without a direct record of floods, and so fulfils an important need.

Simulating possible scenarios of environmental change

A more widespread use of forecasting models is as planning tools. Model forecasts explore the expected implications of alternative plans without the expense of putting each into practice. The model may range from a simple projection of trend to a complex hydrological simulation. If its forecasts are correct, then it allows a rational choice of plan from the alternatives modelled. In the context of these 'what–if' forecasts, a suitable model not only needs to give detailed accuracy but should also be able to take account of a flexible range of possible options.

A model which is widely used in this way is the Universal Soil Loss Equation, developed to forecast rates of soil erosion. The model is based on regression of factors associated with rainfall erosivity, soil erodibility, topography, cropping management and erosion control practice. Within an area for which parameters have been established, farmers can be advised on the most appropriate erosion control practices. Although the physical basis of this model is open to question, it has nevertheless been fairly effective as a means of pinpointing areas of high erosion risk and controlling soil erosion in the eastern United States.

Linking short-term process measurement with long-term forms

Geomorphology and pedology provide evidence of process rates which can be measured over a few years, and of forms (landforms and soil profiles) which may take many thousands of years to develop. There is no way of directly measuring such changes, so models are needed to extrapolate over these time-spans in a sensible way. It is rarely appropriate simply to multiply short-term rates up to these longer time-spans. There are commonly changes in environmental controls, especially climate, and the processes are themselves influenced by the forms they produce. For example, hillslope sediment transport is strongly dependent on gradient and hillslope length. As the hillside erodes, sediment transport rates change in response to the changing topography, so that long-term average erosion rates are generally different from those measured in the short term. A suitable model for hillslope evolution, allowing for changing topography and changing climate, is therefore required to judge whether an observed landform is compatible with measured process rates.

Time–space compression in general

The need to link short-term measurements with long-term evolution of forms is just one example of the need for models which compress time or space scales. This is a common problem in hardware models where cost and laboratory size often demand work on reduced scale models. It is also frequently desirable to speed up the rate of processes to obtain results in a reasonable time. Similarly, in computer models spatial dimensions are generally set by a grid of points at which properties like elevation or temperature are defined. The maximum area which can be modelled is equal to the number of grid points multiplied by the area which each represents, and is in practice limited by available computer memory. The time-span represented by a computer model is similarly related to available computer run times (roughly proportional to length of period modelled and number of grid points). In many cases, computer models are therefore able to describe and forecast in detail over areas and times very much greater than can be instrumented in the field. Perhaps the most extreme models of this kind to date are the General Circulation models used to follow and forecast global weather in 100 ×100 km grid squares.

Developing 'explanations' at all scales

The uses for models described above have a major practical component in them. The final use for models is to improve understanding of the system which the model attempts to describe. This is primarily to advance our environmental science, but has substantial practical pay-offs in improving the range of applicability of models, and the measurability of their parameters. For this purpose, models need to be as well grounded as possible in established principles. Such a model acts as a thought-experiment, pursuing the logic of a set of assumptions to one or more points where its outcome may be compared with experience.

At the simplest level, we may ask what follows if we assume that the moon is made of green cheese, or any other more serious assumption. Putting this new assumption together with other better founded views, we hope to reach conclusions about, for example, the size and shape of lunar craters which may be compared with our knowledge of them, and so help to refute or confirm our assumption. In the same way, we can put together knowledge and assumptions about how any system works to produce a hypothesis in the form of a model. In many cases, we may not know the exact form but only the direction of relationship, but it is still useful to assemble a provisional model and compare its forecasts with observation. One question we will want to ask is how sensitive the model outcomes are to the exact form of the unknown relationship. The answer, obtained by trying different forms, may lead to greater knowledge of the relationship and help to focus future field research along potentially fruitful lines. This kind of model development is a vitally important route towards the development of established laws or theories in environmental science.

TOOLS FOR MODEL-BUILDING

The minimum requirement for any model is that it should be built on the basis of logical reasoning. Traditional models, like those of Davis or Penck in geomorphology, are expressed in words but it is their logical basis which allows forecasts to be made from them. Most other kinds of model do not seek to replace this logical basis but merely attempt to strengthen it, perhaps by relying on the formal logic of mathematics or systems analysis to guarantee consistency, or by relying on the mechanical behaviour of physically or mathematically similar laboratory experiments. Neither approach guarantees a faithful model but each forces greater consistency than can be guaranteed in a model which is stated in words alone.

Logical reasoning

Logical consistency is the only necessary requirement for any model, but it also contains assumptions, deductions and conclusions. The assumptions can be derived from qualitative or quantitative observations, or from a basis of theory. Either theory or observation may be incorrect or only partly correct, in which case the model built from it, however internally logical, cannot reach sound conclusions. It is all too easy to feed data into a statistical model, and to allow the logical consistency of the statistics to mask the unreliability of the input data, but the 'Garbage in – Garbage out' rule applies to all models. Similarly good data or assumptions may be used to infer false conclusions if the model is not logically consistent. The only differences between the many kinds of model are that it is easier to detect false reasoning within formal frameworks like mathematics than when expressed in words.

Scale models and other analogues

Where the model consists of a reduced scale model of a section of the real world logical consistency is provided by the action of the same physical or chemical processes in model and prototype. This reliance on similarity must always be looked at very carefully. Is everything scaled correctly? How should time scales change as linear scales change? Can we distort or ignore scaling for some processes which are negligible in their impact? These questions require a very thorough understanding of the detailed physics and/or chemistry of the processes.

Analogues may operate at levels other than that of strict physical look-alikes. One degree removed are models which rely on physical laws with a common mathematical form. For example, Darcy's law for saturated ground-water flow and Ohm's law for electrical conduction may both be expressed in the form:

$$\text{Flow} = \text{Conductivity} \times \text{Pressure gradient}.$$

An analogue model for flow in an aquifer may therefore be made from a sheet of conducting material of the same shape as the aquifer. Flows can be forecast by measuring electrical currents in the conductor when appropriate voltages (analogous to pressure heads) are applied. At a further degree of abstraction are analogues which share with the prototype only the form of the equations. As the degree of abstraction is increased, the dangers of incorrect analogy in assumptions or processes are more evident than for the scale model, though not necessarily any greater. At the same time, the modeller is increasingly freed from the physical constraints of finding materials with suitable (scaled) physical properties, or laboratories large enough to house the experiments.

Mathematical formulations

Mathematical equations representing the physical or chemical processes of the real world are the most abstract form of analogue. They allow access to techniques which help to minimise, though not eliminate, the risks of logical inconsistency in a model. As with most stages in the process of abstraction, they encourage simplification. Model assumptions must therefore be checked to assess whether the processes included are accurately described and whether all essential processes have been included.

Systems analysis

One way to examine the completeness of a model is through systems analysis, which focuses attention on the presence or absence of linkages between parts of the real world or model systems. The geographical literature is now littered with the box-and-arrow diagrams which form a starting point for systems analysis. They remain a useful step in the construction of an effective model to increase understanding as well as provide forecasts; that is, a model with some physical rather than purely statistical basis. Box-and-arrow diagrams commonly represent only the first stage in the sequence of model development and as such have limited usefulness in themselves. The full use of systems analysis is, however, powerful and widely applicable to the development of models of all kinds.

Computer simulation

Computers are increasingly used to obtain model forecasts, both through the analysis of data by standard statistical models and through the construction of more or less physically based simulation models. They are advantageous in any model which has been abstracted to the level of mathematical equations or formal logic, although they cannot readily handle word models. The advantages of rapid and reliable calculation cannot be overestimated, and cost-effective model building has shifted substantially from the laboratory hardware model to the computer model during the 1970s. Computer models allow a much wider range of conditions to be simulated than most laboratory experiments, and model runs can be repeated with increasing ease, but these advantages should not be allowed to conceal the importance of any model's logical basis. The importance of appropriate assumptions and a logical structure remain as strong as ever.

STYLES OF QUANTITATIVE MODEL

It is rarely effective to express models in words alone, now that quantitative and computer methods and literacy are widespread. Even the most qualitative models can be fruitfully explored by trying out quantitative functions in place of general statements, and discussion here is therefore focused around quantitative models.

The three types of model discussed below are not mutually exclusive categories. Black box models relate output forecasts to input data with no explicit statement of how they are functionally related. All real world systems are incompletely understood, so that models always have some element of 'black box' about them, although in many models the 'box' is in shades of grey because the model physically represents at least some of the internal processes acting. In most models, we are concerned with flows of material and/or energy, and these quantities are conserved within physical systems, so that models with any physical basis are commonly constrained by some kind of mass or energy storage equation. Although the physical and chemical environments are largely deterministic at all geographical scales of interest, we may choose to incorporate a stochastic element into models.

Black box or input/output models

Multiple regression is by far the commonest technique for generating black box or input/output models. Although the assumptions of each statistical model must be met, the method is powerful and is readily applied, usually through one of the many computer packages available. The main disadvantage of a regression model is that it adds very little to our understanding of how the system functions, because its parameters have no direct physical interpretation. As a result, regression models cannot readily be applied to conditions outside their original data range, or to other areas. The level of understanding may be improved somewhat by other multivariate statistical methods. Step-wise multiple regression adds variables only if their addition significantly reduces the differences between observed and forecast values. Principal component and factor analyses create new composite variables which are mutually independent, so that each may be interpreted as a distinct causal influence. These and other methods help to make a link between data and understanding. Although they can never independently generate a causal model or theory, they are a valuable guide in combination with a consideration of the physical or chemical principles involved.

Once any model has been constructed, it may be used as a black box model without an understanding of the internal function of either model or the real world it represents. This is particularly evident for computer models but is equally, if less obviously, true in principle of all models. Even a hardware scale model might be run without an understanding of its internal workings. In many forecasting applications, end users wish to treat the model as a black box, whether it is based on a detailed analysis of the physics or on a simple regression.

Models constrained by mass or energy balance

Except in radioactive reactions, energy and mass are conserved, and mass is conserved not only in total but also for each chemical element. These conservation principles are not necessarily a useful component of models since we are invariably dealing with open systems in physical geography. There are, however, many cases where conservation of mass is a useful system constraint, and a few cases where conservation of energy is of use.

An energy balance is at the heart of many micro-meteorological models to forecast heating of the lower atmosphere or evapo-transpiration, for example through the use of the Penman–Monteith model. In these models the partition of incident solar energy between reflection, long-wave back radiation, heating of air/soil and evapo-transpiration is the essence of the model, so that the energy balance must be maintained. In other contexts, for example plant growth or sediment transport, energy must still be conserved, but only small proportions of the total available energy are used within the model system, while most is effectively 'lost' because it is used in ways which are not included in the model. In such a model, it is useful to monitor the working substance which forms the 'currency' of the model. In the case of a plant growth model, this is essentially biomass, which may be monitored via a carbon balance. In the case of a sediment transport model, the currency is total flow of earth materials so that again a mass balance is central. In general, therefore, it is sensible to use a 'currency' for storages and flows in a model which can be budgeted by relying on the principle of mass or energy conservation which is most relevant for the purposes of the model.

Mass or energy balances may usually be expressed in the form of a storage equation, which is perhaps most familiar in the hydrological context:

Input − Output = Net increase in Storage.

This may be applied for the system as a whole, or for many individual components within it. For example in a catchment hydrological model, storage equations might be written for successive layers down the soil profile; or for separate unit areas of the catchment. The requirement of mass or energy balance makes these models much more robust in forecasting outside their original data range or geographical area than a regression model, for example.

Stochastic versus deterministic models

In many engineering design contexts, we need to forecast not a particular event such as a flood but the distribution of floods, or the size of flood which will be exceeded on average about every fifty years. A model may estimate this distribution for a catchment directly from its physical properties. Alternatively a deterministic model may predict stream runoff from rainfall sequences. A stochastic submodel is needed to draw rainfalls at random from their distribution (which is usually better known than the corresponding flood distribution). Other stochastic elements may also be introduced into the rainfall–runoff model to allow for probable differences in, for example, catchment discharge and soil moisture at the start of each rainstorm.

Stochastic models are also useful where soil or other properties vary over a catchment, but it is impractical to measure values at every point. If sufficient measurements are taken to establish the form of the distribution, it may be adequate to use a randomly drawn value from this distribution for each computation point in place of a measurement. The model can then take account of the effect of the spread of values without explicitly using all the individual values. Here the purpose of the stochastic element is to allow for variation in factors which cannot practicably be measured in detail, but where it is thought that the use of an average value throughout will lead to inadequate forecasts. Where models are run over time, it may also be necessary to use stochastic values to define initial states which cannot be known in any detail after the event.

From these examples it can be seen that stochastic elements in a model may be associated with inputs to it, with its parameters or their spatial distribution, or with the processes modelled. In none of these cases is there an indeterminacy in principle. An increasing number of models contain some stochastic elements despite a consequent slight increase in model complexity. Their main advantage lies in the ability to forecast a distribution of outcomes, which can then be assigned confidence limits to assess their similarity to the real world they purport to model.

THE LIMITATIONS AND POTENTIAL OF MODELLING

Despite their uses, models are far from being a universal panacea and should be used with respect for their limitations, and with an open mind about the best way to overcome them. Perhaps the greatest problem with models is in assessing the goodness of fit of their forecasts. A second, related problem is in assigning values to model parameters and identifying them with physically measurable parameters in the real world. These problems do not arise for regression or other multivariate statistical models. Goodness of fit is inherent in the theory of these models, and no attempt is made to identify parameters (i.e. regression coefficients) with physical properties. For more physically based models, however, goodness of fit is much harder to measure and the physical identification of parameters is important because it is the basis for transferring models from their original test area to others.

Comparison with statistical methods provides some methods for testing the quality of fit of a particular model to a particular outcome. An 'efficiency' can be derived for any model, although it is not always an ideal criterion. For example in a hydrograph model, small errors in the timing of peak flows have a disproportionate influence on the efficiency, even though subjectively the quality of fit is good. Least squares criteria for testing the accuracy of a model also give great weight to large differences in either direction, sometimes for observation which might subjectively be excluded as being in error. Such problems apply to regression models and are not unique to physically based models, but should be borne in mind before accepting any single criterion for goodness of fit.

The main difference between regression models and physically based models lies in the nature of the fitting procedure. In a regression model, all the coefficients may be varied to optimise the efficiency (or whatever other criterion is adopted). In a completely physically based model, all parameter values are derived from field measurement, so that none may be varied to improve the efficiency. Most actual models lie between these extremes, with some parameters which have been measured and others which have not, and so may be varied to optimise efficiency. In practice then, models may usually be optimised for some parameters which represent the more black box elements of the model. For parameters which are measured, but with some error or with spatial variability, it is possible to run a model many times, using parameter values drawn from a suitable stochastic distribution. A distribution or 'envelope' of forecasts may be generated in this way, and we can assess the probability that the observed outcomes might be drawn from this distribution of forecasts.

Models, especially computer models, tend to proliferate parameters, many of them with little physical basis, even in models which claim to be physically based. This trend has been very apparent in flood forecasting models, where many models have thirty or more independent parameters, representing possibly significant flow processes within a catchment. Nevertheless, hydrograph forms are remarkably conservative and may be fairly well described by parameters to describe the delay from peak rainfall to peak runoff, the rate of recession from the peak and the rate of rise to it. The success of the unit hydrograph as a model demonstrates the degree of truth in this assertion. If a hydrograph model has many more than three parameters which are not physically predetermined, then the fit obtained by optimisation is likely to be ambiguous. As a result it will not be possible to say, for example, which of the possible flow processes is dominant, because equally good fits might be obtained for several alternatives. A well-designed model will therefore aim to have a minimum of undetermined parameters.

Models can never replace field observations and laboratory experiments, but can greatly increase their effectiveness in several ways. Any programme of research begins with one or more hypotheses to be tested or compared with each other. These views necessarily form a model which the research is testing. By making this model explicit and formalising it, we can make provisional forecasts which usually help in designing the experiment.

The choice of an appropriate model as part of a research design must always be closely geared to the needs of that research. It is necessary to choose or construct a model which operates at appropriate time and space scales, even if it is designed to be compatible with models at another scale. Similarly, the model used must be related to the techniques and variables which are to be measured, so that they can exchange data and forecasts in a meaningful way.

As research proceeds, there should be some continuing dialogue between experiment and model; experimental data being used to improve or replace the model and the model giving fresh forecasts which are relevant to the field site and data set, to refine the experimental collection programme. As the advantage of cost and time shifts from fieldwork to model building, so the experiment needs to be designed with increasing care; with large amounts of prior testing through models and very thorough analysis of the data at every stage. An extreme example of this is found in programmes of space research. Experimental work in physical geography has not reached this extreme, but the balance of effort is steadily changing, and the pattern of research effort has tended to lag behind.

Models then are seen as having great potential for research and communication in physical geography, both as a means of building useful forecasting models and as a means of developing understanding and theory. Despite their limitations and the technical difficulties in implementing them, they are becoming increasingly important as a component of physical geography, and perhaps the component which is doing most to develop a soundly based theory for our subject.

FURTHER READING

A non-technical introduction to modelling in the context of physical and resource geography with many informative flow diagrams but a minimum of equations and programs is given by:

Frenkiel F. N. and Goodall D. W. (eds) (1978) *Simulation Modelling of Environmental Problems* (Chichester: SCOPE/John Wiley).

The following two texts on modelling approaches and methods span the whole of geography, but with a substantial physical content. They illustrate the areas of common ground, or lack of it, between human and physical geography. The first makes particularly challenging reading.

Bennett R. J. and Chorley R. J. (1978) *Environmental Systems: Philosophy, Analysis and Control* (London: Methuen).

Thomas R. W. and Huggett R. J. (1980) *Modelling in Geography: A Mathematical Approach* (New York: Harper & Row).

Three books can be suggested which exemplify the use of models in three different subject areas. Anderson is at a more advanced level than the others, and represents the current state of the art in hydrology.

Maynard Smith J. (1974) *Models in Ecology* (Cambridge: Cambridge University Press).

Carson M. A. and Kirkby M. J. (1972) *Hillslope Form and Process* (Cambridge: Cambridge University Press).

Anderson M. G. (ed.) (1985) *Hydrological Forecasting* (Chichester: John Wiley).

Finally, two collections of research papers give some idea of the range of models now being developed. The latter covers the wider field, ranging from coasts to channels on Mars!

Ahnert F. (ed.) (1976) *Quantitative Slope Models*, Zeitschrift für Geomorphologie, Supplementband 25.

Woldenberg M. (ed.) (1985) *Models in Geomorphology* (London: Allen & Unwin).

1.4
Remote Sensing – Global and Local Views

John R. G. Townshend
University of Reading

The rate at which data about the biophysical environment are collected from satellites now exceeds that from ground level. Satellite sensors already gather data about the Earth's atmosphere, oceans and continents, and in the next ten years the number of satellites, the sophistication of their sensors and the proficiency with which their data are analysed will increase dramatically. But why are satellites so useful in gathering data? What are the benefits to be gained from taking this distant view of the environment? Most satellites fly in orbits more than 500 km from the Earth's surface and those in geostationary orbit are more than 20000 km from the Earth's surface. The outstanding benefits of this distant view relate to the overview of the resultant images, which provides us with reliable internally consistent data of large areas almost instantaneously. Moreover, the same area can be imaged regularly, which permits us not only to obtain base-line data but also to monitor the dynamic components of the Earth's surface for the first time.

The increasing value of satellite-derived data is a consequence of a number of technological advances, principally relating to the types of radiation that can be sensed; the amount of detail which can be detected; the frequency with which the data are provided; and the use of digital computers for automated extraction of information from the data.

WIDENING THE SPECTRAL VIEW

The electro-magnetic radiation which human beings can sense and from which they derive pictorial representations forms only a small proportion of the radiation which naturally occurs (Fig. 1.4.1). Not surprisingly, the radiation that we can see is the most abundant naturally occurring radiation, but radiation of other types is also of considerable use. In Figure 1.4.2, the proportion of radiation reflected from a healthy plant is shown both for visible radiation and for rather longer wavelengths in the near and middle infrared (Fig. 1.4.1). 'Infrared' in this context has nothing to do with emitted

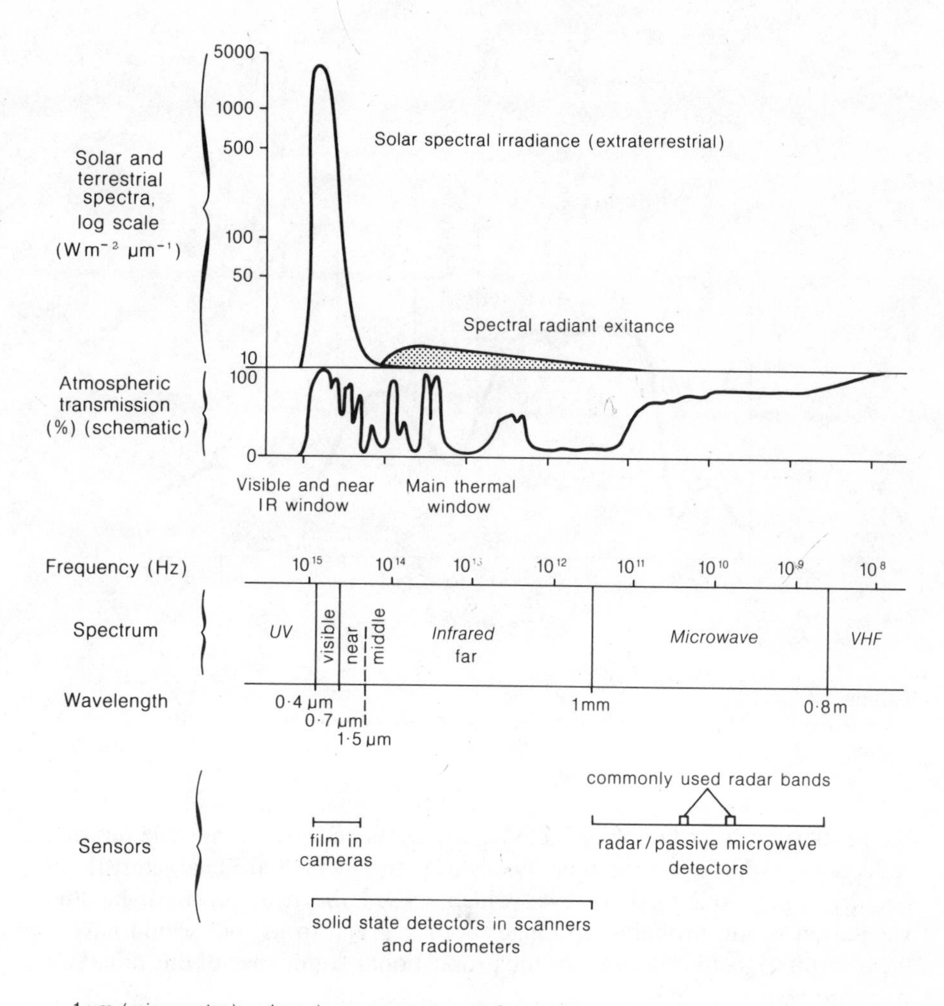

FIGURE 1.4.1
Types of electromagnetic radiation and sensor types. (Hardy, J. R. 1981). Data collection by remote sensing for land resources survey. In Townshend, J. R. G. (Ed.) *Terrain analysis and remote sensing*, George Allen & Unwin, London 16–37).

heat, which will be considered below. Concentrating on the visible part of the spectrum, it can be seen that there is a small peak at around 0.5 micrometres. Radiation at this wavelength stimulates our eye–brain sensor system to perceive green colours, which is why healthy vegetation appears green to us. The prime cause of this greenness is the presence of chlorophyll, which plays such a central role in photosynthesis. However, at wavelengths slightly longer than the red, specifically in the near infrared, there is a dramatic increase in

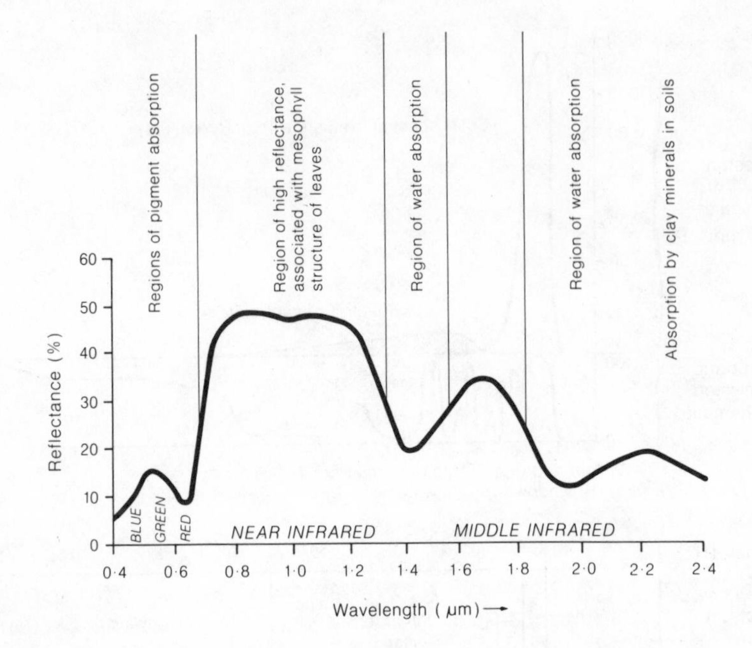

FIGURE 1.4.2
Proportion of radiation reflected from healthy green vegetation for different wavebands.

the proportion of radiation reflected. This arises because of multiple internal reflections within open mesophyll cells of plants, which are characteristic of healthy vegetation. Had we eyes which sensed the near infrared, healthy vegetation would probably no longer appear green to us, but would have a 'near infrared' colour because of the proportional dominance of this radiation compared with the green.

At still longer wavelengths, the spectral response curve (Fig. 1.4.2) takes two dramatic plunges resulting from the absorption of radiation by water within the plant leaves. A sensor which detects radiation at the edge of one of these absorption features permits the monitoring of leaf moisture. One which senses radiation in the centre of one of these absorption features will reveal little or nothing of the Earth's surface, but will show us variations in atmospheric water.

At longer wavelengths still (Fig. 1.4.1) the far or thermal infrared allow representation of heat. Human beings can sense thermal radiation but cannot construct pictures of their environment with such radiation. Figure 1.4.3 shows a thermal image in which variations in tone relate to changes in temperature, along with images taken in the visible and near infrared.

In the microwave part of the spectrum we have to generate our own radiation if we require anything but a general overview of the Earth. Radar

FIGURE 1.4.3

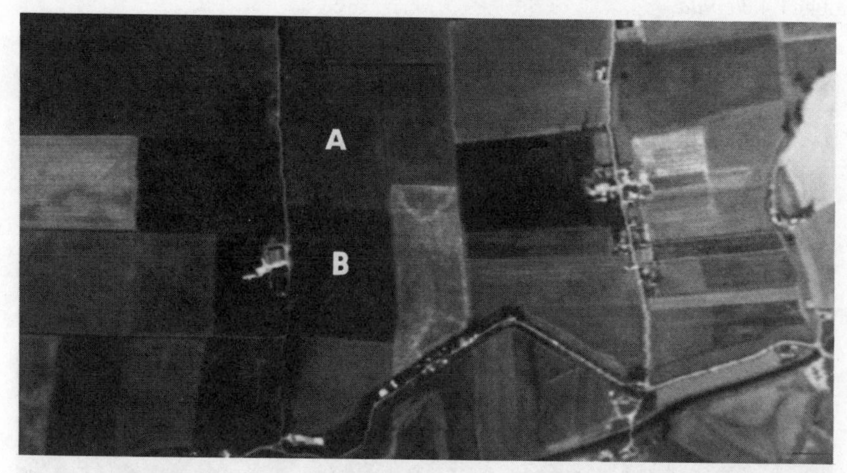

(a)

(b)

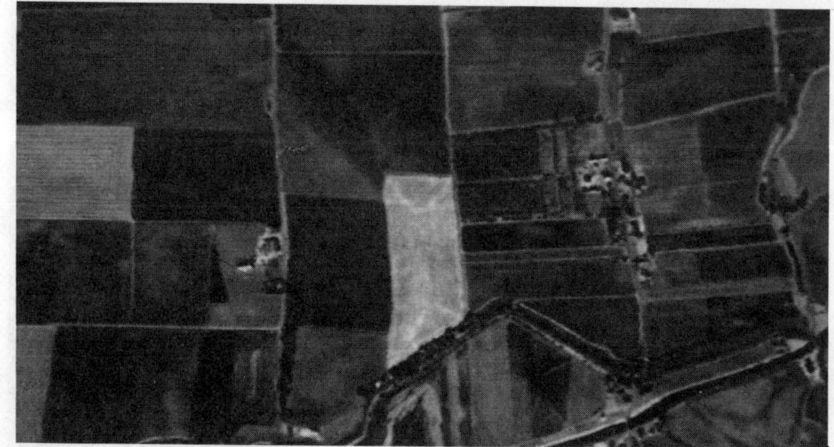
(c)

FIGURE 1.4.3 *contd.*

(d)

(e)

(f)

FIGURE 1.4.3
Images taken from an aircraft of part of the English fenlands near Peterborough. The images represent an area roughly 2.5 km across. The bands shown correspond to those of the Thematic Mapper (TM) onboard Landsats 4 and 5 though the actual satellite has significantly coarser resolution.

(a) blue-green 0.45–0.52 μm (TM band 1)
(b) green 0.52–0.60 μm (TM band 2)
(c) red 0.63–0.69 μm (TM band 3)
(d) near infrared 0.76–0.90 μm (TM band 4)
(e) middle infrared 1.55–1.75 μm (TM band 5)
(f) thermal infrared 10.5–12.0 μm (TM band 6)

Clearly different spectral bands show very different information. It is particularly instructive to compare field A (bare soil) and field B (100% cover of sugar beet). Note that the sugar beet is relatively dark compared with the bare soil in the blue green, red, middle infrared and thermal infrared. Referring to figure 1.4.1 it can be seen that this is related to absorption by pigments in the blue-green and red, and to absorption by leaf moisture in the middle infrared. In the thermal infrared, active evapotranspiration by the sugar beet depresses the temperature and hence the amount of emitted radiation. The sugar beet appears relatively bright in the green part of the spectrum due to reflection by the chlorophyll pigment, but is much brighter in the near infrared, due to strong reflection in the mesophyll layer of the green leaves (based on Townshend, 1984).

---

sensors generate their own radiation and then produce an image using the part of that radiation reflected from the ground. The outstanding advantage of this type of radiation is that it penetrates cloud and hence ensures very regular monitoring even where the frequency of clouds is very high. The first European Earth Resources Satellites (ERS-1) will have a radar sensor to monitor the state of the ocean surface.

Sensors which can detect these different types of radiation greatly extend the quality of data about our environment beyond that available within visible wavelengths. Many of these sensors have two other outstanding characteristics. First, they can detect radiation in several different parts of the spectrum simultaneously. This greatly improves our ability to distinguish between different features on the ground and, in the case of atmospheric sensing, allows us to construct vertical temperature lapse rates through the atmosphere. It is proposed that the Earth Observation System to be placed on the Space Station will be able to sense up to 128 separate bands within the 0.4 micrometre to 2.5 micrometre part of the spectrum (Fig. 1.4.1). This sensor would be of particular value in mineral identification in arid and semi-arid areas. Secondly, the sensor can provide data in a quantitative form which consequently allows computers to aid in their analysis.

In Figure 1.4.4 the ways in which the most common sensors operate are outlined. In the case of the multispectral scanner, radiation is split up, and for each component, set wavelength ranges are detected simultaneously by

FIGURE 1.4.4

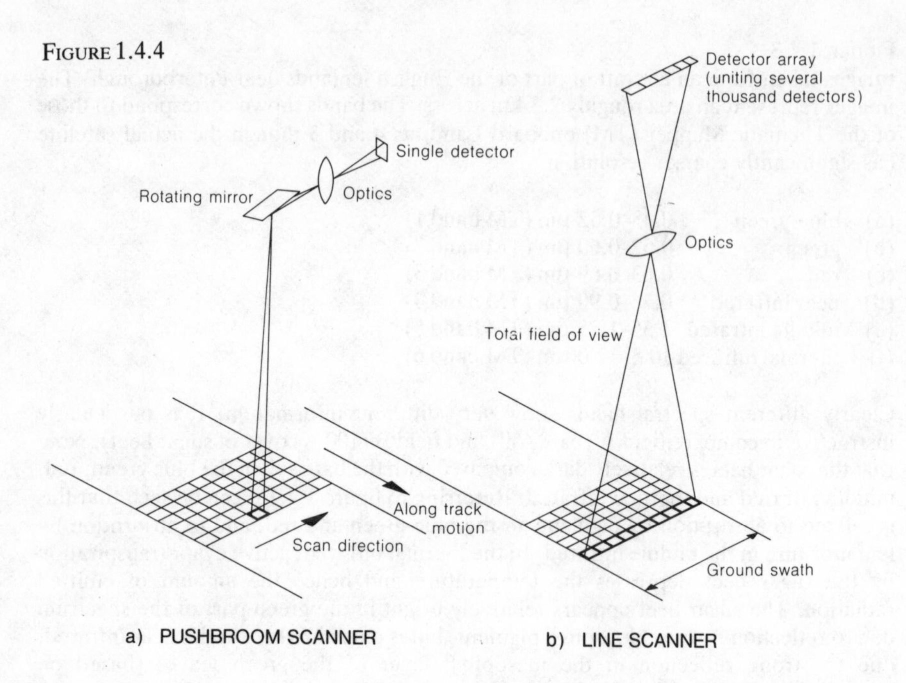
Detector array
(uniting several
thousand detectors)
Single detector
Rotating mirror
Optics
Optics
Total field of view
Along track
motion
Scan direction
Ground swath
a) PUSHBROOM SCANNER
b) LINE SCANNER

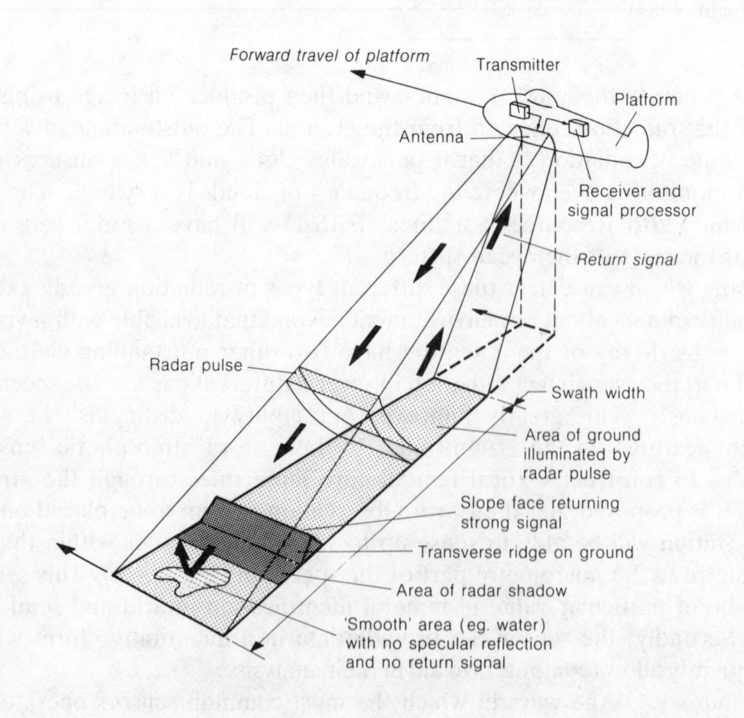
Forward travel of platform
Transmitter
Platform
Antenna
Receiver and
signal processor
Return signal
Radar pulse
Swath width
Area of ground
illuminated by
radar pulse
Slope face returning
strong signal
Transverse ridge on ground
Area of radar shadow
'Smooth' area (eg. water)
with no specular reflection
and no return signal
c) RADAR

FIGURE 1.4.4
Operation of sensors used in aircraft and satellites.
(a) Pushbroom scanner
(b) Line scanner
(c) Radar
Based partly on Hardy, J. R. (1981). 'Data collection by remote sensing for land resources survey' in Townshend, J. R. G. (ed.) *Terrain analysis and remote sensing* (George Allen & Unwin, London) 16–37.

separate detectors. At any one instant a given detector is only looking at one small piece of ground. A moving mirror *scans* the surface building up a line of signals. The forward motion of the satellite allows successive lines to be sensed, the speed of the mirror's rotation being just sufficient so there is no over or underlap between scans. The separate small areas detected at any one moment produce picture elements or pixels, which can be given tones or colours proportional to the amount of radiation detected so that a convenient picture can be produced. In most future satellites, solid state technology will be used with thousands of sensors on a single chip detecting complete lines, without the need for a moving mirror. In the case of radars, the time that the signal takes to return to the sensor is used to determine the ground location and hence to generate a line of pixels. The forward motion of the sensor again allows successive lines to be built up (Figure 1.4.4).

IMPROVING THE SPATIAL DETAIL

One of the main ways in which satellite sensors improved in the first half of the 1980s was in the fineness of detail they could detect. For example, satellite sensors designed for surveying the Earth's surface have improved by better than an order of magnitude in terms of smallness of pixels that are detected. In Figure 1.4.5, a part of the countryside to the north-west of Reading is shown, at 79m and 30m pixel sizes respectively. These are representative of images from the Multispectral Scanner System (MSS) of Landsats 1 to 5 and the Thematic Mapper (TM) sensor found only in Landsats 4 and 5. It is apparent that the first image provides a rather poor representation of the rural scene. Many of the pixels are mixed, containing more than one cover class (i.e. type of terrain). Hence identification of cover categories by visual or by computer-assisted means is unlikely to be very successful. In contrast, the Thematic Mapper image has a much sharper appearance related to the small proportion of the pixels which are mixed and identification of cover types can be carried out much more reliably.

The Thematic Mapper image is, however, not very satisfactory in representing the urban scene because of its high spatial complexity. In the north of the area displayed, an area of residential development is found: individual roads are not resolved and there is a clear limit to the amount of information which can be extracted. Better resolutions are clearly required.

FIGURE 1.4.5

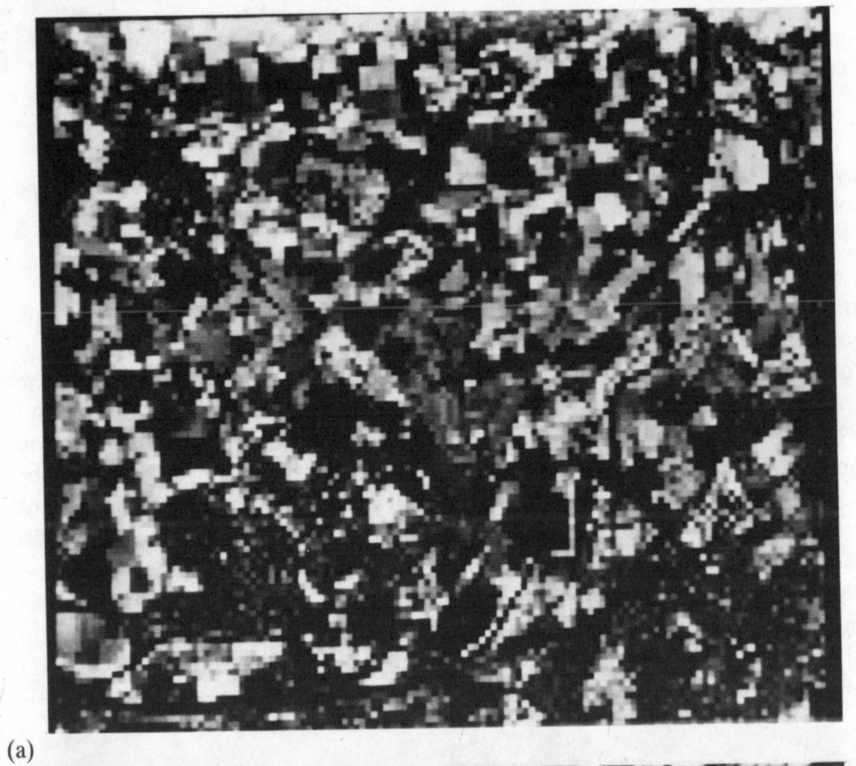

(a)

(b)

FIGURE 1.4.5
Images of the area to the north west of Reading
(a) Multispectral scanner (79 m resolution)
(b) Thematic Mapper (30 m resolution)

The higher resolution imagery which will be produced by the French SPOT satellite provides the necessary improvements. Figure 1.4.6 shows part of southern Reading, on images which simulate the performance of the sensors of this satellite. The pixel sizes are 10m and 20m respectively. The 10m data are particularly impressive, though as reference back to Figure 1.4.1 indicates, these images are only sensed in a single, broad spectral band in the visible part of the spectrum. Consequently the value of applying computer-assisted information extraction techniques is restricted.

(a)

FIGURE 1.4.6

(b)

FIGURE 1.4.6

Simulated images of the sort to be derived from the French SPOT satellite
(a) 20 m resolution
(b) 10 m resolution

Visually, images with smaller pixels and hence higher resolution always appear better. However, if we use automated computer-assisted methods of information extraction, better resolution does not necessarily yield more accurate results, as the graphical plots of land cover classification accuracy illustrate (Fig. 1.4.7). In order to understand how this arises, it is necessary briefly to consider how computer-assisted image classification is carried out. The most common procedure is to choose a 'training set' of sites whose ground characteristics and digitally recorded spectral characteristics are known. As shown in Figure 1.4.8a, each class is found in a distinct part of the plot of data from two spectral bands. Consequently there is a good chance that we can classify the remainder of the scene using remote sensing data alone, assuming that our training set was a statistically valid sample. On the other hand, we might find the situation as shown in Figure 1.4.8b, where

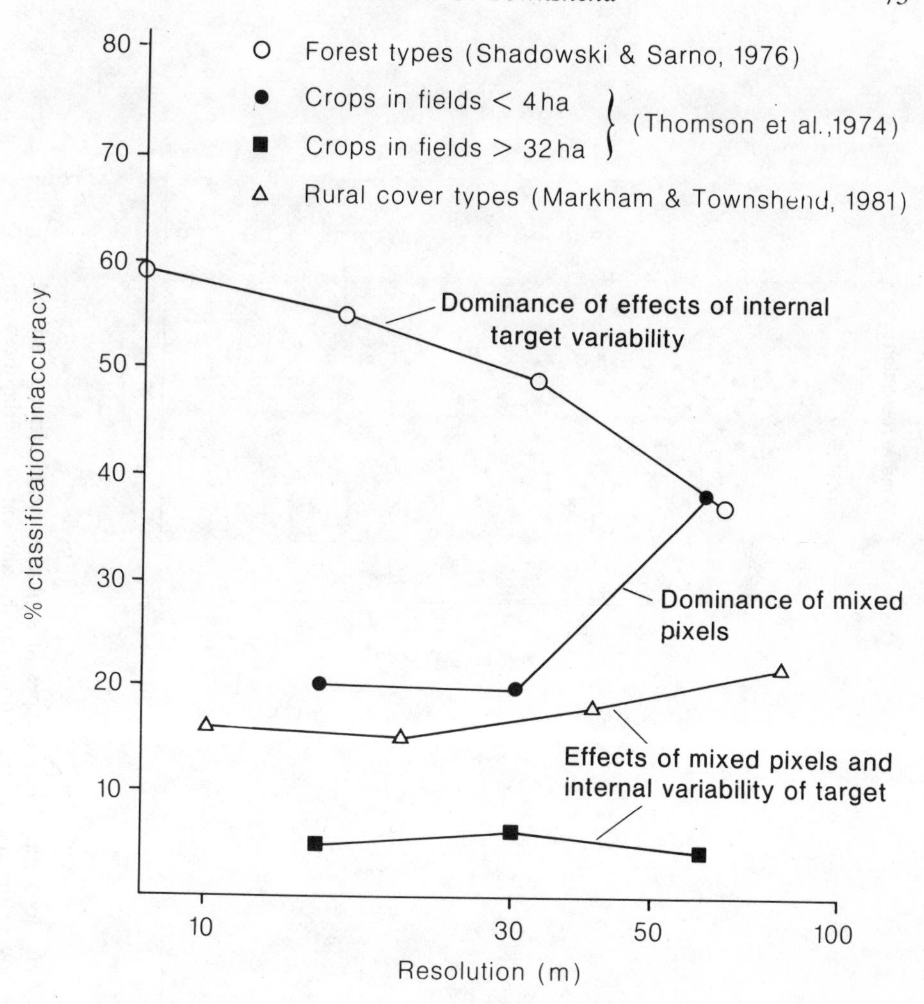

FIGURE 1.4.7
Plots of errors of classification against spatial resolution. Note this is based on the most commonly used automated classifier which only uses data from each individual pixel to carry out the classification. More advanced classifiers would change the observed trends especially for forest types.

there is considerable overlap between classes. In such a circumstance particular classes do not have a unique spectral response and errors will occur if reliance is placed on spectral response alone. As pixel size becomes smaller, there is often a marked tendency to move from a situation like that in Figure 1.4.8a to that in Figure 1.4.8b. In other words, the variability of the spectral response for a given class increases with coarsening resolution. Explanation for this lies in the fact that as a given object is viewed in ever more detail, then

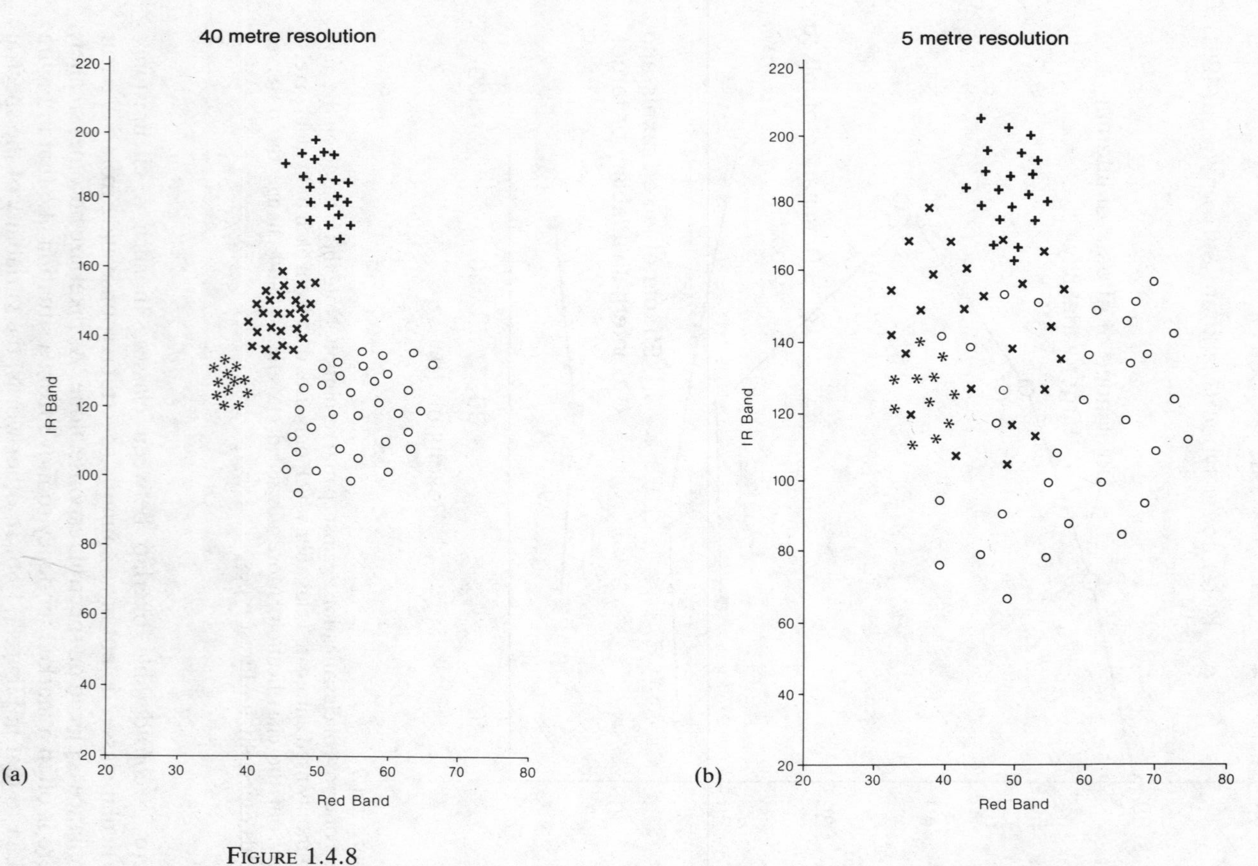

FIGURE 1.4.8
Representation of the location of cover types in a plot of near infrared and red reflectance. Note how at the coarser resolutions (a) the amount of overlap between categories is much smaller than for the finer resolutions (b).

internal variations become more and more apparent. Within a woodland, for example, internal variations due to the crowns of trees being differentially illuminated will occur; within a field of a single crop, internal variations due to different soil characteristics may appear; in a residential area, individual roads, roof tops, patches of woodland and gardens will be resolved. If we are relying solely on the recorded signal from each individual pixel then the distribution of a cover class becomes more like that in Figure 1.4.8b, and hence misclassifications will occur.

Conversely as resolution becomes finer, then the proportion of pixels which contain more than one class of the phenomena under consideration, will inevitably decrease. Visually this can be appreciated from Figure 1.4.5. Mixed pixels are inherently more difficult to classify and thus in this respect classification accuracies should improve with small pixel size. There are thus two counteracting tendencies present as pixel size declines – one tending to worsen the accuracy of classification and the other to improve it. Depending, therefore, on the spatial frequencies present with the area being analysed, it is possible for finer resolution to lead either to improved accuracies, poorer accuracies or virtually no change at all across a wide range of resolutions. The capability of a satellite sensor cannot therefore be divorced from the spatial characteristics of the terrain being sensed.

Two important caveats to the above conclusion must be made. First, it is possible to use not only the response of individual pixels, but also to use textural variables which describe the spatial variation of spectral response of groups of pixels. Use of such variables may well lead to improved classifications at higher resolutions. Secondly, the success of classification depends strongly on the task for which the information is being extracted. For example, consider a residential area with its complex assemblage of cover types. If our concern is to map the overall distribution of cover types within the urban area, then a lower resolution image with bigger pixels may give us better results than a finer resolution image. But if we are interested in the hydrological response of the catchment containing the residential development, then the most significant parameter may be the proportion of the catchment covered by impermeable surfaces such as houses and roofs. In such circumstances small pixel sizes would be a distinct advantage.

IMPROVING TEMPORAL FREQUENCIES

Reliable monitoring of the environment relies on regular data collection. Geostationary satellites, whose speed of rotation around the Earth matches the speed of rotation of the Earth's surface, are particularly well suited to this role. Meteorological satellites provide information of a large proportion of the world every fifteen minutes, and it is this frequency which allows production of the time-lapse moving images which are now so familiar on the weather forecasts on television. Apart from their meteorological use, these data are being used for applications such as the monitoring of soil moisture in semi-arid areas.

There are also a number of meteorological satellites which have so-called polar orbits, which move rapidly relative to the surface of the Earth and provide a more detailed but more infrequent set of images. The best known of this type are the American NOAA (National Oceanic and Atmospheric Administration) satellites. The resolution of the main sensor is much coarser than that of Landsat or SPOT, the pixel size being 1.1 kms. Until relatively recently, the only use made of such data was for meteorological purposes but now they are being applied to the monitoring of the Earth's vegetation, since although they provide a very coarse overview, such data are available for every day compared with a frequency of once every sixteen or eighteen days in the case of Landsat. Furthermore, the large pixel size is actually a positive benefit, by reducing the data quantities to a manageable size for analysis when continent-sized areas are being monitored.

The data used for monitoring vegetation undergo two or three important stages of manipulation to make them amenable to analysis. Instead of using digital numbers from individual spectral bands, the ratio between two bands is used: specifically the ratio between reflective radiation in the near infrared (NIR) and Red bands. The ratio used is called the normalised difference vegetation index (NDVI) and is calculated as follows:

$$NDVI = (NIR - Red)\ (NIR + Red)$$

As mentioned earlier, an open mesophyll layer of cells in healthy green vegetation is associated with high near infrared reflectance. Conversely such vegetation tends to have depressed red reflectance because of absorption by chlorophyll. Thus the *NDVI* will tend to be high where the photosynthesis is high and the vegetation is healthy, and conversely will tend to be very low where there is bare ground and where water or clouds are present. Calculating this ratio also tends to correct for variations between distant locations as a result of differences in the height of the sun above the horizon.

The second important stage of data manipulation is to make the images obtained at different times spatially registered, so that they can be compiled and compared pixel by pixel and the differences between them directly estimated. Thirdly, it can also be very profitable to carry out a procedure of 'temporal compositing' (i.e. aggregating through time). This is especially useful when large areas are being analysed, since completely cloud-free scenes are in such circumstances often impossible to obtain. This procedure involves for a given time period the allocation to each pixel of the highest daily *NDVI* value from the period of compositing. Thus even if only one cloud-free day out of seven occurs, a valid pixel value will be obtained. Since the procedure is done on a pixel by pixel basis, almost cloud-free images are now regularly generated for the whole globe by NOAA with a resolution of approximately 20 km. The reduced resolution of 20 km compared with the original 1.1 km is performed by on-board and ground-based sampling and averaging.

The movement of a continent-wide 'green wave' of vegetation growth across Africa is shown clearly on such images (Figure 1.4.9), as a result of the seasonal northward movement of the rains associated with the migrating Intertropical Zone of Convergence. Such data are being used for tasks including biomass estimation for rangelands in Senegal and Botswana; monitoring rain forest destruction in Amazonia; and estimates of land cover for whole continents. The latter is being carried out utilising the distinctive variability of the *NDVI* through the year for a given cover type. Taking examples from South America (Figure 1.4.10) the rain forest shows relatively little variation through the year; cerrado (savanna) areas show marked seasonal changes as do the arable lands of the humid pampas, which show a somewhat different shape. Semi-arid areas display a lower range of values, and in deserts we find consistently low values throughout the year.

With the improved spatial detail visible and the improved frequency of imaging, a larger and larger range of environmental data requirements can be met through remote sensing. However, this does not mean that all those requirements can be met with current sensors. In Figure 1.4.11, applications are represented within a graphical plot of temporal and spatial capabilities. Over the last two decades, the dashed lines of current capability have been moving progressively towards the origin. Nevertheless, to satisfy fully the requirements for monitoring crop growth in the intensive lands of south-east Asia will require substantial improvements in spatial resolutions (pixel size), and for almost all locations it will demand improved frequency of imaging by higher resolution sensor systems. The latter is likely to be solved either by cloud-penetrating radars or by pointable sensors such as those on SPOT. On the latter, the sensors are capable not only of pointing vertically but also to either side, thus improving the chances of obtaining cloud-free images.

EXTRACTING INFORMATION FROM REMOTELY SENSED DATA

In the previous sections, the importance of a quantitative approach to remote sensing has been implicit throughout. Effective use of remotely sensed data is increasingly dependent on the use of equipment based on computers. In Figure 1.4.12, the main components of a typical image processing system are shown. Data input is usually from computer compatible magnetic tapes which record data from the satellite, and the images are then held on a hard disc. The latter usually has to have considerable capacity because of the large size of images. (A single Thematic Mapper image contains 330 megabytes of data.) The operator controls the image processing from a keyboard, usually by response to a menu of options displayed at the terminal. Additionally, and most importantly, the image being processed is displayed on a monitor. The

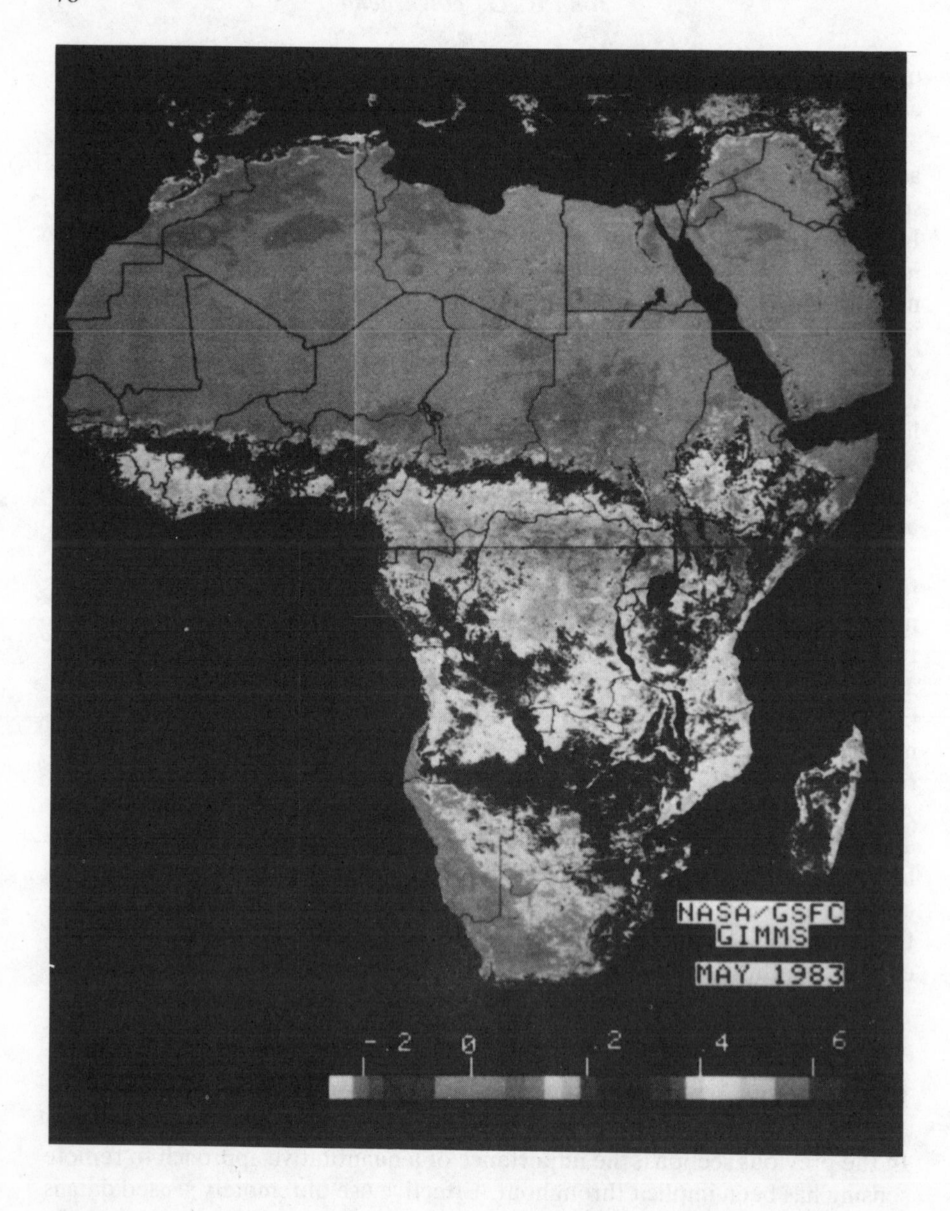

FIGURE 1.4.9
Images of the Normalized Difference Vegetation Index of Africa. This is black and white representation of an original colour image. However the northward movement of the continent-wide green wave in the Sahel is clearly visible associated with the

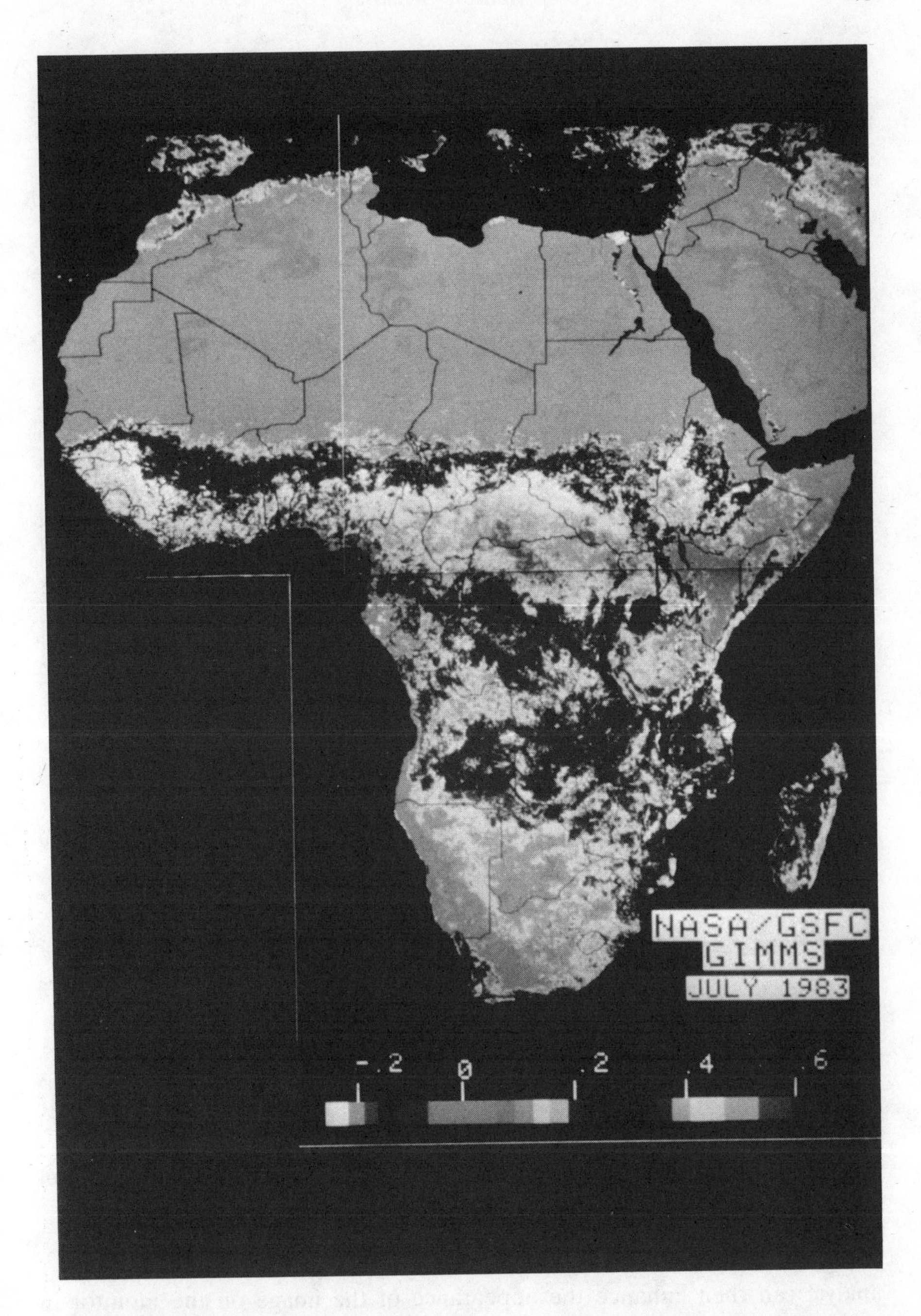

northward movement of the inter-tropical convergence zone and the resultant precipitation (Justice, C. O., Townshend, J. R. G., Holben, B. N. and Tucker, C. J. 1985. Analysis of the phenology of global vegetation using meteorological satellite data. *Int. J. Remote Sensing*, 6, 1271–1318.)

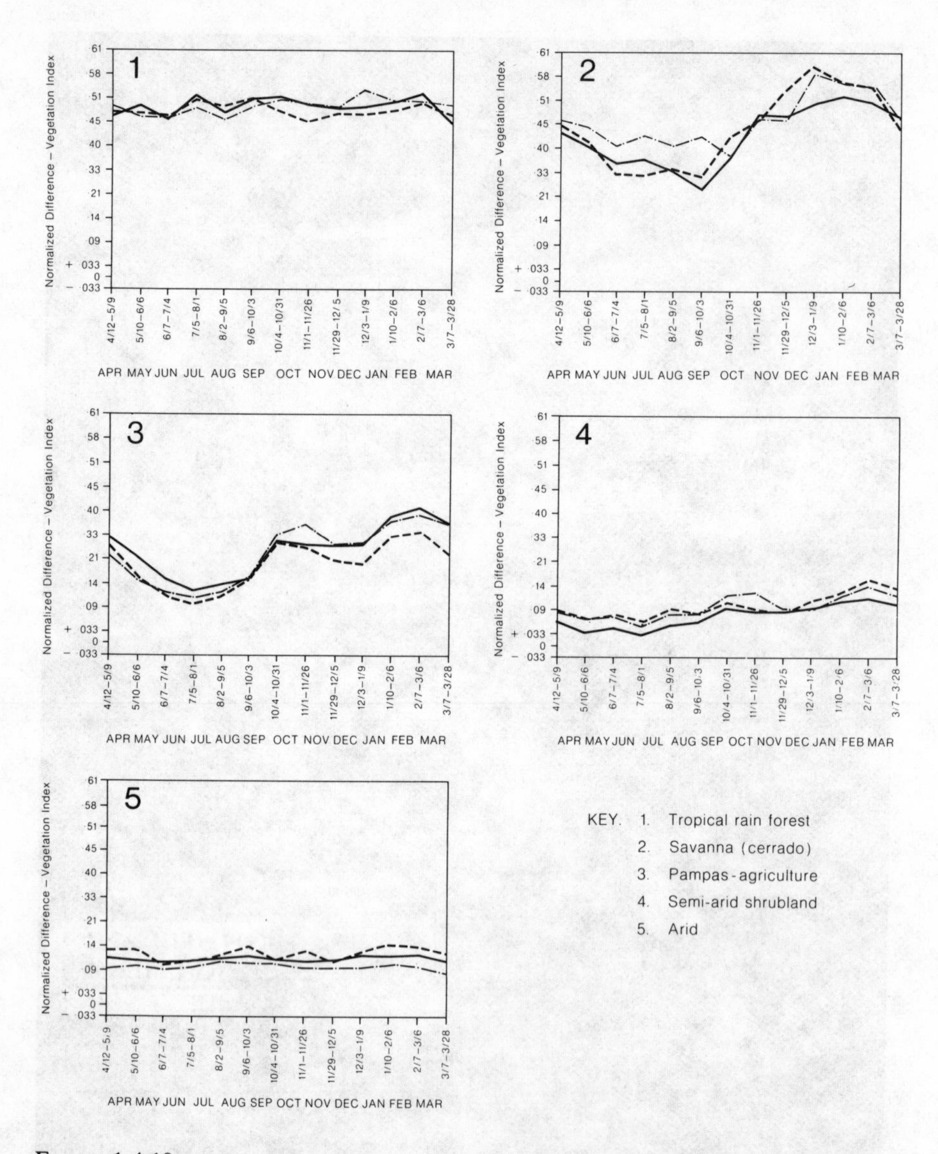

FIGURE 1.4.10
Plots of the Normalized Difference Vegetation Index for sample 3 by 3 pixel areas of cover types from South America.

analyst can then enhance the appearance of the image on the monitor in various ways to aid visual interpretation. Additional uses of the monitor are for tasks such as selection of training sites for automated classification.

Remotely sensed data gain greatly in value when they are combined with other data sources. One vital element of the system shown in Figure 1.4.12 is

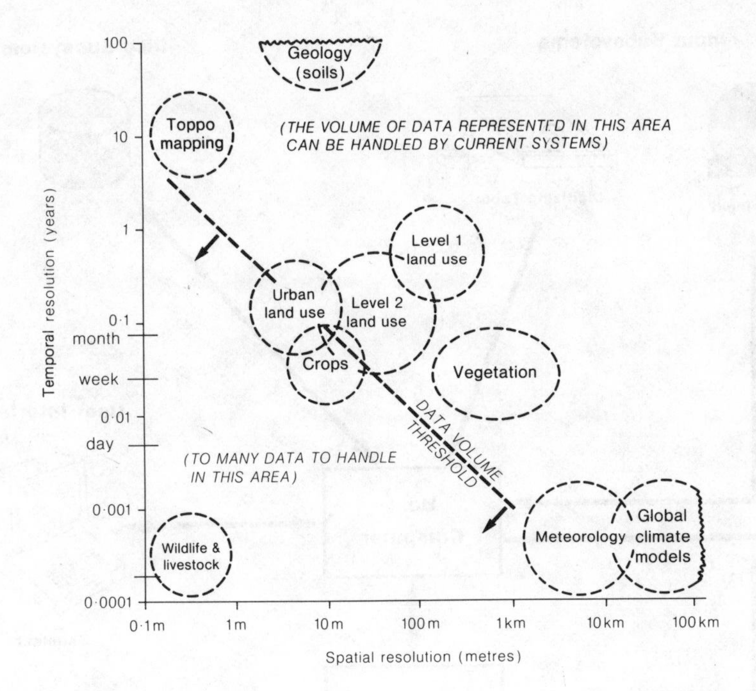

FIGURE 1.4.11
Relationship between frequency of imaging (temporal resolution) and the fineness of detail depicted (spatial resolution) in relation to various inventory and monitoring tasks (based on Allan, J. A. 1984. The role and future of remote sensing. Satellite remote sensing – review and preview. Proceedings 10th Anniversary International Conference, Remote Sensing Society, Reading, 23–30.) Note how the Data Volume Threshold is shown as moving as information technology improves.

the digitisation of other spatial data sources and their integration and manipulation with remotely sensed data. By digitisation, we refer to the representation of points, lines, and areas data by a series of numbers, often in the form of a set of *x* and *y* co-ordinates. It then becomes feasible to relate remote sensing data to characteristics such as relief information, soil boundaries, and administrative boundaries. Numerous possibilities then become feasible. Comparisons can be made directly with urban limits as shown on existing maps and those shown on more recent remotely sensed data; classification of land cover and land use can be carried out using not only remotely sensed data but other environmental properties such as relief, which improve the accuracy of classification. Vegetation amount estimated from satellite data can be related to meteorological parameters such as rainfall to improve understanding of vegetation response in semi-arid areas of the world.

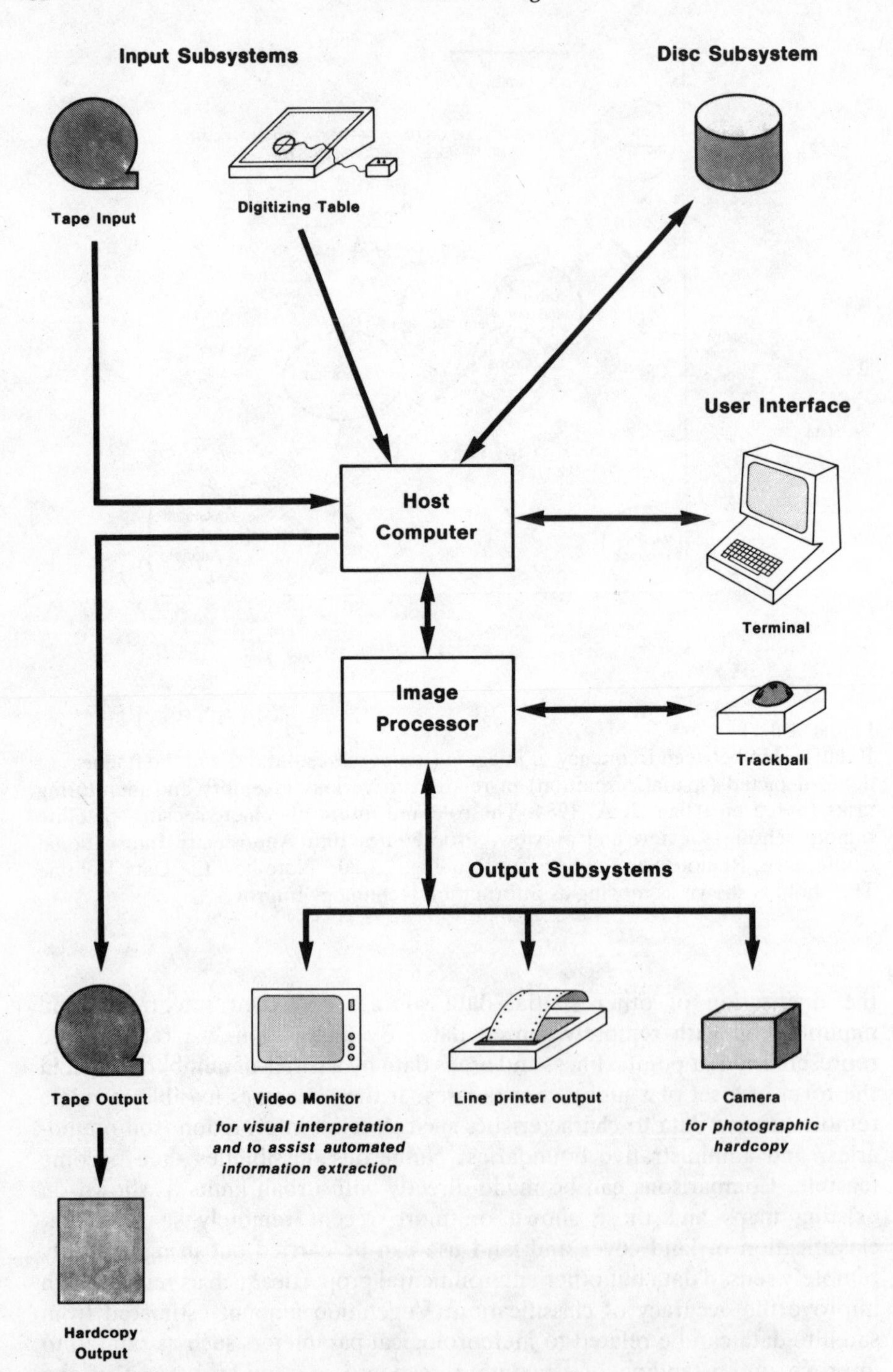

FIGURE 1.4.12 *Elements of an image processing system*

Such tasks are characteristic of Geographic Information Systems, where in this context the term 'Geographic' refers primarily to spatial location. Computer-based Geographic Information Systems ideally allow the analyst readily to manipulate the data sources available, and to examine and analyse their spatial interrelationships. Moreover, these attributes and interrelationships can then be used to make forecasts of future spatial distributions and to make estimates of characteristics such as suitability of specific land use activities.

CONCLUSIONS

The use of remotely sensed data and the related area of automated Geographic Information Systems are the most potent examples of the influence of information technology on the study of geography. The sophistication of the hardware and software involved may well appear to make the use of such techniques the exclusive playthings of the rich 'western' research worker. This view has been expressed to the writer in various forms many times and in most respects it is a mistaken view. The data sources and techniques outlined above are being used operationally by more and more users. The information they gain gives such users increasing power both nationally and internationally. If these methods are viewed as little more than technocratic 'hype' by developing countries, then the technological dominance of the western world will be increased. Moreover, by virtue of improved environmental knowledge, the potential of the developed world for economic and social dominance will also be increased. In this writer's opinion, this situation in the long run will be mutually disadvantageous. In fact, a number of developing countries have recognised the significance of these techniques and have made major investments in order to use remotely sensed data: China is a recent notable example.

Some of the techniques described are not very costly to implement. The simplest of ground receiving stations for weather satellites can be purchased by anyone for less than £2000. Even one which is useful operationally need not be enormously expensive compared say with a Landsat ground receiving station, since the rate at which data is transmitted from the satellite is low and hence the receiving dish size and size of computer equipment for data storage can be correspondingly small. Manipulation of remotely sensed data can tax the largest computer, but useful and interesting forms of data analysis can be performed on even a modest sized microcomputer.

The increasing power of computers and improving abilities of sensors on board satellites give us the power to observe, monitor, model and make predictions about our environment. Their effective use will prove a major challenge to workers from numerous disciplines for many years to come. It is incumbent on the geographic community at the interface between the human and physical worlds to contribute widely to this effort in the coming decades.

FURTHER READING

A revised edition of a very successful introduction to both principles and applications is:

Barrett E. C. and Curtis, L. F. (1982) *Introduction to Environmental Remote Sensing* (London: Chapman and Hall).

For a readable up-to-date account with an emphasis on principles rather than applications, see:

Curran P. J. (1985) *Principles of remote sensing* (London and New York: Longman).

Two collections of essays focusing on the principles and applications of remote sensing of the land surface and of meteorology and climatology respectively are:

Townshend J. R. G. (1981) *Terrain Analysis and Remote Sensing* (London: Allen & Unwin).

Henderson-Sellers A. (1984) *Satellite Sensing of a Cloudy Atmosphere: Observing the Third Planet* (London: Taylor & Francis).

The remaining references assume a more practical and detailed interest in the subject. The first is a do-it-yourself introduction (at a price), whilst the second is a useful introduction to digital image processing requiring only a modest level of mathematical ability:

Short N. M. (1982) *The Landsat Tutorial Workbook* (Washington DC.: NASA).

Schowengerdt R. A. (1983) *Techniques for Image Processing and Classification in Remote Sensing* (London: Academic Press).

Finally, a large compendium of principles and applications which is, perhaps, beyond the needs of the average reader but has become the standard reference source is:

Colwell, R. N. (1983) *Manual of Remote Sensing*, 2 vols (Falls Church VA: American Society of Photogrammetry).

References cited in figures

Allan J. A. 1984. 'The role and future of remote sensing. Satellite remote sensing–review and preview', *Proceedings 10th Anniversary International Conference*, Remote Sensing Society, Reading, pp. 23–30.

Justice C. O., Townshend J. R. G., Holben B. N. and Tucker C. J. 1985. 'Analysis of the phenology of global vegetation using meteorological satellite data', *Int. J. Remote Sensing*, 6, pp. 1271–318.

Hardy J. R. 1981. 'Data collection by remote sensing for land resources survey', in Townshend J. R. G. (ed.) *Terrain Analysis and Remote Sensing* (London: Allen & Unwin) pp. 16–37.

Markham B. L. and Townshend J. R. G. 1981. 'Land cover classification as a function of sensor spatial resolution', *Proc. Int. Symp. Remote Sens. Environ.*, 15th, Ann Arbor, Michigan, 1075–1090.

Sadowski F. A. and Sarno J. 1976. 'Forest classification accuracy as influenced by multi-spectral scanner spatial resolution'. Report No. 109600-71F (Ann Arbor, Michigan: Environmental Research Institute).

Thomson F. J., Erickson J. D., Nalepka R. F. and Weber F. 1974. 'Final report multi-spectral scanner data applications evaluation'. Vol. 1. User applications study. Rep. No. 102800-40-1 (Ann Arbor, Michigan: Environmental Research Institute Michigan).

1.5
Mechanisms, Materials and Classification in Geomorphological Explanation

W. Brian Whalley
The Queen's University of Belfast

Present day geomorphological research is the explanation of landscapes and landforms: what they are, how they function and have developed. This includes a variety of approaches some of which are shown in Figure 1.5.1. Research is a much more complex undertaking than giving names to landforms or even the A-level study of 'processes', both in terms of what we mean by 'process', and how 'explanations' are approached. It is worthwhile examining some reasons for this, as it shows the increasing complexity of the subject and leads to ideas of how geomorphology can be studied.

There is a need to examine carefully both what is taught in geomorphology at school and its relevance beyond. There appears to be a disparity between the necessity to teach the basics about landforms and develop landscape awareness, and the desire to introduce the latest material percolating down from research frontiers. This second trend may be an attempt to attract students to career training in physical geography; it may even be a subconscious desire to show that the subject is 'scientific'. Such a conflict is not new but it may have peculiar results.

Project work is a significant part of many syllabuses and this can place an undue emphasis on 'process' studies which neither pupil nor teacher may fully appreciate. Much 'process' work is merely monitoring effects (even at research level) and not strictly explaining what is happening in the landscape. Elementary hydrology and fluvial geomorphology tend to become case studies because events happen relatively quickly and can be measured easily. Similarly, beaches show rapid changes, but these are studied by means of simple sediment sampling and levelled profiles. Such projects rarely develop into studies of the dynamics and mechanisms of landscape behaviour, particularly where changes are slow. Furthermore, school 'process studies' may concentrate upon the very small scale with little appreciation for the landscape as a whole, and perhaps upon restricted aspects of geomorphology. Significant advances are being made in all branches of geomorphology but

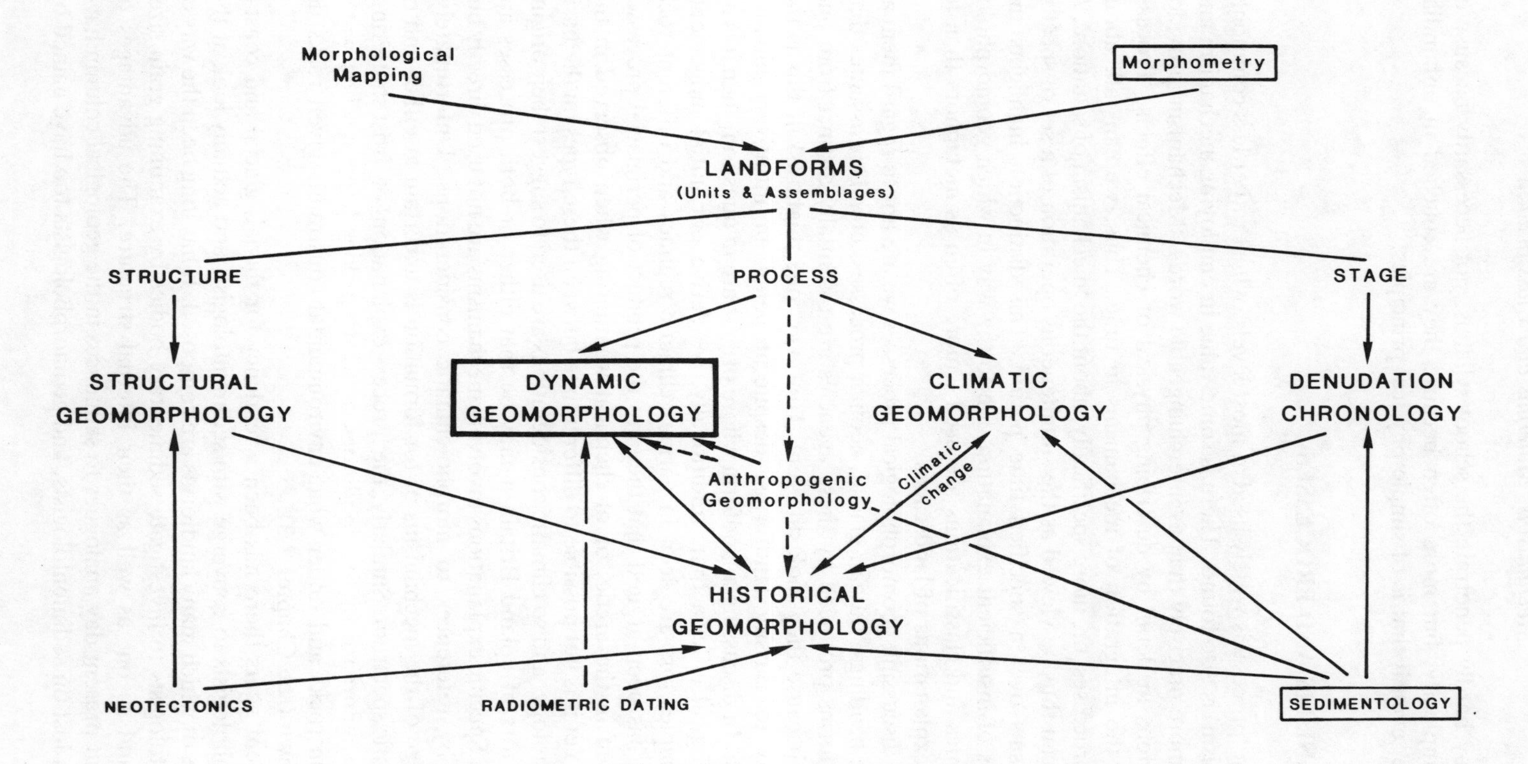

FIGURE 1.5.1 *A view of the component parts of geomorphological study; modified from Jennings (1972)*

these may hardly penetrate the school syllabus, not necessarily because of their complexity, but more often because they are couched in unfamiliar language, even if that is of simple physical principles.

MECHANISMS AND PROCESSES

The term 'process' is widely used at most levels, whilst rather less commonly 'mechanism' is also found. There is some value in employing mechanism and process more precisely than as interchangeable words. 'Mechanism' provides explanations in detail by describing physical or chemical effects. 'Process' signifies the integration of mechanisms in time. Thus, we can talk about 'fluvial processes' or, more specifically, about the 'mechanism of saltation'. A process can thus be viewed as the simultaneous operation of a set of specific mechanisms over a specified time period. This distinction highlights the problems of insufficient explanation due to the way in which geomorphological names and classifications are used – more often as metaphors than as specific explanations of landscapes.

When discussing geomorphological processes we are apt to regard them as textbook headings; fluvial, glacial, aeolian processes, etc. This assumes that the landforms produced by these agencies are essentially distinct from one another because they look different. However, one might ask if this is the only way to classify landforms, despite it being an intuitively obvious approach. In discussing a landform whose origins are in question, then it is by morphological criteria that, traditionally, we derive our initial judgement. For example, tors are seen in various climatic regimes and different rock types. It has been argued that they are the product of periglacial processes when seen in the arctic, or of chemical weathering when observed in the tropics. Yet it is not possible to differentiate between these types on the basis of morphology, and so climatic considerations are used to suggest their origin. For the tors of upland Britain it may be that either or both processes are possible. Specific explanations involving mechanisms are not used directly but implied by references to morpho-climatic considerations. Unfortunately, knowledge of the mechanisms of tor formation is uncertain at either end of the climatic spectrum. Similarly, the 'process explanation' of 'frost shattering' belies an ignorance of the conditions as to what does happen when water freezes in rocks and under what environmental constraints given rocks do break down (see Chapter 2.3).

In recent years there has been a tendency for fluvial, glacial and coastal geomorphologists to converge with sedimentologists and actually look at the materials of which many landforms are composed. It has long been the way of sedimentologists to investigate sedimentary bodies by examining grain size, shape, and so on, as well as their internal structure. The advantages of looking at present day analogues to sequences in the geological column have not been lost on sedimentologists, and geomorphologists too have gained by detailed interpretation of sediment erosion, transport and deposition mechanisms. In the USA, geomorphologists and sedimentologists share a geological affiliation.

Not all features of interest to geomorphologists can be examined by sedimentology; rock weathering, for example. Nevertheless, investigating the mechanics of processes is still just as important, as it aids an understanding of what is going on. Thus, studies of mechanisms help to improve explanation and add to clues from morphology and distribution. This notion of new approaches developing to supplement, rather than supplant, traditional techniques can be seen in several chapters in this book.

DESCRIPTION AND EXPLANATION

The distinction between description and explanation is sufficiently important to warrant further discussion. Studies of the areal distribution of landforms or phenomena are traditionally geographical in outlook, and generally consist of descriptions of those phenomena. Yet, no matter how many ways you describe a landform such as a drumlin, you cannot thereby explain how it was formed. To say 'by glaciers', takes a naive process approach. The mechanisms of formation, how the glaciers behaved, is left unsaid. We need to know something of the materials of which the drumlin is made, as well as the properties of the glacier. Similarly, a map of drumlin distribution may help explanation, but will not provide that explanation by itself. In this way, morphometry says little about mechanisms and hence about processes. On the other hand, morphometry and cartographic techniques may be used to test ideas about formation or suggest lines of investigation, and perhaps indicate the broader controls upon the mechanisms involved. One must beware, therefore, of reading more into descriptions than is justified. To take an extreme example, if climatic conditions are and have been unsuited to glaciation, it is not much use looking for evidence of relatively recent glacial landforms. On the other hand, where there is marginality of climate for the formation of a specific landform, then this does suggest an interesting area for investigation. Classifications tend to be rigid, but the fuzzy edges of natural sequences provide happy hunting grounds for those geomorphologists interested in explaining mechanisms and distributions rather than just mapping them. Unfortunately, such situations, which may delight a research geomorphologist, generally provide far too complex a task for school-level investigations.

The so-called 'ergodic hypothesis' is sometimes invoked in geomorphological explanations. Its basic idea is that temporal distributions are assumed to have spatial equivalents, so that by examining the latter, a sequence of events in time can be constructed (see Fig. 1.5.1). There are considerable assumptions, and thus dangers, in this approach, yet the more we know about how materials react to different stress changes, the better we will be able to explain landform changes in space and time. Such 'environmental stresses' may be actual stress (as in slope failure), or the complex changes resultant upon man's activity, intense rainfall or climatic changes, etc. Overall, therefore, considerable variability in the appearance and distribution of landforms may result from the combined effects of these 'stresses'. It is necessary to consider basic mechanisms and the various, often changing, boundary conditions of landforms production. If we now link the rigidity of classification boundaries to the variability of landscape features in space and time, it is clear that true explanation in geomorphology is a highly complex process ill-suited to school levels.

Processes may be viewed either as the agency itself (for instance rivers), or the type of stresses which produce changes in that environment. Shear stresses are the predominant type of stress in geomorphology. It is shear stresses which produce movement of sand grains on the bed of a river or across a beach. A simple equation involving average shear stress may give a physical representation of what is happening on a river bed, albeit in a simplified way. It is possible to use theoretical and experimentally determined formulas to show why sediment of a specific size should be transported by a river under some circumstances but not under others. Unfortunately, in the quest for 'complete' explanation it then becomes necessary to provide explanations of shear stress, unit weight, etc. A fundamental question which needs to be raised thus concerns the level at which ideas should be taught, given the prevailing aims of geomorphology in schools. Perhaps it is more important to provide the basic vocabulary and simple explanations of processes than to try explanations in detail of mechanisms which are better left to university or college.

THE SIMPLIFICATION OF GEOMORPHOLOGY?

Like any growing subject, geomorphology has become increasingly complex and diverse. It is difficult for an individual to write an advanced general geomorphology text as it is not possible to keep abreast of the research literature for the whole subject. New ideas do seep through the educational system, until it is hardly possible for all branches of geomorphology to be taught – let alone learned – even at Advanced Level. In this way, geomorphology approaches the position faced by physics and chemistry syllabuses years ago, in that only a portion of what might be considered at a given level is actually taught or examined. Some implications of this are clear. For instance, it is doubtful whether undergraduate courses in geomorphology cover all aspects of the subject, certainly not in any depth. Not only is there a shortage of time, but the breadth of understanding needed in some aspects may preclude all but a superficial coverage until option courses allow detailed study of one part of geomorphology. Questions must therefore be raised about the need to present so much geomorphology in schools, as well as about the necessity to pack in 'process' studies as well.

Is it possible to simplify geomorphology in the light of the complexity and increasing volume of material to be found in textbooks at all levels? One solution is to reduce the amount of thematic geomorphology taught by dropping certain subject areas. One, rather radical, alternative is to use physics and chemistry as well as mathematics, at a fairly simple level, to gain a certain simplification of geomorphology as a way to increase the power of explanations. The trade-off is that these subjects must be learned at some stage. For most school students, geography is taken as an alternative to science and so this route is unlikely to have many adherents. One might argue, however, that instead of learning statistical techniques in school geography that time could be better spent on the rudiments of chemistry and physics.

If advanced school physical geography is to provide a better understanding of processes and simple mechanisms, then showing the way materials react to 'systems stresses' is one way to simplify. This could be likened to the way in which valency theory or a knowledge of carbon bonding makes sense in chemistry. The strictly 'reductionist' view of the physical, chemical and biological basis for physical geography may have to be left for courses beyond school, although it does offer an easier understanding of geomorphology – provided that the supporting scientific skills have been mastered. The old method of teaching geomorphology through names of landforms can undoubtedly be boring unless leavened with some judicious fieldwork. Conversely, it must be shown that a project which seems to need measurements of all the pebbles on Chesil Beach relates to proper explanation of a phenomenon. Student motivation needs to be mixed with geomorphological purpose.

The address of Professor S. Gregory to the Geographical Association in 1978 on the 'Role of Physical Geography in the Curriculum' shows one of these views:

> It may be possible to *describe* the physical characteristics of the earth's surface without these [physics, chemistry, biology and mathematics], either in words or in diagrams, pictures or maps, but to arrive at an *explanation* and an understanding of exactly how and why these characteristics are as they are – and how they will react or change in the future under various influences and pressures – is not possible without such basic scientific knowledge.

Gregory was at pains to point out that he was referring to all levels of education and not just at universities:

> much physical geography has been 'explained' not in terms of an answer to the question 'why?' but rather in answer to the question 'how?'. An

account is given that from some initial state this event was followed by that event, etc. until the present state has been achieved – thus an historical succession of specific events replaces a scientific law as the basis for explanation.

Hence, the application of physical laws provides simplification in terms of more general explanations. Although we have already said that this is a route best suited to levels beyond school, it is worthwhile exploring some ideas to see if a middle road might be found.

How can scientific principles be applied to geomorphology? Could they simplify things in practice? What would a 'simplified' textbook look like? Such a text would be a general science text with geomorphological examples. It would be necessary to outline concepts such as basic units, mass, length and time and then build, via forces and friction, through stresses and strains to ideas of strength, elasticity, plasticity and flow. It would then be possible to view mechanisms of sediment entrainment and deposition as a more unified whole with many aeolian, fluvial, coastal as well as glacigenic landforms resulting from the interplay of such ideas. The mechanisms that need to be understood in geomorphology follow from the knowledge of how materials behave. This progression is akin to that employed in teaching physics or chemistry. Strahler was propounding very similar ideas in 1952:

> To place geomorphology upon sound foundations for quantitative research into fundamental principles, it is proposed that geomorphic processes be treated as gravitational or molecular shear stresses acting upon elastic, or fluid earth materials to produce the characteristic varieties of strain, or failure, that constitute weathering, erosion, transportation and deposition.

In Britain, the approach to geomorphology by way of stress–strain relationships of materials had little effect until recently. The quantitative research hoped for by Strahler initially produced morphometric analysis rather than evaluation of landscapes with respect to materials' properties. Strahler called his paper 'The Dynamic Basis of Geomorphology' and this is a good way to describe the approach; nevertheless, it is his work on stream ordering which has been passed downwards. However, it must be questioned seriously whether stream ordering, and morphometry in general, contributes much to an understanding of landscape formation. Figure 1.5.1 shows ways in which 'process studies' can be complemented in general geomorphological explanation. Which combination is taken depends upon the question in hand. Teaching the subject presents its own problems because of this diversity of possible approaches. To follow the morphometric route because it happens to employ numbers or statistical techniques is being unnecessarily restrictive.

Another quotation from Gregory's address is helpful when we consider the problems of teaching geomorphology that arise as the technology and complexity of research becomes progressively divorced from the basic vocabulary learned in schools:

> Just as it is useless to teach about physical geography in terms of concepts for which the particular students or pupils do not possess the basic science, so it is equally wasteful and intellectually unsound to avoid – or fail to utilise – the basic science that they *do* possess. I wonder how many teachers of geography, at all levels, give time to finding out exactly what scientific concepts are being taught by others to their geography class . . . reply to the question 'what shall I tell them?' . . . 'tell them the truth; or as much of it as they can be expected to understand'. In this way, simplification (which is essential at all levels) need not lead to falsification.

It is most unlikely that such a view will be widely implemented within the current framework of school physical geography teaching. Nor is it hard to see that by far the majority of teachers may find it difficult to pass on some of the modern ideas of landform development that they have learned because the present curriculum may given them little opportunity to do so. Teachers might like to give their more receptive pupils a glimpse of what is in store if they continue their studies, but the impact can only come by keeping the presentation simple. It should be possible to tell a valid story about landscapes by synthesising the way in which 'environmental stresses' operate in landscapes on specific landforms. However, to do so does require some knowledge of basic science.

PROPERTIES OF MATERIALS

Geomorphological knowledge can help alleviate many environmental problems concerned with soil erosion, flooding, road-building as well as natural hazards of many kinds. There can be little doubt that, as geomorphologists learn more about the behaviour of materials they should be better fitted both to understand landforms and to converse with engineers and conservationists. Let us consider a 'simplified' approach to geomorphology in the light of materials' properties in a little more detail. Some examples will show just how the desired simplification can be achieved and point towards a better understanding of mechanisms and processes. Uniformitarianism, the idea that 'the present is the key to the past' provokes particularly pertinent thoughts. The first is that the nature of both materials and the agencies which produce geomorphic features have not changed over geological times: that is, the laws of physics and chemistry have not changed. However, it is not true to say that the rates of activity have not varied. One of the present strengths of geomorphology is an increased understanding of the rates of activities and the means to investigate them. Interest is now being paid to 'magnitude–frequency' concepts to ascertain just how much landforms might have been different at time intervals which stretch from human life spans to geological time.

As far back as 1945, W. H. Ward of the Building Research Establishment argued to geographers that soil mechanical principles could be used to help the understanding of slope behaviour. It is only surprising that ideas from geotechnics did not really find much space in research papers in geomorphology until the early 1970s and in textbooks until the mid 1970s. Although there is a pure research side to the use of mechanics in geomorphological explanation, it is also true that the application of geomorphology to human problems actually becomes quite easy with such an approach.

In 1966, the Japanese geomorphologist Eliju Yatsu published his book *Rock Control in Geomorphology* which looked at the way materials behaved in a geomorphological context. Whilst lip service was paid to this book by other researchers, it was largely neglected. One may thus ask whether the time is now more opportune to reconsider the advantages of a geomorphological study of materials' properties. After the somewhat hesitant start just mentioned, there is now clearly a move in this direction. An example will suffice to show that, not only does it work in its own right, but that it can help solve problems in conjunction with traditional techniques.

The thermal properties of glaciers are now fairly well understood, and it is possible to look more critically at processes of glacier erosion and till deposition than was the case thirty years ago. This has only been feasible with a knowledge of the physics of ice and the mechanical properties of tills. For example, it has been suggested that using the idea of a glacier frozen to the bed in one area and not frozen (i.e. having meltwater present) in another, coupled with ideas of pore water pressures in soils, models of subglacier debris entrainment can be produced. It is clear from such simple models that Late Quaternary ice masses over the British Isles were not behaving in the same way as most of the present Antarctic ice sheet. The study of materials has produced a general and far-reaching conclusion, as can be seen by reference to Chapter 2.3.

In slope studies, the usefulness of soil mechanical principles has now been established. For example, engineers with a geomorphological bias have shown how the clay landscapes of much of southern England have been moulded by periglacial solifluction and shallow failures in the period since deglaciation. The significance of this example is that it shows how the morphology has been created, not just the end-product of a vaguely conceptualised 'process'.

In modelling, it is necessary to provide the basis on which the model will be built, and thus the mechanisms operating in a specific process are needed. This is obvious in the case of deterministic modelling: for instance, a 'flow law' for ice must be used if glacier flow is examined. The naive view of a glacier being a 'river of ice' fails because the mechanics of water and ice flow are quite different in most respects. That there may be several models of flow, applicable under different stress conditions, does not matter as long as an appropriate one is chosen for the task in hand. Without this knowledge, it is not possible to start modelling, however simple; properties of materials provide a link with reality. If, for example, a computer model of slope failures through time is to be made, then mass movements such as creep, flows and slides at different rates and responding variously to environmental changes are involved. It is absolutely necessary to know how the materials will behave under different pore water pressures, slope changes, etc. Such a model might resemble a more complex and time-dependent version of the slope failure models employed by engineers to determine the likelihood of failure.

It should not be thought that geomorphology must always begin with materials and end with mathematical models. The discrete boxes used in Figure 1.5.1 do not mean that there are exclusive paths to an explanation. Strahler was well aware of the necessity to incorporate a study of mechanisms with the historical aspects of the subject. Similarly, Gregory was not deriding historical aspects of geomorphology as such, but the divorce of historical explanation from physical principles.

CLASSIFICATION AND GEOGRAPHY

Do we look carefully enough at landscapes? There are several reasons why we may not be doing so, one of which is that there is a significant tendency to rely too much upon classification. Inevitably, in a world full of complexity, there is a need to apply a systematic examination and classification of what is seen to provide a basic vocabulary. This, however, leads to preconceptions about how things are ordered or how they work. In short, an explanation is implied.

There are distinct problems of scale of investigation which can be seen clearly when we look at classificationary approaches to physical geography. Köppen's climatic classification finds its equivalent in climatic geomorphology where there is a rough equation between the climate of an area, the 'processes' presumed to operate there and hence the landscapes seen. Climatic geomorphology is a potentially useful approach, but because it operates at a rather general level it does not take much account of relict landforms. Such relicts would be, for instance, landforms developed under a different set of climatic conditions than at present, and would thus not fit the climatic classification. Furthermore, geological controls are an extra complication not easy to incorporate into a climatic geomorphological classification. Thus broad (world or continental) views of denudation are often only approximations to a global classification.

Slope classifications, either morphological or of slope failure type, are similar products of generalisation, but at a smaller scale than that generally assumed in climatic geomorphology. At a still smaller scale, the classification of micro-weathering forms on rock surfaces can be extended down to detailed descriptions of micro-topography on individual sand grains. One can validly ask which scale, as well as which type of classification, is the most appropriate for geomorphological understanding? This point is addressed in the next section.

Classifications can be seen as restrictive by examining a textbook on physical geography. Chapter headings such as 'weathering', 'slopes', and 'fluvial processes' show how classifications rule our geomorphological lives. Classifications are useful as a response to real world complexity, but can lead to several problems which, if they cannot be avoided totally, must be recognised and their more insidious effects minimised. First, classifications tend to consolidate ideas. If such models are substantially 'correct' then they may be helpful in trying to understand landscapes. However, there may be flaws in the classification. Landforms may not have a single, causal explanation and a classification may thus consolidate unhelpful views. A good example is that of the Davisian 'methodology'. The embracing classification of 'normal' landscapes presumes 'normal' processes, whilst the 'arid cycle' presumes arid processes, implying that normal and arid landscapes are quite different and were produced in different ways. It is unlikely that Davis thought that they were totally different, but there is a tendency for the classification to suggest that they are. Not that the Davisian view is necessarily wrong, but it has to be recognised that other concepts of landform development may be more useful. Does our more modern 'process approach' harbour similar pitfalls?

As a further example, a size-morphological classification of glaciers (e.g. glacier versus ice sheet) neglects the very important influence of temperature effects on the properties of ice and glaciers, yet the traditional view of glacier studies is to look at them in terms of size. This is not wrong, but it is restrictive. In many cases multiple classifications are needed simultaneously, and the distinction between 'cold' and 'temperate' ice masses has been of great value in the interpretation of both ancient and modern glacigenic sequences at all scales of investigation. The aspect of greatest importance is how ice as a material behaves and, in turn, how the till beneath responds to the imposed stress field. Traditional classifications of glacial deposits can be rather misleading in this explanatory context.

A second, related, tendency is that of neglecting other important attributes in a classification. For example, function may be subservient to morphology, especially where landforms are classified. There is a recognition that different processes may produce similar-looking landforms (the idea of 'equifinality' or 'landform convergence'). Thus, we might ask whether tor-like forms have several origins according to where they are found, or whether they are the product of one distinctive 'process' in an area.

Thirdly, classification necessitates the giving of names to class members. There is then a subtle tendency to assume that this nomenclature is sufficient to explain what is going on. This overlooks the fact that classification is an aid to explanation, but is not explanation in itself. As an example, 'frost shattering' is assumed to be both a mechanical weathering process and the mechanisms of volumetric expansion of water on freezing. Recent research has shown the latter to be a simplification (there are several possible mechanisms), but it does not mean we inevitably have to abandon the classificatory term. A researcher in this field may appreciate the context and thus avoid misusing the term, but school texts might not. This is only one illustration of the gulf between the apparently neat formulation at school and the complexities revealed by a detailed study of mechanisms.

SCALES OF EXAMINATION

Geomorphological explanation may involve a complexity of study and examination which would deter the physicist or chemist (who might well prefer a simpler system) or the engineer (who would perhaps want a quick and cheaper answer). For the schoolteacher, moreover, there is little time to delve into complexity because the level of understanding needed in a particular curricular context may not demand it. However, this does not mean that the problems should be ignored. Richard Chorley highlighted some of this conflict when he suggested that, with respect to weathering:

> Instead of boldly concentrating on the immanent characteristics of the products of weathering and their implications for slope development, or on observed or inferred rates of weathering, or on the geometrical features and geomorphological implications of the weathered mass, geomorphologists have too often assumed the role of inept biochemists, and become preoccupied with process in a field so complex that no competent biochemist would have the temerity to attack it at that scale.

We can see that this view is pertinent to some of the general problems already remarked upon. However, if it is necessary to investigate the mechanisms of geomorphological processes, how can we avoid examining them in detail? The biologist may be more sympathetic to the plight of the geomorphologist, as that subject too is characterised by diversity and complexity. Yet biology has gained enormously from collaborations with physics, chemistry, mathematics and engineering. This is not to say that geomorphology has not had such imports, but such help is only now on its way in a manner which will help broaden and deepen understanding of what is going on in a landscape. The introduction of ideas from soil mechanics to slope stability in the short term, as well as landscape in the long term, has had a significant bearing in landscape explanation. We are now asking questions about landscape behaviour at all scales.

Implicit in Chorley's statement about weathering is the belief that the geomorphologist need not, and indeed should not, investigate the microscale and that the preferred investigation should be at the mesoscale. Although appealing, such a viewpoint is restrictive. The central core of geomorphology consists of basic processes, landforms and landscapes. However, to understand how they change in time (other than merely describing these changes), a study of mechanisms and materials is required as one of the routes to 'explanation'. There is, therefore, a case for the investigation of mechanisms at all scales, and this invites collaboration with scientists from other disciplines or borrowing of techniques. It is, of course, necessary for the 'small-scale geomorphologist' to be aware of problems at all scales. Without such combinations of scale there seems, all too often, to be explanation of the handwaving 'because it is there' type which has little regard for the geological processes involved. Geological interpretations range widely in scale from microscopic examination of rocks, through interpretation of sedimentary or metamorphic features ranging from those requiring a hand lens, the unaided eye in the landscape to a study of maps and aerial or satellite photographs. If the geologist finds this scale range of examinations essential, why does this not apply to the geomorphologists?

For instance, scree consists of rocks which have fallen from a cliff, yet is is by no means clear how long it has taken for the scree to build up and what conditions were responsible. Climatic change may mean more than an increase in the number of freeze–thaw cycles the rocks have been subjected to; water conditions are also responsible. Moreover, screes can form even where rocks are apparently not subjected to frost shattering. Thus, to explain the significance of screes in a landscape involves some considerable knowledge of the weathering conditions to which the cliff has been subjected. The appropriate scale of investigation depends upon the purpose of the study. Yet even explanation of the geomorphology of a tract of land is likely to gain as the result of detailed investigation of mechanics and materials properties. Figure 1.5.2 shows how one item in an urban landscape, a church, can be examined at increasingly smaller scales. The building style and construction as well as the nature of the masonry plays a part in the way the building is appreciated as a functional part of the landscape. Similarly, a cliff, as part of a mountainscape, benefits from study at increasingly smaller scales. Although individual aspects can be studied separately and in detail, when added together the result is informative about the wider scene.

Small-scale investigations can help explain very large features. The residual weathering deposits collectively referred to as duricrusts (e.g. laterite and the siliceous cemented rocks known as silcrete) cover large areas of present day semi-arid areas. They have also been important in the past, as is shown by the way that silcretes form residual capping rocks in the Paris Basin and are seen as the sarsens of southern England. The elucidation of various models of formation can only come from a detailed examination of the materials and mechanisms involved and, when coupled with information on occurrence, help to present a picture of landscape evolution in southern England. Other studies of duricrusts using microscopic, stable isotope and chemical methods do relate back to the larger scale and can be used to answer a host of questions about landforms and landscape development.

FIGURE 1.5.2

(a)

(b)

FIGURE 1.5.2 *contd.*

(c)

(d)

FIGURE 1.5.2 *The three views of a church and its components roughly match the view of a cliff and the scree fallen from it*
In both cases, changes in scale of examination produce different appreciations of the

FIGURE 1.5.2 *contd.*

(e)

(f)

whole structure. An extension to the microscale is similarly valid as it is then possible to see, for example, how both cliff and building react to 'environmental stresses' such as weathering. Changes at the micro-level may ultimately be visible at the largest scale.

CONCLUSIONS

In stepping back to view the spectrum of geomorphology teaching over the past eighty years it is as well to remember that, at one time, Davisian (and Penckian) thinking was at the research frontier; it was then passed down and subsequently taught in schools. Apparently, the Davisian cycle has now been displaced from some schools altogether. This shows that research ideas do percolate downwards and can be replaced, and that it may not be fanciful to predict a science-based approach to geomorphological materials being exploited in the next twenty years.

A knowledge of the way materials behave is central to many problems of explaining landforms. Geomorphological materials – rock, soils, water or ice – react to environmental conditions in a way which, in theory, can be replicated and modelled. A frequent problem is that the conditions within which this behaviour takes place may not be well understood. Knowledge of materials can best provide explanations when landforms are viewed at a variety of scales, the most appropriate of which will depend upon the problem in hand. Geomorphology, without this knowledge, is restricted. However, the needs and methods of the research geomorphologist are now at considerable variance with geomorphology taught at schools. The aims of both are still to explain the landscape and individual landforms, but their audiences are now quite separate. Perhaps we should be aware of the potential dangers of teaching too much 'process' before the basic mechanisms are appreciated.

Acknowledgements

I thank my colleagues Mike Clark, Anne Gellatly, Julian Orford and Bernard Smith for discussions and constructive criticism during the preparation of this chapter.

FURTHER READING

Reading can profitably be directed towards two general themes, the first dealing with broad issues and the second introducing specific details of a materials-based approach. The following three references open the discussion by raising a variety of methodological matters:

Chorley R. J. (1978) 'Bases for theory in geomorphology', in Embleton C., Brunsden D. and Jones D. K. C. (eds) *Geomorphology, Present Trends and Future Prospects* (Oxford: Oxford University Press).

Gregory S. (1978) 'The role of physical geography in the curriculum', *Geography*, vol. 63, pp. 251–64.

Jennings J. N. (1972) '"Any milleniums today, lady?" The geomorphic bandwaggon parade', *Australian Geographical Studies*, vol. 11, pp. 115–33.

Next, various aspects of materials-based explanation are considered, starting with two overall statements of the potential:

Whalley W. B. (1976) *Properties of Materials and Geomorphological Explanation* (Oxford: Oxford University Press).
Pitty A. (1985) *Themes in Geomorphology* (Beckenham: Croom Helm).

Finally, the following sources introduce a variety of detailed examples of the significance of materials at various scales, starting with weathering and moving on to erosional and depositional environments:

Goudie A. S. and Pye, K. (1983) *Chemical Sediments and Geomorphology* (London: Academic Press).
Wilson R. C. L. (1983) *Residual Deposits: Surface Related Weathering Processes and Materials* (Oxford: Blackwell Scientific).
Brunsden D. and Prior D. B. (1984) *Slope Instability* (Chichester: John Wiley).
Caine, N. (1983) *The Mountains of Northeastern Tasmania* (Rotterdam: Balkema).
Boulton G. S. (1975) 'Processes and patterns of subglacial sedimentation: a theoretical approach', in Wright A. E. and Moseley F. (eds) *Ice Ages: Ancient and Modern* (Liverpool: Seel House Press) pp. 7–42.

Section 2
A Role for Environmental Process

2.1
Hydrological and Fluvial Processes: Revolution and Evolution

D. E. Walling
University of Exeter

CHANGES AFOOT

Down yonder green valley there flowed a meander
that drifted out onto a wide peneplain.
But our Ken Gregory he was not contented
to see such simplicity within his domain.

If my memory serves me correctly, this was the opening verse of a satirical jingle presented at the Annual Dinner of the University of Exeter Geographical Society in 1964. It points to the undergraduates' perception of notable events over the previous years, and more particularly to changes in the approach to, and teaching of, fluvial geomorphology associated with the arrival of a new member of staff – one of the editors of this volume. This was a change which would have been mirrored in many other university departments in the early 1960s. It reflected the movement away from the traditional Davisian approach (which although paying lip service to 'processes' did so in an essentially qualitative and frequently misleading way), towards an increasing awareness of the need for precise quantitative information on the nature and rates of operation of these processes. This movement was closely linked with a shift in emphasis away from the study of landform evolution and denudation chronology to interest in the functioning of the present day landscape and the rate of operation of contemporary processes, which had been stimulated by the work of such North American geomorphologists as Arthur Strahler and Luna Leopold. Some would argue that such a change in focus was not without drawbacks, because the overall aim of understanding and explaining landform evolution was to become subordinate to the study of contemporary processes in their own right. However, it is possible to reconcile these two attitudes, at least partially, by suggesting that the subject had reached a stage where it was necessary to develop new understanding and skills before returning to a new assault on the long-term objective. There would clearly be a need for some to devote all their efforts to developing this new understanding, whilst others pondered the long-term strategy.

Parallel with this upsurge of interest in contemporary fluvial processes, university geographers began increasingly to view the subject of hydrology as offering a worthwhile focus for geographical study. Dr John Rodda and Professor Roy Ward, whose names have been closely linked with the development of hydrology in the UK, commenced their doctoral research in the Geography Departments at Aberystwyth and Reading in the late 1950s and hydrology was, for example, introduced into the undergraduate syllabus at Exeter in 1965. Hydrology is a wide ranging subject, but it was natural for physical geographers to direct their attention to hydrological processes and the branch of the subject frequently referred to as physical hydrology. There was a clear need for improved understanding of the various mechanisms and pathways associated with the movement of water through the hydrological cycle. The study of these processes represented a logical extension of the physical geographer's concern for the working of the natural landscape, and afforded a useful link between the geomorphological and the meteorological and climatological aspects of the subject. The relevance and value of such work was also emphasised by the launching of the International Hydrological Decade (IHD) by UNESCO in 1965. The primary objective of this international programme was to develop an improved understanding of the science of hydrology and it placed considerable emphasis on elucidation of the physical processes involved, through the setting up of instrumented basins.

Although the parallel development of interest in contemporary geomorphological and hydrological processes can be viewed as essentially a chance occurrence, their convergence provided an important impetus for subsequent developments and a clear example of synergism. The fluvial geomorphologist and the hydrologist had a common interest in such processes as surface runoff, infiltration, streamflow and sediment transport, and it can be suggested that geomorphological investigations benefited greatly from the adoption of a wide range of hydrological measurement techniques. Equally, hydrologists can be seen as having profited from the appearance of a group of like-minded researchers with whom they could co-operate and exchange ideas.

In many respects the distinction between process geomorphology and physical hydrology was blurred at this time, but whilst some would have claimed interests in both, others would have readily defended their primary objectives as relating to one or the other. It has been suggested that many fluvial geomorphologists emphasised the hydrological aspects of their work in order to exploit financial resources for research which were more readily available to the latter, but it is questionable if such a clear distinction existed in the minds of the Research Councils.

The empirical assault

The great burst of activity in the study of fluvial and hydrological processes that resulted from these developments and that characterised the latter years of the 1960s and the early 1970s has been referred to as the 'measurement movement'. With converts responding to the dictum 'go ye forth and measure', instrumented drainage basins sprang up in many parts of the country. In addition to drainage basins, river reaches, meander bends, gullies and small plots were also instrumented in order to provide information on the nature and the rates of operation of contemporary processes. All represented a marked reduction in the scale of study when compared to the more traditional approaches of the physical geographer. Similar activity was apparent in many other countries as evidenced by the establishment of the International Geographical Union Commission on Present Day Processes in 1968.

It is easy to criticise this period as one where preoccupation with 'techniques' meant that the need for carefully defined objectives was sometimes overlooked, where there was a preoccupation with site specific studies, and where the availability of novel statistical methods encouraged the collection of vast amounts of data in the hope that these methods would in some magical way reveal the answer, and perhaps even the question! However, such criticism ignores the fundamental contribution made by these investigations. In the first place, they permitted the testing and development of a wide variety of measurement techniques which were essential for future progress. Secondly, they provided valuable new information on the likely rate of operation of a variety of fluvial processes. In some cases the data obtained were of limited accuracy, but even as 'order of magnitude' estimates they played an invaluable role in developing our knowledge. Thirdly, they permitted an improved understanding of the basic nature of many processes, and even recognition of others hitherto unappreciated. For instance, much of the more recent work on subsurface throughflow owes its origins to simple empirical studies which involved digging trenches and monitoring the seepage of water from the individual horizons, thereby demonstrating its existence.

Mechanics, models and methodology

Contemporary dissatisfaction with the measurement movement can, however, be seen as having prompted an increasing awareness of the need to synthesise and consolidate the results of site-specific field measurements into more generally applicable models. The need to consolidate empirical data inevitably necessitated a search for underlying principles and mechanisms, and these were to be found in the theoretical and experimental work of the physical scientist and the engineer in the fields of thermodynamics, fluid mechanics, hydraulics and soil mechanics, to which the geographer increasingly turned. Professor Carson's book *The Mechanics of Erosion*, published in 1971, provides a clear example of an attempt to stress the need for a knowledge of the relevant aspects of basic physics in the study of erosion processes. Work on river channel development can, for instance, be seen as having benefitted greatly from an improved understanding of the hydraulics of open channel flow and the physical constraints governing channel adjustment and development.

Contacts with physical scientists and engineers were promoted by the appearance and acceptance of Environmental Science as an interdisciplinary umbrella for studies of the physical environment which embraced both geomorphology and hydrology. These contacts also encouraged the development of laboratory-based studies of geomorphological and hydrological processes involving both prototypes of field situations and more theoretical abstractions. Common to both this laboratory work and more traditional field investigations, there was an increasing trend towards the study of the processes themselves rather than surrogates relating to their effects or end products. For example, simple measurements of rates of river bank recession were replaced by investigations of the flow patterns impinging on the banks, the associated shear stress and the response of the bank material to changing moisture conditions and to undercutting.

Cascading systems, process–response systems, mathematical models, computer simulation models and the many attendant concepts and techniques also provided a timely basis for synthesising and generalising much of the disparate information collected by the measurement movement. It would, however, be wrong to suggest that there were always close links between the field investigator and the modellers and theoreticians. The former frequently distrusted and criticised the generalisations emanating from the armchairs of the latter, whilst they in turn stressed the need to extend the findings of site-specific studies beyond a particular basin or channel. Nevertheless, in retrospect, a synergistic effect can be discerned. Synthesis was clearly essential for progress, but many of the models relied heavily on the findings of field measurement programmes for identifying the processes and linkages involved. Acceptance of this complementarity and the merging of the two approaches was an important milestone on the way forward.

Concern for mechanics and models inevitably prompted a reappraisal of the overall approach to field investigations of fluvial and hydrological processes. Much of the early work in this field can be seen as essentially *inductive* in approach, involving the measurement of what appeared to be happening and the drawing of conclusions from the data collected. This contrasted with the *deductive* approach advocated by the philosophers of scientific method. Here a study of fundamental principles provides the basis for formulating a hypothesis which is subsequently tested by a carefully designed programme of measurements. The merits of these two contrasting approaches, and the perceived scientific respectability of deductive reasoning, provided scope for considerable debate. A sceptic might deny the existence of clear-cut alternatives, pointing to instances where measurements were an essential prerequisite for defining the system to be studied, or where a hypothesis had been conveniently formulated at the end of a study in order to justify the measurement programme employed. However, the increased attention to objectives, hypotheses and method that inevitably emanated from these debates can only be seen as an important advance. In this context it is significant that the IGU Commission on Present Day Processes was succeeded in 1976 by the Commission on Field Experiments in Geomorphology, reflecting an increasing awareness of the need for carefully formulated measurement programmes and experiments in order to provide improved understanding of contemporary processes.

Application and management

The great advances in the understanding, measurement and modelling of fluvial and hydrological processes that were clearly apparent by the end of the 1970s also provided an opportunity for the physical geographer to contribute to the more applied problems of environmental management. Thus an understanding of the processes of soil erosion and sediment yield, the dynamics of river channel development and the functioning of floodplains, could be, and indeed was, applied to problems of soil degradation, reservoir sedimentation, river training and flood control and floodplain management. In practice, however, the geographers' input to such problems was severely limited by a lack of appreciation of their potential contribution by the engineers traditionally involved. In most cases it was necessary to embark upon a lengthy programme of proving and marketing these skills. Success in this field was, however, not always seen as marking progress. Some argued that these developments represented a deflection from more important aims and objectives. This sentiment was particularly relevant in the case of fluvial geomorphology, for here it could be suggested that the need to link studies of contemporary processes to longer-term landform evolution was easily forgotten.

Achievements

This brief outline cannot embrace all the developments, debates, and potential interpretations relating to the growth of interest in hydrological and fluvial processes over the past twenty years, and its subjectiveness must be emphasised. Few would deny the great burst of activity in the study of these processes, that (physical) hydrology consequently emerged as a clearly defined area of geographical study and that in the field of fluvial geomorphology the historical paradigm was replaced by a dominant concern for contemporary processes. A useful reminder of the magnitude of these shifts in outlook is provided by Figure 2.1.1 which depicts in simplified form some of the changes in approach to the study of those meanders referred to in the introductory verse. Traditionally, the treatment of these features focused on planform development and their role in the evolution of river valleys, with scant regard for the processes involved. Initial 'process' measurements focused on rates of planform development as a surrogate for the processes involved, but increasing concern for mechanics and models has now generated a greatly improved understanding of the hydraulics of these features, even if the factors ultimately responsible for their formation remain uncertain.

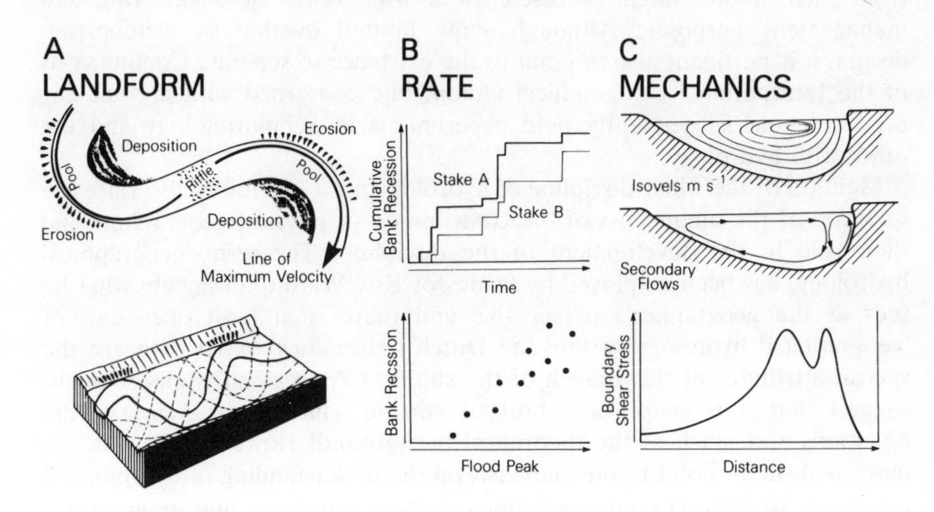

FIGURE 2.1.1 *Stylised examples of changing approaches to the study of river meanders*

CURRENT STATUS

Hydrology or geomorphology?

In the early days of interest in processes, the distinction between the fluvial geomorphologist and the physical hydrologist was blurred, and the close links between workers in the two fields has already been emphasised. To some extent this overlap of interests has been perpetuated and the term 'hydrogeomorphology', suggested by Professor Ken Gregory as an appropriate descriptor of process-oriented fluvial geomorphology, could perhaps be used to denote a unified subject. Certainly, in the school syllabus, such unity is frequently implied. However, any such suggestion of integration would deny several important distinctions between the two subjects. Perhaps most importantly, it must be recognised that the ultimate goal of the physical hydrologist is to understand and to model the processes comprising the hydrological cycle. These processes embrace both quantity and quality dimensions, and hydrologists can therefore justifiably include considerations of sediment and solute transport within their sphere of interest. In many respects, the interests of the fluvial geomorphologist start where those of the hydrologist finish, because the former is primarily concerned with the geomorphological effects of these processes on the landscape. Furthermore, it must be accepted that the study of hydrological processes represents only one aspect of the wider subject of hydrology, and that the knowledge gained from such studies might be used for a wide range of engineering and management purposes. Although some limited overlap of membership occurs, it is pertinent also to point to the existence of separate Commissions of the International Geographical Union, one concerned with present day processes and subsequently field experiments in geomorphology and the other with hydrology.

Mention of the wider discipline of hydrology must inevitably also raise the question of the uniqueness of the contribution of physical geographers and their role in the development of the discipline. The term 'geographical hydrology' has been employed by Professor Roy Ward to designate what he sees as the geographer's perspective and there is at least one chair of 'geographical hydrology' within the Dutch universities. What then are the special attributes of this branch of the subject? A sceptical engineer might suggest that it is simply an abridged version which omits many of the equations and much of the theoretical background! However, it would be more realistic to point to the emphasis on the understanding of hydrological processes, to field-based investigations of these processes, and to an awareness of their spatial variability at scales ranging from that of a small drainage basin to that of the various physiographic and climatic regions of the globe.

The expanding technological base

Studies of fluvial and hydrological processes have benefited greatly from the numerous technological developments that have characterised the past twenty years. Many of the significant advances in understanding can be closely linked to the availability or development of appropriate hardware on which to base field and laboratory investigations. Improvements in data recording equipment afford an excellent example of this expanding potential. In early studies, the possibilities for continuous recording were limited by the paucity of appropriate sensors and even when recording was possible, as in the case of a simple float and counterweight-activated water level recorder or an autographic rain gauge, this was invariably achieved through the use of a drum chart and pen. Such charts provide a valuable visual record, but their temporal resolution is frequently limited and considerable effort may be required to reduce the graphical plot to a digital record. Subsequent developments in multichannel electrical strip chart recorders paved the way for high resolution and synchronous recording of several parameters, and the introduction of magnetic tape loggers and more recently microprocessor controlled solid state loggers provided even greater scope for automated data collection, since the records could be transferred directly to a computer for processing without the need to convert analogue charts to a digital form. Furthermore, the low power requirements for these loggers meant that they could be operated for long periods in remote locations from small batteries.

However, all these advantages have been marked by an increasing requirement for financial resources to purchase the necessary equipment. Whereas, in the early days useful information might have been derived from equipment improvised from inexpensive components and materials, the need for solid state loggers and related sensors could necessitate a budget running to several thousands of pounds. If the funds available were an order of magnitude greater than this, it would even be possible with current technology to link the monitoring equipment to a satellite data transmission system so that the data could be sent straight to the investigator's office without the need to visit the field installation!

In any assessment of the current impact of technological advances on process studies, it is important to reflect on whether the introduction of new techniques merely permits existing geomorphological or hydrological problems to be investigated in a more intensive manner, or whether it provides the stimulus for the recognition and formulation of new problems. The distinction may admittedly be blurred, but it is relatively easy to point to examples of the former situation. However, instances of the latter may also be identified. Satellite remote sensing has, for example, opened up many new possibilities for documenting spatial patterns of vegetation cover, soil moisture conditions and erosion and deposition features, and for sequential studies of such phenomena as floodplain inundation and sediment plumes in lakes and reservoirs. The development of mineral magnetic studies by Professor Oldfield and his colleagues (Chapter 1.1) and the use of Caesium-137, a naturally occurring radioactive isotope associated with the fall-out from atomic bomb tests, to study erosion and sediment redistribution, must equally be seen as opening up many new possibilities not embraced by more traditional techniques. Furthermore, the major changes in both the temporal and spatial resolution of field and laboratory measurements associated with technological advances in sensing and recording equipment have frequently necessitated reassessment of the significance of specific processes. For example, the use of repeated airborne remote sensing and of automatic recording tensiometer networks – to record runoff and soil moisture conditions within a drainage basin – has emphasised the dynamic nature of the areas contributing to storm runoff and therefore of sediment and solute sources.

The example of material transport by rivers

Current studies of material transport by rivers can be used to provide a useful example of the nature, scope and potential application of recent work on fluvial processes by hydrologists and geomorphologists. It is easy to recognise that measurements of the transport of material in solution, in suspension, and as bed load are of fundamental importance to the geomorphologist, because they provide invaluable information on contemporary rates of denudation within a drainage basin. Many simplifying assumptions are involved and the exercise should therefore be approached with caution, but nevertheless it is possible to convert a value of mean annual sediment and solute load to an estimate of the rate of land surface lowering by fluvial denudation. A combined sediment and solute load of 250t km^{-2} $year^{-1}$ could thus be seen as representing an annual rate of lowering of the order of 0.1mm $year^{-1}$, assuming a rock density of 2.5g cm^{-3}. Equally, hydrologists have an interest in river load data, since they represent an important facet of the quality dimension of the hydrological cycle and they are central to studies of material transport to the oceans and of biogeochemical budgets.

Studies of material transport are being undertaken at a variety of scales ranging from the global overview to investigations of small drainage basins, and they highlight the wide range of scales embraced by fluvial process studies Recent increases in measurement activity in various parts of the world have afforded a greatly improved picture of the degree of variation of suspected sediment yields. These range from less than 10t km^{-2} $year^{-1}$ in areas such as northern Europe and part of Australia to in excess of 10 000t km^{-2} $year^{-1}$ in certain regions of the world where conditions are especially conducive to high rates of erosion. These locations include the island of Taiwan, South Island New Zealand and the middle Yellow River Basin in China. In the two former cases, steep slopes, high rainfall, and tectonic instability are major influences, whilst in the latter case the deep loess deposits and the almost complete lack of natural vegetation cover are important. Rates of land surface lowering associated with this component of fluvial transport can be seen as varying over more than three orders of magnitude from less than 0.004mm per year to in excess of 4.0mm per year. The broad pattern of global suspended sediment yield depicted in Figure 2.1.2 reflects the influence of a wide range of factors including climate, relief, underlying geology, vegetation cover and land use.

Information on the dissolved loads of rivers at the global scale is less complete than that available for suspended sediment, but existing knowledge points to loads ranging from less than 1.0t $km^{-2}year^{-1}$ to approximately 500t km^{-2} $year^{-1}$. These values are somewhat lower than those associated with suspended sediment transport, and if total material transport from the land surface of the earth to the oceans is considered, the value for suspended sediment, which is of the order of 15×10^9t $year^{-1}$, exceeds that for dissolved material (*c*4×10^9t $year^{-1}$) by nearly four times. As a global generalisation, the relative efficacies of mechanical and chemical fluvial denudation could therefore be seen as being in the ratio 4 to 1.

At the other end of the scale, detailed studies of small drainage basins are being undertaken to elucidate the various processes contributing to the suspended sediment or dissolved solids yields. Whereas, in many early process

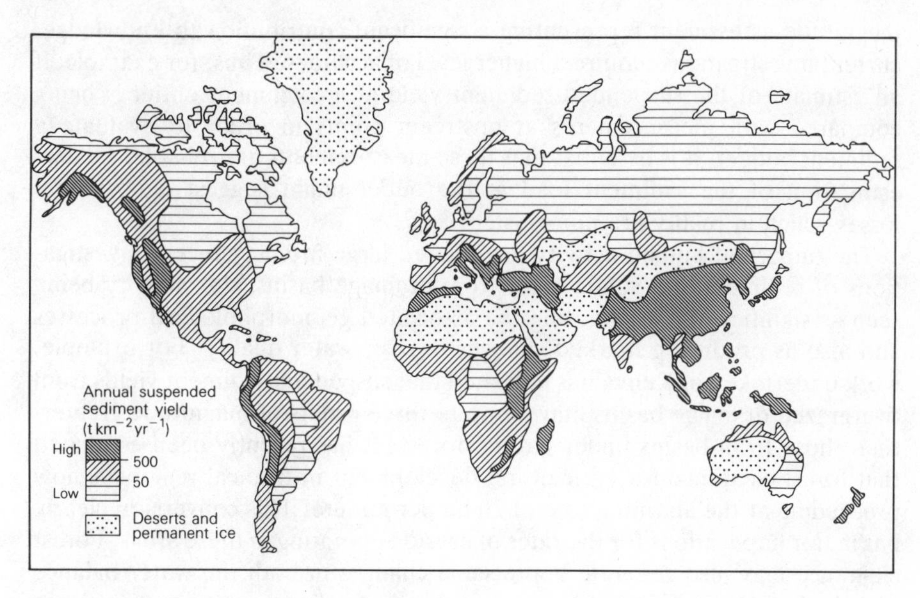

FIGURE 2.1.2 *A generalised map of the global pattern of suspended sediment yield from intermediate-sized drainage basins (i.e. c. 10 000 km²)*

studies, attention was focused on documenting the load at the catchment outlet, the questions now being posed demand information on the precise sources of the sediment and solutes within the basin, and on the pathways involved between source and outlet. In the case of suspended sediment studies, the sediment budget concept provides a valuable framework for assessing the various sources of particulate material and the potential sinks involved in the movement of sediment to the basin outlet. Figure 2.1.3 provides a hypothetical, although typical, example of the major components that might be included in such a budget. In this case the sediment leaving the drainage basin represents only a small percentage of that mobilised by erosion. A similar approach has been adopted in many investigations of the solute yield from drainage basins. Figure 2.1.4 shows how it is necessary to understand the various pathways of water movement and the biogeochemical processes associated with these pathways if temporal variations in river water quality are to be evaluated in a meaningful way, and if the solutes which are associated with chemical weathering processes and are of prime interest to the geomoropholgist are to be distinguished from those of atmospheric and biotic origin. Geomorphologists are now beginning to investigate patterns of solute generation by chemical weathering on hillslopes with a view to determining whether rates of weathering vary according to position on the slope profile, and therefore if slope development associated with these processes is characterised by parallel retreat or rectilinear recession.

Whereas relatively little attention was given to the accuracy and precision of load measurements in early process studies, with even an order of magnitude assessment representing a significant contribution to knowledge, current investigations require a higher level of accuracy. Thus, for example, if an estimate of the suspended sediment yield at a catchment outlet is being compared with measurements at upstream points in order to evaluate a sediment budget, it is essential that these measurements are reliable. Underestimation of the sediment load at the outlet might suggest depositional losses which in reality are non-existent.

The impact of human activity has figured large in many recent investigations of sediment and solute yield from drainage basins, this activity being seen as significantly accelerating the associated geomorphological processes and also as producing marked changes in river water quality. For example, work undertaken in Kenya has indicated that suspended sediment yields from overgrazed drainage basins may be up to three orders of magnitude greater than those from basins under natural forest. It has recently been estimated that forest clearance for agricultural development in tropical regions is now proceeding at the alarming rate of 20 ha per minute! This conversion clearly has major implications for the rates of erosion operating in these areas. Forest clearance may also generate appreciable changes in both the water balance and the nutrient budget of a drainage basin, leading in turn to significant changes in stream solute concentrations and solute yields. Work in small catchments in Western Australia has, for instance, shown how decreased evapotranspiration losses have resulted in rising groundwater levels which have mobilised saline deposits in the soil and caused the chloride loads of the streams to increase by up to four times. In countries such as Britain, urban development similarly exerts a marked influence on fluvial transport. Construction activity may cause sediment yields to increase up to thirtyfold Stormwater drainage from established urban areas may also represent an important input of material to rivers, much of which may be toxic.

Concern for human impact on fluvial processes is closely related to many of the practical applications of process studies that have been demonstrated in recent years. Increased rates of erosion and suspended sediment yield can rapidly lead to problems of reservoir sedimentation, and meaningful prediction of the potential impact of land use change must be seen as an important requirement for effective water resource development in many developing countries. Growing awareness of the problems of non-point source pollution has also inevitably generated a need for an improved understanding of the processes involved in the movement of sediment and solutes from the slopes of a drainage basin, into and through a river system. Although Figures 2.1.3 and 2.1.4 were originally introduced as reflecting current approaches to fluvial process studies, they must equally be seen as emphasising the potential value of the improved understanding of sediment and solute movement which is being obtained. It is well known that many agricultural pesticides are readily adsorbed on to soil particles when applied to farmland, and knowledge of the location and nature of the major sediment sources within a drainage basin and of the pathways and sinks associated with subsequent

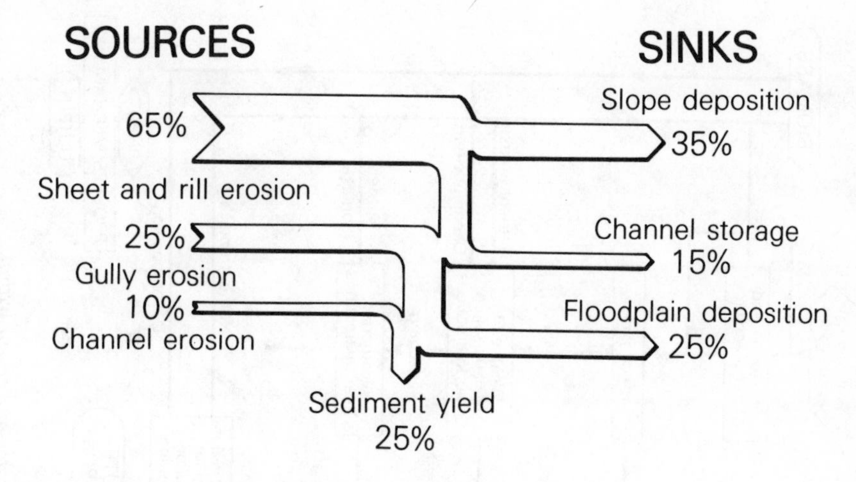

FIGURE 2.1.3 *A hypothetical sediment budget for a drainage basin*

sediment movement must be seen as fundamental to any attempt to predict the fate of these toxic chemicals. Likewise, assessment of the potential impact of acid rain or of increased rates of fertiliser application on stream water quality demands an understanding of the various process linkages depicted on Figure 2.1.4.

THE FUTURE

Much has happened since the awakening of interest in fluvial processes that characterised the early 1960s, and the closer integration of field measurement programmes with the demands and aspirations of model builders and theoreticians has gone a long way towards producing a rational and effective approach to the study of fluvial processes. The geomorphologist must still ponder the question as to whether the study of contemporary processes can be justified in its own right or if such work must have the understanding of landform evolution as its long-term goal. This potential dichotomy is, however, already becoming somewhat indistinct. The results of process investigations are now being applied to problems of landform development, as is evidenced by recent work on palaeohydrological reconstruction and the formulation of models of the evolution of fluvially eroded landscapes. Awareness of the significance of high magnitude low frequency events and geomorphic thresholds in process studies has also encouraged a longer-term perspective for such work. The time may now be ripe for a renewed assault on the long-term objective. Recognition and exploitation of the increasing potential for practical application of the results of recent process investigations may well deflect the geomorphologist from this objective, and priorities

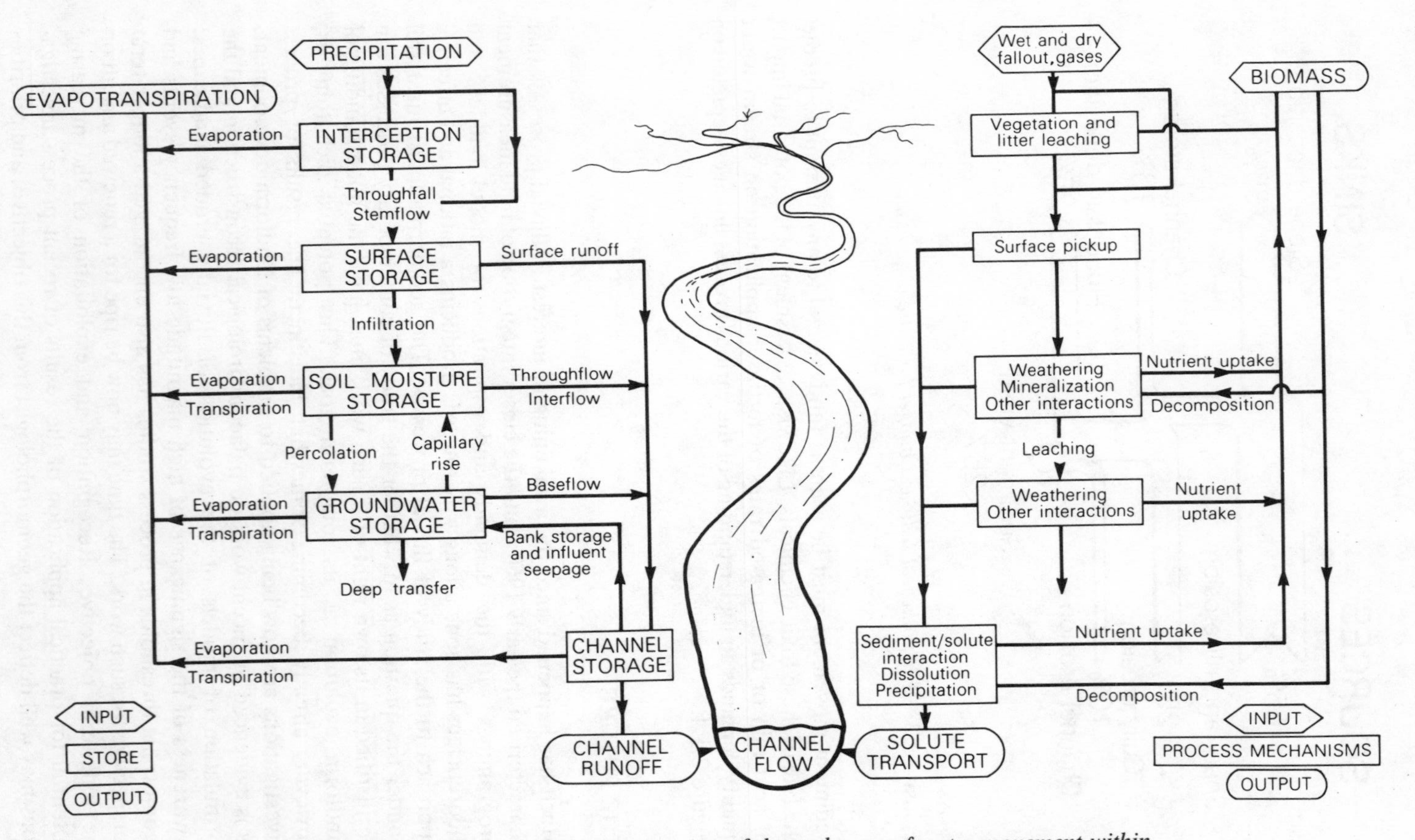

FIGURE 2.1.4 *A simplified representation of the pathways of water movement within a drainage basin and associated mechanisms governing solute levels in streamflow*

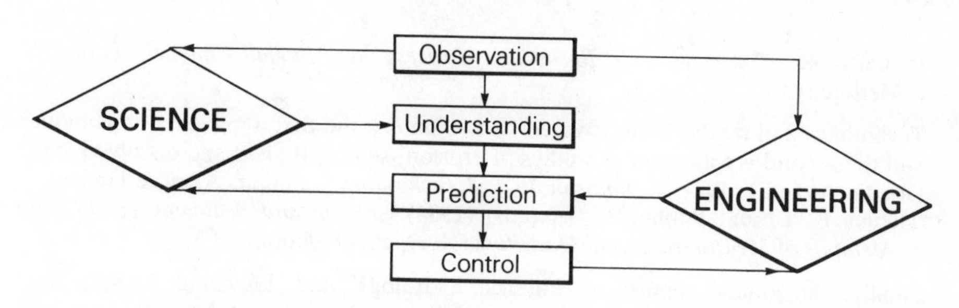

FIGURE 2.1.5 *Dooge's view of the contrasting roles of the scientist and the engineer in environmental management*

must be debated. However, the hydrologist, if not also the geomorphologist, should seize these opportunities and strive to strengthen the role for the physical geographer in dealing with an increasing body of environmental problems which demand a detailed understanding of fluvial processes. In this context the physical geographer is well qualified to provide a scientific input. The approach of a *scientist* to environmental management differs fundamentally from that of an *engineer*. As Professor Dooge, an eminent engineer has noted, the scientist is concerned with *understanding* the processes, whereas the engineer must adopt a more pragmatic approach, being primarily concerned with prediction and control (Figure 2.1.5). Future improvements in predictive and control capabilities will depend heavily upon improvements in understanding the processes involved.

FURTHER READING

An overview of the nature of process studies undertaken in Britain during the late 1960s and early 1970s can be obtained from:

Gregory K. J. and Walling D. E. (eds) (1974) *Fluvial Processes in Instrumented Watersheds*, Institute of British Geographers, Special Publication No. 6.

Gregory K. J. (1978) 'Fluvial processes in British basins', in Embleton C., Brunsden D. and Jones D. K. C. (eds) *Geomorphology: Present Problems and Future Prospects* (Oxford: Oxford University Press), pp. 40–72.

These reviews can usefully be compared with a more recent compilation of the results of drainage basin studies in Britain and overseas:

Burt T. P. and Walling D. E. (eds) (1984) *Catchment Experiments in Fluvial Geomorphology* (Norwich: Geobooks).

Two further references exemplify the increasing emphasis on mechanics that characterised process studies in the 1970s, and provide valuable treatments of the basic physics involved, whilst a third gives a detailed treatment of the hydraulic principles underlying the behaviour of river channels:

Carson M. A. (1971) *The Mechanics of Erosion* (London: Pion).

Embleton C. and Thornes J. (1979) *Process in Geomorphology* (London: Edward Arnold).

Richards K. (1982) *Rivers: Form and Process in Alluvial Channels* (London: Methuen).

Techniques and methods are covered by two books, the first a manual of techniques and the second a collection of studies of erosion, sediment yield and dissolved load:

Goudie A. (ed.) (1981) *Geomorphological Techniques* (London: Allen & Unwin).
Hadley R. F. and Walling D. E. (eds) (1984) *Erosion and Sediment Yield: Some Methods of Measurement and Modelling* (Norwich: Geobooks).

Finally, the annual reports of 'Physical hydrology' and 'Fluvial geomorphology' published in *Progress in Physical Geography* (London: Edward Arnold) since 1977 refer to many of the contemporary issues and debates, and provide useful sources of references to recent work.

2.2
Atmospheric Processes: Global and Local

B. W. Atkinson
Queen Mary College, University of London

Physical geography is concerned with the natural environment, and includes study of the solid and fluid constituents of that environment together with the life forms that exist within those constituents. The core of enquiry into the natural environment is the question 'How does it work?' and this immediately throws the attention on the investigation of processes (here considered to be changes in time and movements of things, such as heat and water). Understanding of the atmosphere has emerged gradually over the past three centuries through observation and the derivation of theory. From the rudimentary beginnings of the mid seventeenth century we now observe the global atmosphere from about 7000 surface stations, about 700 upper air stations, polar orbiting and geostationary satellites and, at smaller scales and for more particular purposes, we use radar, lidar, sodar, instrumented towers, and constant level balloons among other techniques. Theoretical study of the atmosphere also has its origins in the mid seventeenth century but took on a distinctive meteorological flavour only in the mid to late nineteenth century. The major developments have occurred in this century.

Geographers entered this pursuit of understanding about sixty years ago. Although there had been efforts in the preceding half century, it was largely books such as that by A. A. Miller that stamped climatology as the 'proper' branch of atmospheric science for geographers. The emphasis was on observation, organisation of data and regionalisation of climate. In the UK these elements, in a local climate context, became the important contributions of senior figures such as G. Manley, P. R. Crowe, T. J. Chandler and S. Gregory. This tradition still flourishes, but since the beginning of the 1970s applicable and applied studies have emerged. Whilst virtually all modern geographers would shun the epithet, it cannot be totally rejected that applied studies have a deterministic flavour. To this small extent, therefore, they exemplify a closing of a methodological circle that we initiated almost a century ago. Whilst in no way undervaluing the recent applied studies this chapter, by being concerned with atmospheric processes, takes a pure science approach. Good application of science can result only from sound understanding of fundamental processes.

SOME BASIC PROPERTIES

The atmosphere comprises a stable mixture of gases that forms a fluid envelope about 100km deep around the rotating earth. All weather and climate is fundamentally related to the way in which air moves. For example, the distribution of radiation is dependent upon cloud amount and type which in turn results from vertically moving air. Temperature at any location is as much a function of advection as of the local radiation budget. Humidity is a function of evapo-transpiration and advection, and the former is itself a function of wind speed. It follows that if we understand how air moves we have deep insight into the nature of weather and climate. It is for this reason that so much of the research in atmospheric science has been in dynamical meteorology, the greatest strides being made since 1945.

If the atmosphere were this simple, its behaviour would be far easier to predict than it actually is. The difficulties arise because additionally the atmosphere is compressible (greater densities near the ground than at high altitudes), unevenly heated, three-dimensionally turbulent and has a very complicated lower boundary surface (the surface of the earth). Nevertheless, the facts that the atmosphere is a fluid, and that its motion is fundamental to weather and climate mean that its behaviour has been intensively examined with the aid of fluid dynamics. Hence, the vast amounts of observational data have increasingly been complemented, and to a large degree 'explained' by theoretical analyses of air motion. Understanding of weather and climate is currently progressing in the time-honoured way of physical science – the combined attack with both observational and theoretical approaches.

Recently the literature of physical geography has been strongly flavoured by the term 'process-response'. This term is but another way of expressing part of the well-known Davisian dictum of 'structure, process and time'. Given an initial state of a 'structure', if 'processes' operate on or in it through a period of 'time', clearly some other state (or response) will emerge. The essence of virtually all scientific enquiry is to establish the nature of processes. In some subjects this is particularly difficult. For example, in earth science the processes are frequently so slow that they cannot be measured directly. Hence processes are frequently inferred from observations of form. In other subjects, particularly atmospheric science, it is indeed possible to observe processes such as air movement, and hence our direct knowledge of important processes is significantly increased. But the problem is not resolved at this stage: indeed it is really only being posed in a reasonably satisfactory manner. Given that we know the configuration of airflow at a particular time, the fundamental questions are how and why that configuration came into being. These questions explain the necessity of dynamical concepts in the examination of 'process'.

Much fundamental behaviour in the atmosphere can be described by six equations, all derived from classical physics. The first three are forms of Newton's second law that relates forces and accelerations, and usually apply to the latitudinal, longitudinal and vertical components of air velocity. The fourth equation expresses the conservation of mass – the continuity equation. The fifth is the Boyle–Charles equation of state, and the last is the First Law of Thermodynamics. In these equations the variables air velocity, density, pressure, temperature and the rotation of the earth are all related in a rigorous way. It is possible to add terms to these equations explicitly to incorporate the effects of radiation, evaporation, condensation and turbulent transfer. But the more terms (and therefore processes) are added, the more difficult it becomes to solve the equations, that is to predict a value of any particular variable that can then be compared with reality.

Examination of these equations has continued for virtually a century – a testimony to their great power as descriptions of the multitude of processes in the fluid environments. A major approach using the equations is to make approximations, by ignoring some terms, in the search for significant atmospheric mechanisms. Thus it was already established some forty years ago that the large-scale atmosphere was essentially geostrophic, hydrostatic and adiabatic. Despite their great significance these properties are frequently understated in textbooks, only the geostrophic wind being explicitly treated. In contrast, the conceptually much simpler notion of a budget or balance has wide prominence in most geographical texts. It is particularly useful in analysing the disposition of heat, water and momentum between earth and atmosphere. Heat and water budgets are commonplace in the geographical literature; momentum budgets less so.

By about 1960 observation (largely, by operational meteorological networks) and theory had produced a preliminary, rather rough, picture of the global atmosphere with a similar level of understanding of flows at and larger than the cyclone scale. In the past twenty-five years tremendous advances have been made in both observational and analytical techniques with the result that many new 'forms' have been discovered, old ones better described, and even new mechanisms comprising particular combinations of basic processes have been suggested.

NEW OBSERVATIONAL TECHNIQUES

In common with most sciences, meteorology has benefited immensely from the advent of new observational techniques. The traditional mode of observation and one which is still used today involves relatively cheap instrumentation and simple procedures to monitor at the surface, temperature, humidity, precipitation, wind speed and direction, visibility, cloud amount and type together with other atmospheric characteristics. Upper air conditions have been monitored with radiosondes since the 1930s. Global networks of surface and radiosonde stations provide the bedrock data on atmospheric behaviour. These data are collected primarily for routine forecasting and as such have a limited value for research purposes because they are geared to the synoptic rather than to any other scale.

Over the past quarter of a century many new tools have become available. Radar, first used for meteorological purposes immediately after the Second World War, became a far more refined and widely used device, particularly in North America, Japan and Europe. Nowadays ranges extend to 400km, echoes are colour-contoured, composite images of separate radars and video sequences may be constructed, and spatial resolution is as high as 5km. In addition to conventional ground-based radar, surface and airborne Doppler radars provide detail on echo intensity and wind speed and direction at a resolution of a few hundred metres.

The very small-scale structure of the atmosphere, usually lying in the planetary boundary layer (the bottom 1 km or so), is observed with the aid of instrumented towers, constant pressure balloons, omnidirectional anemometers, and accoustic sounding, among other methods. At the other extreme synoptic and planetary-scale systems have been more closely observed over the past two decades by satellite sensors. Both digital and pictorial data are available from polar-orbiting and geostationary satellites. In addition to visual images, infra-red pictures allow earth and atmosphere temperatures to be calculated and the distribution of water vapour can also be established. Satellites now play a major role in both operational and research meteorology.

A final significant addition to the observational armoury is the meteorological buoy. Technological development of buoys has a history of about thirty years but significant problems remain – not least that the buoys have a distinct tendency to capsize. Nevertheless, much progress has been made and there is now a real possibility of using free-floating or tethered ocean-going buoys, to be linked to satellites, in the 70 per cent of the earth's surface covered by water.

Whilst most of the above devices are appropriate for particular atmospheric characteristics, an observational programme using many different types clearly is of major significance. Such a programme was the First GARP (Global Atmospheric Research Programme) Global Experiment (FGGE), which observed the whole atmosphere for one year (1 December 1978 – 30 November 1979) on a daily basis with a horizontal resolution of about 1500 km and a vertical resolution of about 1km. In this way the mass, thermal, humidity and motion fields were described in unprecedented detail, and such a vast data base now provides a vital touchstone for numerical models of the global atmosphere. At smaller scales, but still of substantial size, were the GARP Atlantic Tropical Experiment (GATE), the Monsoon Experiment (MONEX) and the Alpine Experiment (ALPEX) among many others. These smaller programmes had more restricted aims than FGGE – the above cases, for example, establishing the processes operating in the tropical atmosphere (including the monsoons) (GATE and MONEX) and those operating over high mountainous regions (ALPEX).

NEW ANALYTICAL TECHNIQUES

Of equal importance to the investigation of atmospheric behaviour are good analytical techniques. Prior to 1940 when meteorology was largely, but not totally, an empirical science, the data were analysed either climatologically or synoptically. The former approach involved the familiar statistical methods required to generate means, variances, frequencies and so on – a mainstay of 'geographical' climatological work throughout the history of geography as a university discipline. Indeed it was this firm quantitative foundation of climatology which long preceded and provided a base for the wider so-called Quantitative Revolution within geography. By contrast, synoptic analysis comprises the preparation of a temporal sequence of instantaneous 'snapshots' of the weather over a specified area, fundamental to weather forecasting from its origins in the late nineteenth century to about 1965. Whilst sequences of maps are still produced, the manner in which they are produced reflects a profound change in the analytical techniques employed over the past two decades. In fact they are deeply rooted in the theory of atmospheric motion.

Theoretical analysis is of two kinds: analytical and numerical. Purists may argue, with some justification, that numerical experimentation is not truly theoretical but such a distinction need concern us no further here. Analytical study essentially means the probing of the equations that describe fluid flows and other atmospheric behaviour to gain insight into such behaviour. This kind of analysis has a long pedigree having been a main stream of classical mechanics for over three centuries. In atmospheric terms it has led to such major discoveries as geostrophic behaviour, the Rossby waves and the baroclinic instability of the extra-tropical atmosphere.

Whilst this type of analysis continues with impressive results, it has been most effectively complemented by numerical modelling, which in some cases requires the use of the most powerful computers in the world. This type of modelling is primarily concerned with solving the six fundamental equations outlined earlier, so that the temporal rate of change of the air velocity and temperature can be evaluated at a series of points. Together with observed 'initial' values of velocity and temperature at these points, the solutions allow the simple prediction of future values at those points. The time periods over which one set of calculations is valid are usually of the order of a few minutes; so to make a useful prediction the procedure must be repeated many times. This is one reason why such large and fast computers are necessary.

The above procedure is conceptually not very difficult, but putting it into practice is extremely difficult, largely because the equations 'resist' solution by numerical methods. Many types of numerical technique exist, all aimed at giving results of acceptable accuracy by avoiding computational instabilities (calculations that 'runaway' to give nonsensical answers). Much current research is concerned with the development of better numerical techniques as well as the application of well-tested techniques to various atmospheric phenomena.

In addition to numerical modelling, hardware modelling has been increasingly used over the past two decades or so. This involves the construction of scaled-down models of atmospheric flows so that controlled experiments can be run in the laboratory. The most important of the early models of this type were constructed in the University of Chicago just under forty years ago. They were the famous 'dish-pans' which crudely, yet remarkably effectively, simulated the very large-scale flows that exist in the tropical and extratropical atmosphere. The technique now tends to use rotating annuli and the aims of the experiments are more fundamental – to increase understanding of rotating fluids. Hardware models have also been used effectively on the analysis of lee-waves, tornadoes, urban heat islands, airflow around buildings, cellular convection and small-scale turbulence.

NEW FORMS AND PROCESSES

The tremendous advances in observational and analytical techniques outlined above have led to greatly increased knowledge of forms and processes in the atmosphere. To a large degree the different forms are an indirect expression of processes. It is true that the fundamental processes of mass heat and water transfer underpin the whole edifice, but it is the many ways that these fundamental physical processes generate and interact within the atmosphere to give distinctive regimes of weather and climate that is the central problem of meteorology. These regimes are largely dependent upon the type, size, duration and frequency of systems of air motion – the atmospheric 'forms' mentioned above. The most convenient way of handling the multitude of forms is to classify them by size. 'Size' is determined by horizontal extent or duration, the two being closely related. Vertical extent is not a good discriminant. The classification can be done both non-dynamically and dynamically. Non-dynamical scaling emerges from experience with synoptic and climatological maps and radar and satellite imagery, where the analyst gains a 'feel' for the typical sizes of circulations. Dynamical analysis is a more rigorous approach based upon basic physical characteristics of the atmosphere and planet Earth. Thus the vertical distribution of temperature and density imposes a certain duration (tens of minutes) on small-scale vertical motion in the atmosphere. Secondly, the rotation of the Earth leads to geostrophic flows and synoptic systems with durations of the order of tens of hours. Thirdly, the curvature of the Earth's surface causes a variation in the effect of the Earth's rotation on fluid flows and hence influences systems with durations of tens of days. Given these critical threshold durations, the typical wind speed (about 10 ms^{-1}) clearly defines different scales, as shown in Table 2.2.1. Whilst these scales are rationally defined by the fundamental atmospheric frequencies, their threshold values are not rigid. Nevertheless, they do mean that on Earth we do not find cumulus clouds 1000km across or Rossby waves 5km across.

TABLE 2.2.1 *Scales derived from fundamental frequencies within the atmosphere*

Scale	*Duration*	*Horizontal extent (km)*
Small	$\leqq$10 min	$\leqq$1
Meso	10 min to 17 hr	1 to 100
Synoptic	17 hr to 1 week	100 to 1000
Planetary	1 wk to 3 months	>1000

Small-scale forms and processes

At this scale, the significant processes are vertical transfers of heat, momentum and water, the first by radiation and all three by turbulence. The nature and mechanisms of turbulence *per se* are unsolved problems but it is possible to gain a feel for its nature and effects through the notion of an 'eddy' – a recognisable configuration of fluid flow, exemplified by the vortex of water surrounding an outflow in a bath. This idea was suggested some seventy years ago and has been remarkably resilient. Throughout these seven decades, and particularly within the last two, it has proved possible to make detailed, accurate measurements of the very small (fractions of second in duration) gusts in the bottom few hundred metres of the atmosphere. In turn, special analysis of these data has allowed the discrimination of eddy sizes and an assessment of their abilities to transfer heat, water and momentum.

Much of the analysis of turbulent transfer relies on the flux-profile relationship. This simply means that the transfer (or flux) of an element, such as heat, is directly related to the gradient of the element (in meteorological parlance a profile is a vertical gradient). Hence the stronger the gradient, the greater the flux and vice versa. The direction of the flux is nearly always down the gradient. The measure of proportionality between the flux and the profile is known as the 'eddy diffusivity' or 'coefficient of eddy viscosity' and is traditionally represented by K. Hence:

$$\text{Flux} = K.\ \text{Gradient}$$

If K is known, this is a remarkably simple way of establishing fluxes by measuring spatial gradients. Unfortunately K is frequently neither known nor constant. It is in fact a function of the very flow that it is helping to describe. As the flow configuration changes so does K. Hence, much research over the past three decades has been concerned with the better specification of the eddy diffusivity.

For both practical and scientific reasons, virtually all work on turbulent transfer takes place in the planetary boundary layer (PBL) – the bottom kilometre or so of the atmosphere. Investigation of the PBL is important for two main reasons: it can strongly influence the behaviour of the remainder of the atmosphere and it is the part of the atmosphere occupied by people. It is now appreciated that the PBL has very different structures in day and night. In daytime the PBL comprises convective eddies capped by a thin inversion of temperature at a height of about 1.5km. This results in vertical distributions of temperature, humidity and wind that clearly define two layers within the PBL: first, a surface layer, usually above 50m deep, within which vertical gradients are very large and vertical fluxes are virtually constant with height; and secondly, an overlying mixed layer, about 1.25km deep, within which there is no vertical variation in temperature, wind speed and humidity. In night time, stability is the primary determinant of the character of the PBL. The layer is shallower than in the day time, eddy activity is diminished, the top of the layer takes on a wave form and radiative heat transfer becomes of comparable importance to turbulent heat transfer.

Meso-scale forms and processes

Study of the meso-scale structure of the atmosphere has been a major growth point over the past two decades. Using the criteria in Table 2.2.1 and remembering the flexibility of the limits, we see that this scale can include the individual cumulus cloud at the small extreme and circulations affected by the geostrophic force at the large extreme. Within this range a whole suite of systems exists, some strongly influenced by topography, others more dependent upon free atmospheric characteristics.

At the cloud scale, the past two decades have seen a shift from investigation of the microphysics of clouds to the examination of airflows within clouds. This in turn, has increased understanding of the interactions of the microphysical and cloud-dynamical processes. For example, entrainment of relatively dry, cool air into clouds has two significant effects on their behaviour. First, in familiar fair-weather cumulus, the entrainment strongly reduces the buoyancy of cloudy air. Secondly, in larger precipitating cumuliform clouds, the mixing process tends to alter the spectrum of droplet sizes and hence to affect the coalescence process. In the largest cumuliform clouds, cumulonimbus, entrainment probably plays a minor role, and research over the past twenty years has shown that vertical instability and strong vertical shear of horizontal wind favour their existence. These two factors also strongly influence the particular configuration of airflow within cumulonimbi.

Topographically forced meso-scale features are usually larger than the recognised 'cloud-scale'. Typical systems are lee waves, föhn and bora winds, sea and land breezes, urban-induced circulations and the closely related 'urban climate'. In all these cases a vast literature has emerged over the past few decades. Lee waves are waves to the lee of the obstacle that perturbs the airflow. They are one type of gravity of buoyancy wave. The atmosphere has a vertical temperature structure such that in neutral and stable conditions any parcel of air perturbed in the vertical, will oscillate up and down with a period of about ten minutes. If at the same time the vertically oscillating parcel is moving horizontally with the gradient wind, clearly the trajectory of the parcel will be a wave. The vertical oscillation is due to the action of gravity or buoyancy on the parcel – hence the description of the waves.

Föhn and bora are distinguished by their temperatures and for over a century this difference has been accepted as implying different mechanisms for the two types of wind. In fact, the temperature difference between the winds is as much a function of the temperature of the points that experience the winds at the foot of the leeward side of the hills. So, the chinook is 'warm' in the Albertan winter because before the onset of the wind lowland Alberta is very cold. The bora is 'cold' in the Adriatic coast because this coastal area is quite warm. In both cases descending air is warmed adiabatically, but the resultant temperature of the sinking air can be either warm (föhn) or cold (bora) relative to the air temperatures prior to the onset of the wind. Hence the fundamental problem now becomes, what makes the air sink? There is presently no clear cut answer, but it is highly likely that a mechanism similar to that of lee waves is involved, whereby the downward part of a wave-like flow, hugging the leeward ground surface, represents the föhn or bora.

Sea-land, slope and mountain valley and urban breezes all result from the same basic process. These breezes are essentially thermally driven, the pressure gradient forces that directly induce air movement themselves being due to density and temperature differences over different types of surface. In contrast to many earlier text book explanations, these breezes do not result because 'heated land air being less dense, rises so that a low pressure area forms and therefore sea air moves in'. It is true that density of the heated land air decreases, but this has the effect of creating a relative high pressure at the top of the heated layer (due to the hydrostatic relationship), that is at a height of about 1km. Hence air at this height moves out to sea and it is this export of air from over the land that results in a surface low pressure. In association the import of the land air by the marine atmosphere results in a high pressure area at the sea surface. The couplet of high pressure at the sea surface and low pressure over the land ensures a sea breeze. The reversal of this process at night gives the land breeze.

The same basic mechanism operates over irregular terrain to give slope winds and over urban areas to give the very gentle centripetal flows found therein. Closely associated with airflows over urban areas are most of the other facets of urban climates. The distribution of temperature (heat/cool island), humidity and pollution in urban areas are probably primarily determined by the appropriate surface budget. For example, the heat/cool island is simply a result of the urban surface having a different heat budget from the surrounding rural surface. Nevertheless, part of these budgets is the movement of heat, water and mass by airflow (known as the advective effect) and the complex interplay of airflow and other elements continues to be investigated. More clearly related to airflow is the urban precipitation anomaly – usually manifest as a slight increase in mean summer convective precipitation over urban areas. In this case cloud dynamics are of paramount importance, at one and the same time being influenced by the underlying anomalous conditions of the urban surface and also influencing the efficiency of the microphysical processes that produce precipitation.

Whereas the study of topographical influence upon meso-scale airflows has a history of over a century, free-atmosphere circulations have been closely examined only over the past two decades. The nature and mechanisms of severe local storms, gravity waves, jets, shallow cellular convection and frontal structure have all been substantially clarified in this period. In many cases, these systems were indeed 'discovered' by the new observational tools mentioned above, and the investigations of their mechanisms by both analytical and numerical methods proceeds apace. Examples of these free-atmosphere features are the meso-scale structures within cyclones, more particularly within fronts. These structures frequently are about 50km by 200km and are manifest as 'rainbands'. They themselves comprise smaller systems about 50km across, features which are the basic building blocks of the meso-scale structures to be found in cyclones.

Synoptic scale forms and processes

It is probably fair to say that to the majority of people the synoptic scale means that of extra-tropical cyclones, and cyclones in turn comprise the greater part of the public's perception of meteorology. This is probably so because the model of the frontal cyclone is some sixty-five years old, far older than any other currently accepted meteorological model with the possible exception of the Hadley circulation. The essentially descriptive qualitative model of the 1900s was put on a firmer dynamical, quantitative footing in the 1940s. In its essentials this model is still accepted, not least because it married so well with later ideas of development and the theories of planetary waves.

Most of the recent developments at the synoptic scale have emerged from the recent closer inspection of the tropical atmosphere. Large cloud clusters have been observed by satellite, and tropical cyclones have been extensively observed by aircraft, radar and satellite and have also been studied theoretically. It is now very clear that the release of latent heat is vital to the existence of these storms. Squall lines – lines of severe convective storms, usually tens of hundreds of kilometres long – have also recently been better observed in the tropics. Whilst such features are morphologically similar to their counterparts in the extra-tropical atmosphere, their internal airflows are quite different. A major problem is to explain why this is so.

The long-standing importance of the synoptic scale to weather forecasting has meant that for much of the first half of meteorology's twentieth-century history it received the lion's share of research attention. In turn this resulted in a relative neglect of the smaller and larger scale features. In common with meso-scale phenomena, and in slight contrast to the synoptic scale, planetary-scale features have received more attention in the past two decades.

The planetary scale

Attention has concentrated upon the investigation of the general circulation of the atmosphere *in toto* and on its largest constituent parts. In the latter category, the nature and mechanisms of the trade-wind flows, the monsoons, the high-level jet streams, the quasi-biennial oscillation – all tropical features – have been under intense examination. In many cases, satisfactory descriptions are only beginning to emerge. In the extra-tropics the planetary-scale atmosphere is quite well observed even by the operational networks. The northern hemispheric maps available several times per day to the weather forecaster contain a vast amount of both surface and upper air information in the mid-latitudes with Rossby waves and jet-streams clearly delineated.

Despite the great progress in observing substantial parts of the atmosphere, it became increasingly clear in the 1960s that a respectable knowledge of the general circulation of the whole atmosphere could ensue only from a set of good observations of the whole atmosphere. Such was the aim of FGGE, which served two main purposes: first, to allow empirical investigation of the general circulation with data from an attempt (albeit not successful) to be complete and accurate; secondly, to provide observational bench-marks for general circulation models.

This realisation that it had finally become feasible to observe the whole atmosphere also profoundly influenced the views of the nature of climate and climatic change. Instead of being seen as essentially a statistical description of the elements of most immediate importance to mankind (e.g. temperature, precipitation), climate and its variability emerged as a result of the workings of the general circulation – concerned with the fluxes of energy, water, mass and momentum. In more recent years it has become fashionable to think of the 'climatic system', an entity that includes not only the atmosphere but also the oceans, the cryosphere and the biosphere. It is increasingly appreciated that atmospheric behaviour on the time-scale of weeks and months owes much to the nature and behaviour of the ocean and ice surfaces. The bases of this long-term dependence are the vertical transfer mechanisms. For example, if the sea-surface temperature of a given area becomes anomalous, then the transfer of heat, water and momentum between atmosphere and ocean will become anomalous. In turn, this affects the large-scale pressure and hence motion field in the atmosphere, which result in anomalous weather for weeks, possibly for months, in some areas. These 'teleconnections' between areas of sea-surface temperature (SST) anomaly and areas of anomalous weather have been intensively investigated in both the UK and USA since the term was introduced in 1969. It appears that the brilliant summer of 1976 and, the poor winter of 1981–82 in the UK were closely related to anomalous SSTs in parts of both the Atlantic and Pacific Oceans.

The change in the view of climate has been further encouraged by the capability to construct numerical models of the whole atmosphere – now frequently called 'climate modelling'. The first attempt to model the general circulation was in 1956 in a classic paper by Dr N. Phillips. It was, however, only in the early mid 1960s that reasonably realistic models began to emerge, notably from the laboratory under the direction of Dr J. Smagorinsky in the USA. The past two decades have seen a plethora of models, some of the atmosphere, some of the oceans, some including both; some with no orography, some with a realistic distribution of land and sea and orography, and so on. Their aim is to simulate the general circulation on time-scales ranging from days to years. The weather forecaster has a strong interest in results in terms of days, possibly even two to three weeks. Crossing the threshold of one month leads to the simulation of climate and seasonal changes. Accurate forecasts on this time-scale could be of inestimable value for long-term planning. At longer time scales it may be possible to simulate climatic changes. The prospects are clearly very exciting. It is already possible to simulate the climate in a glacial period of 18000 BP, the possible changes that could ensue from the increasing content of carbon dioxide, and the diversion of the Siberian rivers. No doubt the near future will see some models used to assess the effects of acid rain and nuclear winters. Attractive as these experiments are, it is vital to appreciate that the results are only as good as the models. This is why many research groups are active world wide. Nevertheless, considering that this difficult work has a history of only twenty years or so the achievements to date are indeed impressive.

This section of the chapter has adopted a scale hierarchy which rests upon the basic physical laws that apply to the universe: it is the particular combinations of processes allowed by these laws that results in atmospheric behaviour at any one scale. All features at all scales result from the triggering of some form of instability, negative feedbacks ensuring their eventual demise. Within this framework two powerful instabilities are evident. At the small scale, strong vertical wind shears and buoyancy conspire to create the many forms we find. The resultant potential for growth is called Kelvin–-Helmholtz instability. At the synoptic and planetary scales, baroclinic instability is a major cause of motion systems. There is no well-recognised intrinsic meso-scale instability, but the search is active and its nature and mechanism will probably be clear by the end of the decade.

CONCLUSIONS

It is clear that present day study of the atmosphere is a far cry from the climatologies of the inter-war years. Perhaps their common facets are their quantitative natures. But whereas 'form' and statistical description and analysis were once paramount, it is theory, modelling and appropriate observational support which now occupy the literature. The models are not merely schematic diagrams; they are predictive devices which are used for both operational and research purposes. The test of their quality is the most severe of all – does the predicted value accord well with nature? So successful a technique has this been that there seems to be little doubt that it will continue to be widely used in the next decade or so. Plans are even now being completed for a major international observational and theoretical onslaught on the whole 'climate system'. This truly massive programme is one of the most ambitious experiments ever. In ten years time it may well be possible to claim that we have a respectable predictable model for the global, fluid environment. It is a most exciting prospect.

FURTHER READING

Two books combine to provide a rigorous survey of dynamical, observational and theoretical aspects, the former in particular using a minimum of mathematics, and the latter providing a comprehensive review of meso-scale atmospheric circulations:

Atkinson, B. W. (ed.) (1981) *Dynamical Meteorology: An Introductory Selection* (London: Methuen).

Atkinson B. W. (1981) *Meso-scale Atmospheric Circulations* (London: Academic Press).

More specialist in coverage, but still entirely readable with only a modest mathematical requirement despite being rigorous, are the following treatments on micrometeorology and atmospheric behaviour:

Munn R. E. (1966) *Descriptive Micrometeorology* (London: Academic Press).

Wallace J. M. and Hobbs P. V. (1977) *Atmospheric Science: An Introductory Survey* (Academic Press: New York).

A largely non-mathematical review of the atmosphere in the boundary layer, that is below a height of about one kilometre, is:

Oke T. R. (1978) *Boundary Layer Climates* (London: Methuen).

Two journal articles can be consulted for a consideration of the nature, limitations and values of mathematical general circulation models, and for a mathematical model applied to the simulation of ice-age climate:

Gilchrist A. (1979) 'Concerning general circulation models', *Meteorological Magazine*, vol. 108, pp. 35–51.

Gates W. L. (1976) 'The numerical simulation of ice-age climate with a global general circulation model', *Journal of Atmospheric Sciences*, vol. 33, pp. 1844–73.

Finally, a concise and easily read case in support of the need to analyse the ocean and atmosphere as a single system is given by:

Perry A. H. and Walker J. M. (1977) *The Ocean–Atmosphere System* (London: Longman).

2.3
Geomorphological Processes in Cold Environments

Clifford Embleton
King's College, University of London

Progress in studies of glacial and periglacial processes has, in the last thirty years or so, been more rapid than in any other similar period since the advent of the glacial theory itself approximately 150 years ago. In the field of techniques, the ability to drill deeply through the ice (to more than 2km in Antarctica), to obtain ice-cores and measure englacial temperatures, has revolutionised knowledge of the nature of ice-sheets and glaciers. Photogrammetric and satellite surveys are yielding increasingly accurate maps of glaciers, patterns of former glaciation and the complex physiography of frozen ground areas. New dating techniques are establishing a chronology for the growth of ice-sheets and for pinpointing climatic changes recorded in ice-cores.

Ice and permafrost today occupy about 20 per cent of the land surface of the globe, but in the Quaternary these cold regions included about one-third of all the land. These are without question areas of exciting geomorphology, where unique groups of processes are at work, producing distinctive assemblages of forms. Another compelling reason for the scientific study of cold environments has been a growing appreciation of their economic and strategic potential. In both the North American and Soviet Arctic mineral wealth, especially gold, was the magnet first attracting economic development and settlement to the frozen regions; more recently, oil and natural gas have been in the forefront. Developments of towns, transport systems, even reservoirs, have been possible only through detailed studies of the engineering properties of the permafrost and the processes associated with it.

Interest in the cold environments also springs from the evidence they furnish on environmental change. Ice-sheets, glaciers and permafrost are not static, but possess a dynamic history of large-scale fluctuations whose consequences for man are immense. The world's ice is part of an extraordinarily complex and sensitive climatic–hydrological system, characterised by numerous feedback loops that make exact prediction of the effects of change in any one variable almost impossible. Research suggests that quite minor climatic changes (possibly including man-induced changes) can affect the stability of the major ice-sheets, and that the environmental consequences may be far-reaching – a field explored by David Sugden in Chapter 3.2.

Research into the processes of cold environments has, then, both pure and applied aspects, but it has been the need to solve practical problems that has often triggered major research activity. The resulting volume of publication is huge. Between 1975 and 1978, for example, the US National Report to the International Geophysical Union refers to over 900 papers on snow and ice, while the World Data Center A for Glaciology in Colorado included 4400 entries in its comprehensive bibliography for 1978–1982!

PROCESS DOMAINS

In the cold regions, many processes may be distinguished: glacial, nival and permafrost-related processes form the chief groups (Fig. 2.3.1). They are not mutually exclusive: for example, nival processes are also often active in the glacial domain, and permafrost exists beneath cold glaciers. The processes also vary in intensity, depending on relief and climate. Thus, there are important differences in the dominant processes between the areas of continuous, discontinuous and sporadic permafrost. In the glacial domain, there are likewise major contrasts according to glacier thermal regime (Fig. 2.3.2). Cold-based glaciers frozen to their beds cannot move by sliding but only by internal flow, a situation which minimises bedrock erosion and entrainment of debris. Warm-based (temperate) glaciers, on the other hand, can slide over their beds leading to possibilities of bedrock erosion. Entrainment of debris is favoured by thermal conditions at the bed in which meltwater is refrozen to the base of the glacier, trapping rock particles. The varying thermal regimes in glaciers and permafrost are essentially related to present day climate (though some areas of permafrost in southern Siberia may be relict from the Pleistocene). Relief conditions are the other external control on process intensity: the steeper the slope, the greater the rates of glacier flow, other things being equal, and the greater the mass movement when the permafrost thaws.

GLACIOLOGICAL PROCESSES

It is impossible to understand how ice erodes, transports and deposits without understanding its properties and behaviour under stress in different conditions.

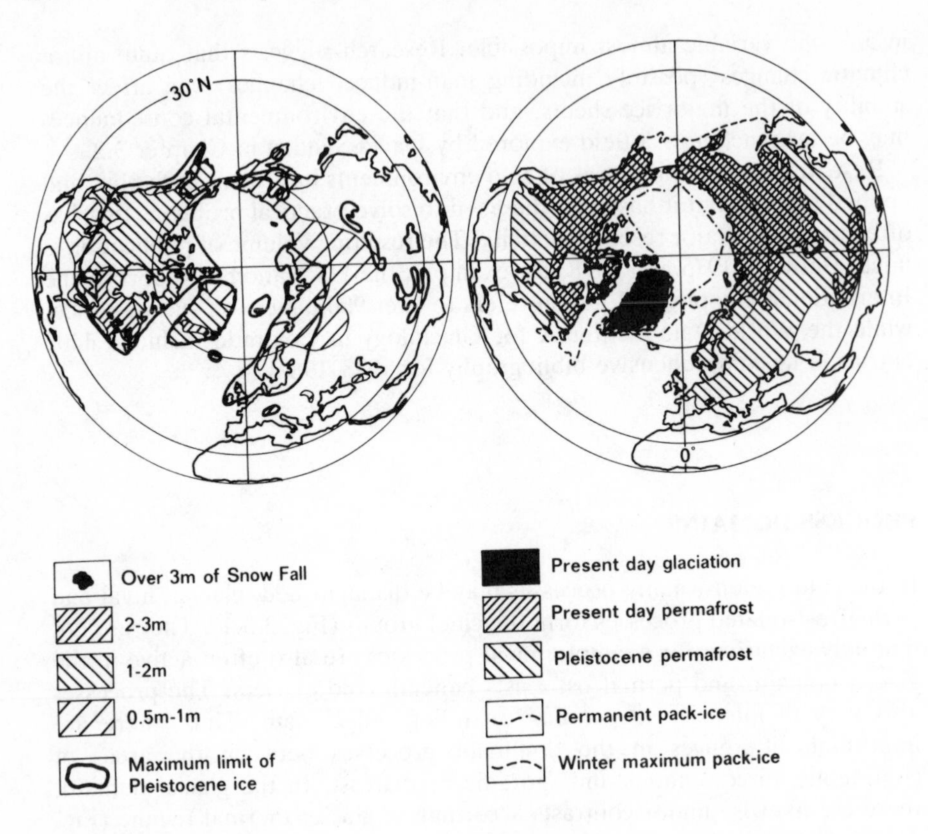

FIGURE 2.3.1 *Cold regions of the northern hemisphere*

The flow of ice

The first simple flow laws defined glacial ice as a sort of plastic which could support small stresses without deforming, but which would deform rapidly at higher stresses. This approximation was abandoned in the 1950s in favour of Glen's power flow law, in which the deformation of ice is proportional to the *n*th power of the applied stress, *n* being greater than one and usually lying in the range of 2 to 4. As more data were acquired, more elaborate mathematical formulations of glacier flow became necessary to reflect the complex behaviour of ice. It is now clear that ice temperature, for instance, is a major factor in the rate of deformation, 'cold' ice deforming more slowly than 'warm' ice close to pressure melting point.

Further complicating the flow law are the discoveries that the rate of response of ice to stress changes with the stress level, and that the debris content of the ice is very important. Near to a glacier bed, ice usually contains debris; once the debris content exceeds about 10 per cent, it progressively

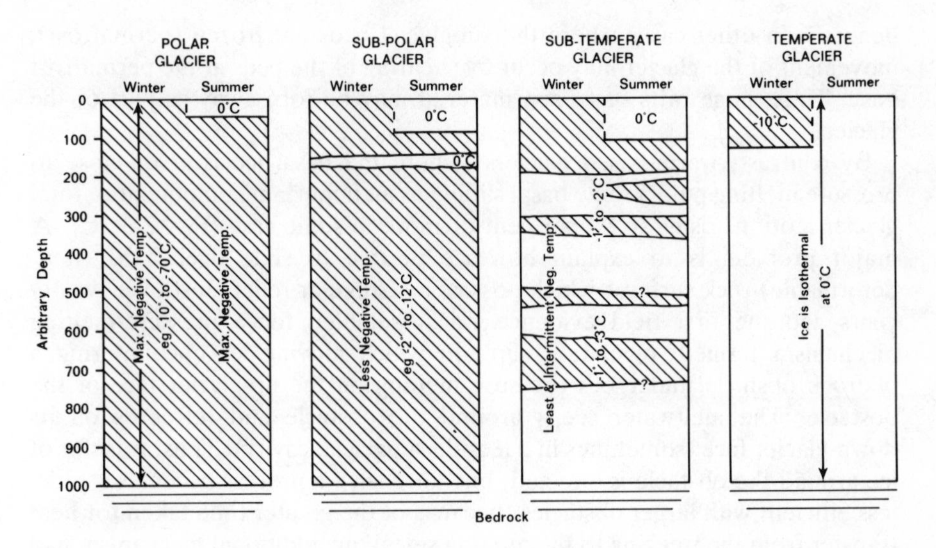

FIGURE 2.3.2 *Thermal classification of glaciers (M. M. Miller, 1973)*
[M. M. Miller, 'Entropy and the self-regulation of glaciers in Arctic and alpine regions', *Research in Polar Alpine Geomorphology*, Proceedings of the 3rd Guelph Symposium on Geomorphology (1973), ed. by B. D. Fahey and R. D. Thompson, pp. 136–58. Published by Geo-Abstracts, Norwich. The diagram appears on p. 137 as Figure 1.]

'stiffens' the ice. In addition, ice with a strongly oriented crystal structure (typical of the ice near the bed of a moving glacier) deforms much faster in the direction of crystal orientation than in any other. The effect of crystal size is not yet fully understood, but it is clear that glacier ice is so variable that a uniform all-embracing flow law is no longer adequate.

Basal sliding

While flow laws represent reasonably satisfactory *approximations* to the actual behaviour of glacier ice, understanding of what happens at the bed of a sliding glacier is still relatively poor. Although the nature of glacier ice is variable, the range of conditions at the glacier bed is infinitely more complex. The distinction between cold-based and warm-based glaciers is a useful one, though somewhat arbitrary. At temperatures below pressure melting point, basal ice will be frozen to bedrock and unable to slide over it. The lowest layers will show rapid shearing and deformation, decreasing upwards, and such glaciers move by internal flow (Figure 2.3.3A) unless the bedrock is itself deformable. It should not be assumed that glaciers rest on impermeable and immobile beds of hard rock: in Figure 2.3.3B a major contribution to the total motion of the ice has been the deformation of unconsolidated materials beneath. In other cases where the subglacial beds are frozen (permafrost), movement of the glacier may occur by shearing of the beds at the permafrost base. Thus, huge rafts of frozen material may be torn away as part of the glacier.

By contrast, warm-based glaciers (where the basal ice is at or close to pressure melting point) show basal sliding as an important component of total glacier motion, rising to 90 per cent or more in some case (Fig. 2.3.3C). A major problem is to explain how ice is able to slide over hard (non-deformable) rock surfaces which possess an irregular relief. Just over twenty years ago the first field evidence was assembled for a pressure-melting mechanism termed 'regelation slip', in which moving ice encountering a bedrock obstacle undergoes pressure melting on the up-glacier face of the obstacle. The meltwater seeps around the obstacle and refreezes on its down-glacier face (sometimes in a leeside subglacial cavity). Thus, transfer of ice around the obstacle is effected, but since such a process becomes rapidly less efficient with larger obstacles (because of the greater time taken for heat transfer from the freezing to the melting sides), an additional mechanism had to be sought.

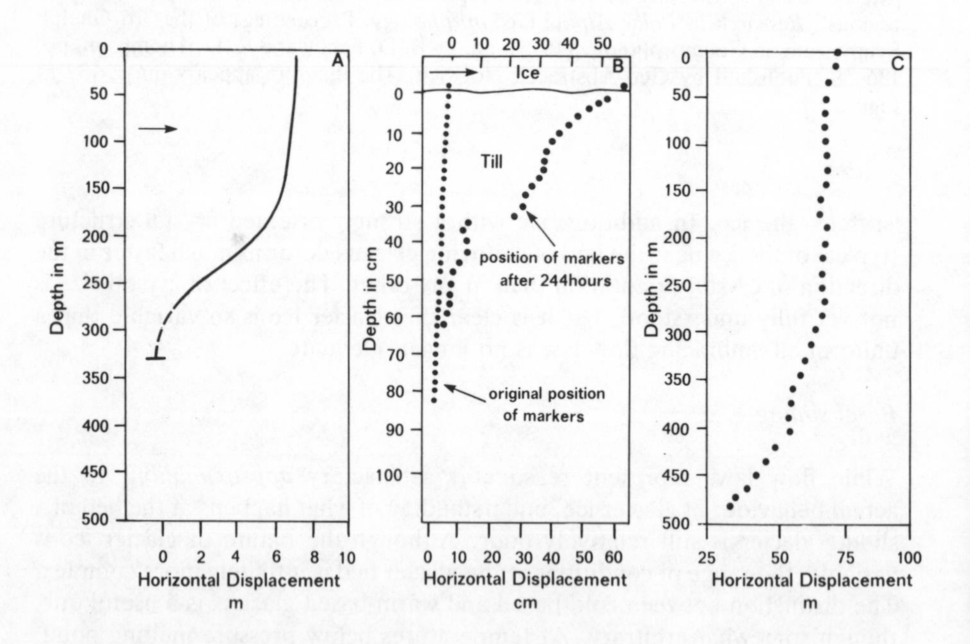

FIGURE 2.3.3 *Modes of glacier flow:*
(A) Velocity profile for a drill-hole in cold ice, Law Dome, Antarctica. The arrow shows ice-flow direction. Extrapolation below 280m assumes no basal sliding. Period of measurement is one year (D. S. Russell-Head and W. F. Budd, 1979).
(B) Deformation of till below Breifamerkurjökull, Iceland. At this site, 88 per cent

of the forward motion of the glacier occurs through deformation of the substratum (G. S. Boulton and A. S. Jones, 1979).
(C) Velocity profile for a drill-hole in temperate ice, Salmon Glacier, British Columbia. Basal sliding over one year amounts to 34m at this site, or 45 per cent of total glacier movement at the surface (W. H. Mathews, 1959).
[References: D. S. Russell-Head and W. F. Budd, 'Ice-sheet flow properties derived from bore-hole shear measurements combined with ice-core studies', *Journal of Glaciology*, 24, 90 (1979), pp. 117–30. The diagram is adapted from Fig. 7 on p. 127).
G. S. Boulton and A. S. Jones, 'Stability of temperate ice caps and ice sheets resting on beds of deformable sediment', *Journal of Glaciology* 24, 90 (1979), pp. 29–43. The diagram is adapted from Fig. 1 on p. 30.
W. H. Mathews, 'Vertical distribution of velocity in Salmon Glacier, British Columbia', *Journal of Glaciology* 3, 26 (1959), pp. 448–54. The diagram is adapted from Fig. 2 on p. 451.]

A few years previously, it had been suggested that regelation slip could be combined with plastic flow involving deformation of the ice due to stress concentrations. This associated plastic flow should be more effective for large obstacles because of the greater stresses. The rate of sliding allowed by this second mechanism would be controlled by the flow law of ice.

The difficulties of testing such hypotheses are formidable. There is some evidence that regelation does occur in subglacial cavities down-glacier from obstacles, but once such cavities are opened for measurement, an artificial temperature situation is created. Another problem is that of formulating a satisfactory flow law for the basal ice which is a structurally complex mixture of ice and rock debris. The greatest difficulty, however, is in specifying the bedrock relief. The sizes and shapes of bedrock irregularities are infinitely variable and have defied realistic mathematical modelling.

Another problem concerns the area of contact between the glacier and its bed, for if subglacial cavities are extensive, contact and friction will be greatly decreased. Friction, however, is also a function of the pressure exerted by the glacier on its bed. The normal pressure on the bedrock is determined by ice thickness and density, but this may be partially offset by the pressure of subglacial meltwater circulating in a thin layer, in discrete channels, or moving through the bedrock if the latter is permeable. Measurements in boreholes have shown that the pressure of basal meltwater may be considerable, in exceptional cases virtually supporting the glacier. High basal meltwater pressures can thus greatly reduce levels of friction and enhance basal sliding. Such glaciers may become extremely unstable and liable to rapid and catastrophic sliding. There is now much evidence that the movement of warm-based glaciers is closely related to varying subglacial water pressures and storage.

However, assumptions that the basal ice of a warm-based glacier is everywhere at pressure-melting point have been recently questioned. The possibility that small patches may be frozen to the bed could greatly increase basal shear stresses, with important implications for sliding and erosion. Such a condition might help to account for the frequently observed 'jerkiness' in temperate glacier motion as the ice attempts to break free from the bed.

GLACIAL GEOMORPHOLOGICAL PROCESSES

Earlier this century, glacial geomorphologists tended to look mainly at the macroforms of erosion such as cirques and U-shaped valleys. Processes were often ignored or described hypothetically, mainly because of lack of data. This did not prevent the formulation of general models for the evolution of glacial landscapes, such as those of W. M. Davis or E. J. Garwood, but these rested on very insecure foundations, discouraged progress in understanding, and have now been largely abandoned. More recent research has concentrated on the processes themselves, while the most significant forms have turned out to be the small-scale bed-forms such as striations, gouges, pot-holes, and flutes.

Erosion processes

It is useful to distinguish processes largely related to ice from processes related to meltwater. Both often work together subglacially (warm-based glaciers), but meltwater will be considered separately for convenience. Ice-related processes can be differentiated into abrasion (responsible for scratching, scouring or polishing) and plucking or quarrying (in which fragments are removed from the bedrock). A third distinction should be made between erosion of hard bedrock and of deformable unlithified sediments.

Abrasion is typical of hard bedrock over which debris-laden ice is flowing. In a few subglacial sites reached by natural or artificial tunnels, the cutting of striae by rock fragments gripped in the basal ice has been described; scoured and polished surfaces result from progressively finer scratching tools. Because of the pseudo-plastic nature of glacier ice, these smaller sized particles are forced against the bed not directly by the ice but by larger rock fragments in the basal ice. The grinding processes produce rock flour which, under warm-based glaciers, is largely flushed out by subglacial meltwater. Measurements of the silt content in pro-glacial streams give abrasion rates beneath actively sliding glaciers of a few millimetres a year. Experiments to measure subglacial abrasion directly (e.g. by inlaying rock plates of known thickness into the subglacial bedrock) have given similar results. Because abrasion operates over substantial areas, the quantities of rock removed can be enormous; up to $5000 m^3/km^2/year$, contrasting with fluvial erosion which, although locally intense, works along narrow channels only. Warm-based glaciers are a prerequisite: ice frozen to its bed does not abrade. Laboratory experiments have also shown that abrasion results from rock-to-rock contacts, the ice simply acting as a transporting medium.

Crushing and fracturing of bedrock protrusions have been discussed both in terms of theoretical rock mechanics and in terms of indirect field evidence. Associated with extreme pressure environments beneath thick ice, there have not yet been direct field studies of these processes. Crescentic gouges and other 'friction cracks' are typical resulting bedforms and are important process indicators. Existing joints or fractures in the bedrock will obviously facilitate the crushing or fracturing process, and the steep broken leesides of *roches moutonnées* may not be due to freeze–thaw action but to direct subglacial pressure acting on a jointed and unsupported rock face.

Some earlier theories of bedrock erosion by 'plucking' action involving, for example, freeze–thaw processes or dilatation mechanisms, have not been well supported by observation. The idea that meltwater produced at the bed of a glacier might refreeze in bedrock joints, widening them and leading to joint-block removal, has proved remarkably difficult to demonstrate. Refreezing beneath warm-based glaciers can occur in leeside subglacial cavities, but it seems doubtful that the small pressure-induced changes of temperature would cause rock failure. More effective might be the migration of meltwater into a zone of cold-based ice or, on a longer time-scale, climatic changes causing shifts in the position of the junction between cold- and warm-based ice.

Bedrock jointing can have many different origins. In the early 1960s it was suggested that dilatation joints could develop as a direct response to rapid glacial erosion, playing a controlling part in the evolving shapes of glacial macroforms. There is some supporting evidence from joint systems paralleling the outlines of cirques and glacial troughs, and evidence too that joints in general exert a powerful control on landform, but a glacial dilatation mechanism remains an hypothesis.

The relative importance of abrasion and block removal (plucking and gouging) has been hotly debated for more than fifty years. Measurements of the output of debris in glacial meltwater streams suggest that much more is produced as fine sediment (presumably from glacial abrasion) than is accumulating as larger fragments. On the other hand, the presence of large fragments in glacial tills suggests that many grinding tools survive abrasion and that they themselves can only come from plucking.

The role of basal thermal regime is fundamental. Warm-based sliding glaciers are required for abrasion. Cold-based glaciers will protect the bedrock floor, except occasionally if large rock fragments in the lower layers of the moving ice happen to collide with bedrock protrusions. Because the bond between cold ice and a rock surface can resist great shear stresses, plucking beneath cold ice can only occur if the bedrock is already well jointed or unlithified. There have been attempts to establish the basal temperature patterns of former Quaternary ice-sheets and to relate them to erosional landscape features. There is still, however, controversy about the erosional role of ice-sheets. One view holds that they are not dramatic agents of erosion but, in their central immobile zones, mainly protective. A contrasting opinion is that they are responsible for an assemblage of equilibrium erosion forms closely linked to basal temperatures and reflecting steady state conditions over periods of the order of 100 000 years.

Geochemical denudation Until recently, there has been little discussion of subglacial chemical weathering. That situation is now changing; studies of the dissolved loads in emerging meltwater, of chemical exchange processes in subglacial cavities, and of chemical coatings on rock surfaces either now or recently under the ice, show that chemical processes are far from absent. Evidence so far indicates that the processes only operate beneath warm-based glaciers where meltwater descending to the glacier bed flushes out solutes produced by geochemical processes at work during periods of limited basal water circulation. Bedrock coatings can affect the strength and permeability of the bedrock, and dissolved minerals will affect the freezing temperatures of meltwater.

The role of meltwater The more the internal hydrology of temperate glaciers is investigated, the more complex the system appears to be. Interest stems not only from the role of meltwater in geomorphological processes and glacier movement but also from its practical importance (for instance, in hydro-electric power schemes), and in connection with the devastation that can result from the bursting of subglacial or englacial water bodies (as in the *jökulhlaups* of Iceland). Such catastrophic events highlight the instability of the meltwater systems.

Subglacial water travels in two distinct modes. First, the bulk of the water flows through complex channel systems which are liable to change radically even from year to year. Secondly, the basal water film vital to the basal sliding process may be a millimetre or less in thickness. In subglacial and englacial tunnels, the dynamics of water flow and sediment transport are quite different from those of open streams. The significance of debris transport by through-flowing meltwater has been highlighted by the finding that as much as 90 per cent of the debris passing through a given cross-section of a temperate glacier may be carried by meltwater rather than the ice itself. Another field of great interest is the study of the bedforms produced by water moving under pressure in tunnels and at high velocities: the familiar pot-holes and the less familiar forms of cavitation erosion show that hard bedrock can be powerfully attacked. Indeed, it is becoming clear that, in talking of glacial erosion, one must include subglacial fluvial erosion as a major component.

Entrainment Uncertainty remains about several aspects of how debris is incorporated into and transported by glaciers, and especially over the processes by which it might be raised from the glacier bed to an englacial position. Clearly, unless the products of subglacial erosion are removed, glacial erosion will cease to be possible, and whereas meltwater flushes out a great part of the debris from beneath temperate glaciers, there are also direct glacial transporting mechanisms. Sediment is introduced to the glacier both supraglacially (rockfalls on the glacier surface, avalanches, wind, surface runoff) and subglacially (from bed erosion). The supraglacial component can easily become englacial (burial by snow in the accumulation zone, falling or washing into crevasses), though part may travel supraglacially for the whole time; likewise, a part of the subglacial material may remain in a subglacial position. The englacial debris may be either disseminated through the ice, or it may be concentrated in debris-rich bands.

Several processes by which glaciers can entrain basal debris have been discussed. Refreezing of basal meltwater will trap sediment and attach it to the glacier base, especially where basal meltwater is migrating into a zone of basal freezing – again, the importance of basal thermal regime is clear. Similarly, whole rafts of unconsolidated sediment may be frozen on to a glacier sole and detached from their origin. Another less important process is the upward squeezing of water-soaked sediment into basal fissures or crevasses in the ice. A third possibility is that debris may move up into the ice along shear-planes rising from the bed. These commonly form near glacier margins where ice-flow is obstructed, for instance by stagnant marginal ice or moraine. In other cases, structures superficially resembling shear-planes have proved not to be of this origin; nor is it clear how debris moves along a shear-plane, or whether the shear is located along debris-rich layers precisely because such layers provide zones of weakness. Acquisition of thick basal debris seems more likely to be due to on-freezing under a specific basal temperature regime; and as more debris and ice is attached from below, so the older material above it will become progressively raised from the bed.

Deposition processes

If the volume of literature is any guide, there is as much interest in deposition as in glacial erosion. However, the actual processes responsible for depositional forms are often rather poorly understood. Given the difficulty of seeing what is happening beneath modern glaciers, there is great reliance on deducing processes from studies of deposits and forms. This approach, however, raises further problems. There is frequently a lack of good exposures, especially since the deposits may vary rapidly over short distances. A more theoretical problem is that of equifinality: many different processes may lead to the same landform. It is dangerous to make deductions about genesis from external form alone: studies of internal structure and composition are essential.

The major advances in understanding glacial and related deposits have come from several directions. First, sedimentology has made great strides in the past twenty years, and analyses of sediments have provided vital information on sediment genesis. Secondly, much more work has been done on the processes of sedimentation in modern glacial environments. Thirdly, new techniques of dating, separating and characterising sediments have been developed, such as radiocarbon, palaeomagnetic signature, X-ray diffraction, heavy mineral analysis and electron microscopy.

Progress in understanding glacial deposition has necessitated an upheaval in terminology and concept. No longer are erosion and deposition seen as mutually exclusive (e.g. the erroneous idea that erosion belongs to highlands, deposition to lowlands). Glacier advance and retreat are no longer considered only in terms of ice-margin fluctuations, but in three dimensional terms of the thickening or thinning of the ice. Simple classifications based on surface form and supposed origin (moraines, eskers, drumlins, etc.) are being abandoned in favour of process-based models.

There is a fundamental distinction between primary depositional processes, which derive debris directly from glacial ice and are unique to the glacial environment, and secondary processes which rework, transport and redeposit sediment. Secondary processes may also be found operating in non-glacial environments: they include fluvial, mass movement, marine, lacustrine and aeolian processes, which impart non-glacial characteristics to the sediments.

Till studies The most distinctive product of primary depositional processes is till, and till studies have recently become an important subdiscipline. Interest in till stems from the information it provides on subglacial processes; from its engineering properties (in connection with its behaviour in artificial cuttings, under building loads, etc.); and from inferences that are possible about former glacier thicknesses (from the degree of consolidation) and directions of flow (from till fabric analyses).

Within some large areas, the composition and structure of till can be amazingly uniform, yet the term till has also been used for a wide variety of deposits, including not only basal lodgement till but also melt-out till, sublimation till, flow till and waterlain till. The problem of terminology and definition is not confined to till: old-established terms such as moraine, drumlin, esker and kame have been applied in widely different senses and have been given assumed process connotations that have turned out subsequently to be partially or wholly incorrect.

The main controls on depositional processes yielding till are intimately related to the dynamic and thermal characteristics of the glacier. Figure 2.3.4 shows a process classification. Melt-out and sublimation demand specific thermal regimes. Melting at the surface of a temperate glacier can be extremely rapid (as much as 20m a year) and, depending on the debris content of the upper parts of the glacier, a cover of supraglacial till may develop. Sublimation at the surface of a cold glacier is much slower, but has similar results. Melt-out till will accumulate beneath warm-based glaciers if not flushed out by meltwater. Lodgement till accumulates from subglacial processes only. Lodgement is the forcible pressing of rock flour into voids

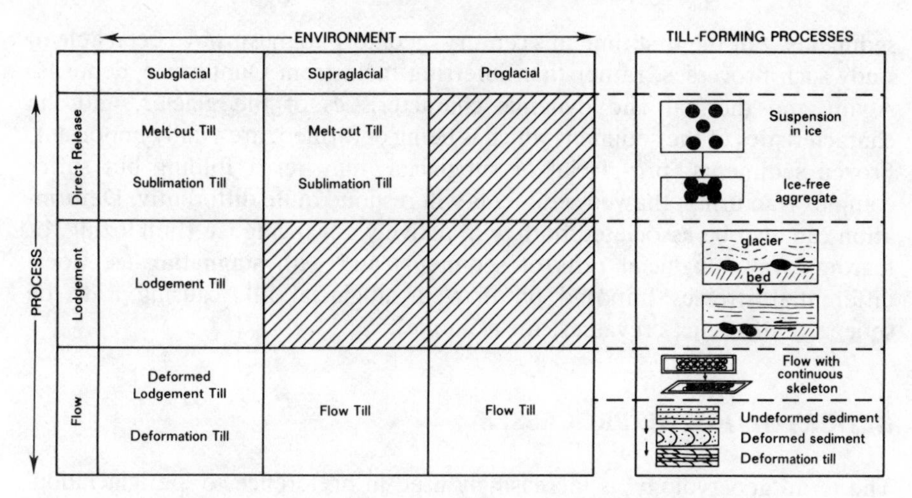

FIGURE 2.3.4 *A process classification for glacial till (G. S. Boulton, 1980)* [Reference: G. S. Boulton, 'Classification of till', *Quaternary Newsletter*, 31 (May 1985), pp. 1–12. The diagram is adapted from Fig. 1 on p. 6.]

between larger particles, the resulting clay till being smeared or plastered on to the bed, accompanied by shearing. Basal sliding is a prerequisite: except where bed obstacles project up into the ice to intercept higher debris-carrying flow lines, it is difficult to conceive how lodgement could occur beneath cold ice. There are complex interactions between the bed and the basal ice carrying debris at temperatures close to pressure melting point. Subglacial melt-out and lodgement will happen simultaneously, but will also vary spatially, so that there may be patchy stagnation of basal ice and patchy lodgement. Continued movement of other ice over and around such patches may be one means by which drumlins, fluted till and other streamlined forms are produced. The number of hypotheses to explain drumlins continues to expand; patchy subglacial freezing, inhomogeneity of basal sediments, dilatancy of till under pressure, squeezing into subglacial cavities, and moulding of a newly thawed bed of till under an advancing glacier are just a few relevant ideas recently discussed.

Flow tills have been considered since the turn of the century when these glacially induced mass movements of water-soaked supraglacial till were first described in Spitsbergen. They are often an important component of glacial deposition whose recognition has revolutionised the interpretation of Quaternary compound till sequences, avoiding the need to postulate multiple glaciations. Flow tills resting on lodgement tills can be explained as the product of the decay of a single glacial episode.

Another rapidly developing field has been the study of glacio-tectonic sediment deformation. One reason for this growing interest is that many glaciers have recently been advancing, pushing or overriding their proglacial sediments. For the first time this century, geomorphologists have been able to study such processes, rather than inferring them from Quaternary deposits. Again, the thermal and dynamic characteristics of the glacier, and the characteristics of the sediment which is being deformed, are vitally important. Frozen sediments, pro-glacial or subglacial, may resist folding but suffer complex fracturing; thawed sediments will respond quite differently. Deformation can also be associated both with actively advancing ice (bulldozing, or tearing away subglacial rafts of sediment) and with stagnating ice (local differential stresses imposed on thawed subglacial till, causing it to be squeezed into relict crevasses, for example).

GEOCRYOLOGICAL PROCESSES

The term 'geocryology' is increasingly used in preference to 'periglaciation' for the study of cold-climate, non-glacial phenomena including some, such as frost weathering, that are not unique to the cold-climate domain. There have been tremendous advances in the understanding of these processes, reflecting partly the general growth in process studies, but also arising from the needs of engineers and others to obtain more and better data.

Frost weathering

This is one of many processes that were once considered simple but have in the last thirty years been totally re-evaluated. The basic idea was that freezing of water in rocks caused their disintegration by volumetric expansion, and there was said to be abundant supporting geomorphological evidence in the form of block-fields, screes, angular tors and other rock faces. Doubts began when field measurements of temperature where freeze–thaw action was supposed to be most effective, failed to show the postulated frost cycles. Many arctic areas, for example, exhibit few frost cycles, often only during the changeover between winter freezing and summer thawing. Further doubts arose when, in laboratory studies, many rocks failed to shatter in the way expected. It was also discovered that at depths of only a few centimetres, under a snow cover, or behind or beneath a glacier, temperatures remained relatively constant.

In the 1950s, many rather inconclusive field and laboratory studies of frost action were undertaken. Hypotheses other than volumetric expansion were proposed – crystallisation pressure in closed systems, crystallisation with water being drawn in, hydraulic pressure due to an advancing freezing plane, and so on – but none has proved completely satisfactory. Since 1965, the Centre de Géomorphologie du CNRS in Caen, France, has undertaken an immense programme of laboratory-based research into the effects of freeze–thaw action. Some aspects of the process have been clarified, but there is still a long way to go. Their experiments showed that more severe freezing does not produce increased breakdown, but rapid freezing does. Rock fatigue can help: disintegration is more likely after several hundred frost cycles. Rocks of low porosity are fairly immune, as are rocks where cracks are wider than 0.5mm. Cracks or pores are obviously needed for water to penetrate in the first place, so that frost action can never take place in a totally closed system; yet if the system is not closed, pressures of crystal growth or expansion can be relieved without cracking. The Caen experiments have shown that rock-type is critical, and suggest that different lithologies may be frost-weathered by different processes. There is a growing feeling that chemical processes, especially hydration, may be involved, and that processes cannot be isolated from one another – for instance, cracks generated by hydration stress may then be exploited by crystal growth.

Nival processes

Since the beginning of the century geomorphologists have debated the role of snow in land sculpture. There is agreement that rapidly moving snow, as avalanches, plays an important part in the geomorphology of many high mountain regions. Avalanches are essentially a type of mass movement. There has been much research on their causes, prediction, prevention or mitigation, especially since 1931 when the Swiss Commission for Snow and Avalanche Research was founded. Their geomorphological effects, however, have been comparatively neglected until the last thirty years. Only a proportion of avalanches contact the ground, and it is only these that are potentially erosive as well as transportational. The velocities attained may be high but do not normally exceed 100km/hour. Although snow–air suspensions may attain two or three times this velocity, their erosive work is negligible, and damage from such avalanches is largely due to wind-blast.

Snow sliding is distinguished from avalanches by much slower movements, measureable but not normally visible. This is akin to the creep and basal sliding of glaciers, for the properties of consolidated snow are not dissimilar. Some evidence has been produced that slow sliding of thick snow beds can be a minor abrasive agent, causing rock surface striation if rock fragments are present in the snow.

The term 'nivation', coined in 1900, is applied to a supposed set of weathering and removal mechanisms associated with stationary snow patches. The basic idea is that intermittent refreezing of snow melt causes rock disintegration beneath or around a snow patch, the weathering products being removed by meltwater runoff or gelifluction. Recent investigations have mostly failed to confirm such a hypothesis. The freeze–thaw weathering process has been questioned, and even thin snow beds efficiently insulate the ground from atmospheric frost cycles. When snow thickness exceeds a metre, ground temperatures remain remarkably close to 0°C, unfavourable to freeze–thaw weathering. More favourable are the margins of snow patches where the ground may be wetted during the day and exposed to freezing at night. Even here, however, any observed rock disintegration may be a hydration phenomenon rather than freeze–thaw weathering.

The possible participation of chemical weathering has also been considered, mainly because the optimum temperature for absorption of CO_2 gas by water is near to 0°C. However, measurements have failed to show that snowmelt contains enhanced amounts of dissolved CO_2, and recent studies in the Arctic have made it clear that limestone solution there is slow compared with temperate or tropical areas. There is no suggestion, moreover, that snowmelt with dissolved CO_2 is a significant factor in the weathering of non-calcareous rocks.

Altogether, the search for a nivation mechanism has produced no clear answer as to what the processes might be. This in turn has led many recently to think that supposed nivation erosion features, such as nivation hollows or cryoplanation terraces, are really structural features, and that the snow banks simply occupy them (as favoured sites for snow collection) without significantly modifying them, except for the action of snowmelt runoff in washing out fines.

Frozen ground processes

Scientific interest in the frozen ground regions commenced in the last century and intensified with their economic development. At the same time, technological advances have triggered a vast increase in the availability of scientific data and have made possible long-term monitoring of physical processes. Traditional and rather vague concepts of a periglacial environment are giving way to more precise geocryological studies.

The classic experiments of Taber on soil freezing, made about 1930, gave the first insights into this process. Current opinion favours a capillary theory in which the pressure differential between soil ice and water results in the migration (by suction) of moisture towards an ice interface. Under the right circumstances (high pore-water pressures, shallow depth, fine-grained materials and slow freezing rates), bodies of ground ice may grow to massive proportions, several metres thick. This is segregated ice whose formation has displaced soil particles and which, on thawing, will liberate more water than can be retained in the soil. The formation of segregated ice is the key to understanding the process of frost-heaving which has immense importance for engineering in permafrost areas. Surface displacements may exceed 15cm in an annual cycle, involving forces of uplift of as much as 140 tonnes per square metre.

The striking forms of patterned ground in Arctic areas have long been the subject of speculation. Although some are now known to be non-cryogenic, most are related to repeated ground freezing and thawing. Sorting of particle sizes takes place both laterally and vertically. In vertical sorting, there is a strong tendency for coarser grades to move upwards and for fines to descend. descend.

Thermal contraction cracking is another long-known phenomenon of severe frost climates, especially affecting ice-rich sediments. Cracks initiated by sudden lowering of ground temperatures fill with ice or fine sediment, depending on the humidity or aridity of the climate; with repeated cracking each winter, the structures widen into ice- or sand-wedges whose width is an indicator of age. It has been shown that mean annual temperatures less than −6°C are required, and therefore fossil Quaternary ice wedges can be useful palaeoclimatic indicators.

As well as the attention paid to processes associated with ground freezing there has been great interest in the processes and results of the thawing of frozen ground, induced either naturally or by man's activities. The term 'thermokarst' was introduced over fifty years ago to describe the features and processes of permafrost degradation, which are of immense practical importance. Because of the high ground-ice content of some permafrost, thawing produces more water than can be held in the soil. Excess pore-pressures are set up, reducing cohesion and shear strength until flow or failure of a slope occurs. The changes take place with exceptional rapidity and cause even low-angled slopes to be unstable. Such thaw consolidation processes are unique to the permafrost environment, and are probably responsible for some of the so-called cryoturbation structures in Quaternary deposits. Involutions and injections, often considered to be the product of ground freezing and ice segregation, may in fact be the product of thaw.

CONCLUSIONS

Ice is a unique substance whose properties and behaviour are now much better understood, mainly as a result of the detailed experimental work of the last thirty years. Understanding of the geomorphological processes related to snow and ice still lags behind, but at least we have now moved into an era when verbal descriptions of assumed processes and forms have largely given way to precise measurement and analysis. Progress has been considerably aided by technological advances, ranging from remote sensing to electron microscopy. Older ideas are being questioned, in some cases abandoned; many processes, such as ground-ice formation, turn out on investigation to be much more complex; 'new' processes, such as hydration weathering in place of freeze–thaw action, and subglacial geochemical denudation, are being considered. On the more practical side, many of the findings of glaciological and geocryological research are of immense importance to man. Studies of snow crystals led to advances in avalanche prediction or prevention; studies of meltwater systems in glaciers have benefited hydro-electric engineering; knowledge of the types and properties of permafrost has paved the way for pipeline and reservoir construction in the Arctic – the list of ways in which applied glaciology has helped in resource utilisation and hazard control is almost endless.

FURTHER READING

Even restricting our review to the last ten years still leaves a huge volume of literature to consider. Three primary sources, the first being indispensable for its introduction to glaciology rather than geomorphology, are:

Paterson W. S. B. (1981) *The Physics of Glaciers* (Oxford: Pergamon Press).
Embleton C. and King C. A. M. (1975) *Glacial Geomorphology* (London: Edward Arnold).
Sugden D. E. and John B. S. (1976) *Glaciers and Landscape* (London: Edward Arnold).

A more recent multi-authored text dealing with glacial land systems and sediments, combined with an engineering approach to their utilisation, is:

Eyles N. (ed.) (1983) *Glacial Geology* (Oxford: Pergamon Press).

A broader geographical perspective of both human and physical geography which places glacial processes into context is provided by:

Sugden D. E. (1982) *Arctic and Antarctic* (Oxford: Basil Blackwell).

For a detailed discussion of the geomorphological processes operating in cold environments, see chapters 3, 6, 8 and 9 of:

Embleton, C. and Thornes, J. B. (eds) (1979) *Process in Geomorphology* (London: Edward Arnold).

On purely glacial processes, a new text is:

Drewry D. (1986) *Glacial Geologic Processes* (London: Edward Arnold).

A comprehensive study of frozen ground areas, their processes and characteristics, is given by:

Washburn A. L. (1979) *Geocryology* (London: Edward Arnold).

Apart from these texts, there are many relevant volumes of conference proceedings, and research papers are published in a wide variety of journals of which the most important in English are the *Journal of Glaciology* (from 1947), the *Annals of Glaciology* (from 1980), *Arctic and Alpine Research* (from 1969) and *Geografiska Annaler* (from 1919).

The best way of locating papers on any particular topic is through the medium of *Geo-Abstracts* (Norwich: Geobooks) and the associated cumulative indices (from 1960). In addition, excellent review articles on glacial and periglacial geomorphology are regularly published in *Progress in Physical Geography* (London: Edward Arnold).

2.4
Theory and Reality in Coastal Geomorphology

William Ritchie
University of Aberdeen

The importance of coastal geomorphology is increasing as a result of a worldwide movement of people to the coast, where there are wide-ranging developments, from agricultural reclamation to sophisticated industrial operations. There is particular interest in tourist and holiday activities, many of which require massive capital and recurrent cash inputs for protective works. Beach nourishment is a common form of coastal engineering, but many other developments are designed to alter coastal landforms to accord with a perceived optimum usage. Since they are unconsolidated and easily excavated, low beach-type coastlines carry the main impact of these changes. The place of coastal geomorphology in physical geography has never been in doubt but the nature, pace and scale of man-made changes, from equatorial to polar areas, have reinforced its relevance and range of applications.

THE SCOPE OF COASTAL GEOMORPHOLOGY

Coastlines fall into two broad groups – the hard and usually high, and the low and almost invariably soft coastlines. Hard, high coastlines offer less to process geomorphologists. With the exception of recent research on biological processes, including the long-neglected geomorphological study of coral and other biogenic landforms, the study of hard rock forms has been concerned with the abrasion and lowering of rock platforms, usually as part of research to determine the origin and age of planation surfaces. Nevertheless, applications for such work exist, such as those concerning the persistence of oil pollution on a rocky shore, but are comparatively rare. Indeed the study of wave processes and erosion rates across these 'neglected coastal features' has been described as one of the main gaps in contemporary coastal research. Nevertheless, this omission is understandable not only as a result of the relative absence of economic developments on hard rock coasts, but also because of the need to shift the time-scale of investigation to that of geology. Essentially, with the exception of the study of rapidly eroding soft cliffs such as occur in boulder clay (Yorkshire) or Crag Formations (Lincolnshire and Norfolk), the interpretation of rock coasts remains dominated by geological factors. Cliffs are frequently inherited forms, and are often the product of land rather than marine processes. Moreover, their shape, position and altitude are a result of the vagaries of sea level change. Nevertheless, there are links with the more dynamic sedimentary environments such as beaches, since cliffs are sometimes primary sediment sources.

Perhaps the role of rock forms in creating boundaries is more important, either as headlands or hinge-points. Until relatively recently, coastal geomorphological literature, particularly of a descriptive nature, made considerable reference to the coastal unit as being some form of bay with headlands, possibly as a consequence of its relative simplicity. More recently, studies of planimetric shape described as a half-heart shaped, log-spiral or zeta curves have reinforced the importance of the headland as the critical boundary in the subdivision of regional coastlines. Another reason for the interest in the headland–bay unit was the past dominance of research in relatively high latitude areas where hard rock forms are more common, and where 'there is a combination of a wave regime dominated by the storm wave, high tidal ranges, and geological variety complicated by the incidence of glaciation' (quotation from Kidson by de Boer, 1973). More recent texts and scientific papers now recognise that the majority of world coastlines, and indeed most of the problems associated with world coastlines, are not concerned with cliffs or rock platforms but with low unconsolidated forms such as beaches and a series of landforms generally described as coastal wetlands. A review of publications over the last ten years shows a dominance of papers from the Gulf and Atlantic coasts of the USA and eastern Australia, with most research derived from beach, barrier island, dune and deltaic environments.

The upsurge of interest in low latitude coastal environments has produced significant geomorphological advances, particularly in quantifying the relationships between sedimentary forms and energy inputs, specifically from waves. The reason for the intellectual excitment engendered by this research is not a product of the use of new instrumentation or technology (for there have been remarkably few developments since the 1960s) but the general applicability of the results; results that were obtained not in coastal areas of Britain, where one has invariably to consider an extended range of factors and the uncertainty of an extreme or catastrophic event, but in more predictable climatic and wave environments. With no disrespect to American or Australian geomorphologists, their outdoor laboratories are simpler, lacking inherited glacial or geological remnants, cyclonic wind and wave patterns and unquantifiable sediment budgets. Geomorphological progress, as in any science, must begin by holding as many variables as possible constant. Once basic models have been developed in these 'laboratory' conditions they can be transferred, tested and refined in more complex situations.

TWO GENERAL MODELS

Most geomorphological research now adopts some form of systems approach, leading to either physical (using hardware) or mathematical/computing simulation. Experimental measurements have produced one of the most useful and widely applicable models in coastal geomorphology; Short's beach-stage model. This is used to illustrate the concept of cyclical development of beach forms and their relationships to the main causal energy input, wave action, and in so doing offers an explanation of the progressive evolution of such associated features as nearshore bars, troughs, cusps, runnels and rip currents. This model can only be described with reference to two diagrams (Figs. 2.4.1 and 2.4.2), and these and the accompanying explanation are condensed and simplified from several papers.

The dissipative extreme has been described as analagous to the 'storm' or 'winter' profile. The beaches as a whole are wide, low gradient with multibarred surf zones. Waves break by spilling and dissipate progressively across the wide surf zone. Fully reflective beaches have typical steep 'summer' profiles. Waves surge up the beach face and collapse with considerable turbulence. The beach often has a 'step' below mid tide level but there is relatively little sand storage below water level. Essentially, this model suggests that beaches pass through a series of identifiable stages that are accretional with falling wave heights and erosional with rising wave heights. A critical level appears to be a wave height of the order of 1 to 1.5m. Stage 1 is fully reflective and Stage 6 fully dissipative. The stages may reverse if wave conditions change. Figure 2.4.1 shows the plot of probability of occurrence (out of 10) of each stage for a given level of wave height. The profiles and planimetry of the main stages are shown in Figure 2.4.2 which provides considerable information on such associated beach features as troughs, rip currents, bars, ridges and runnels and introduces the concept of rhythmic (longshore) morphology.

The value of this beach-stage model to coastal geomorphology relates to the manner whereby the relationship between wave energy input and changes in beach profile are established. Further, additional morphological elements such as the onshore–offshore movement of sand bars are integrated into the model. As a teaching device it is not difficult to envisage a field situation where a group of pupils or students could compare an actual area of beach with the appropriate stage of the model. Questions concerning differences between the model and reality would be engendered, and would inevitably lead to a discussion of other possible variables. Some assessment of wave heights could be made over a subsequent period of a few days to test the validity of the proposed sequence of morphological change. Admittedly,

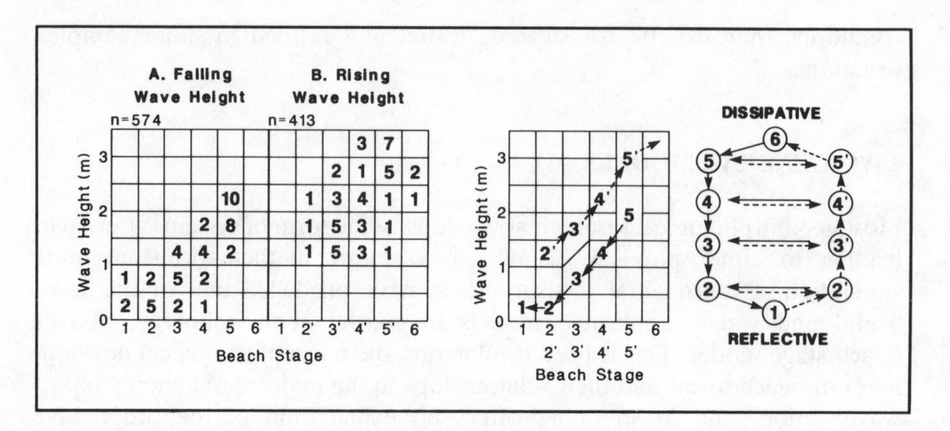

FIGURE 2.4.1 *Plot of probability of occurrence (out of 10) of a specific beach stage* (1 being reflective and 6 being dissipative) – see Figure 2.4.2 – for a given wave height as it falls (A) and rises (B) (left-hand diagram).
Right-hand diagram shows location of model beach stage for levels of wave height under rising and falling wave conditions. Solid arrows indicate falling wave height; dashed arrows indicate wave height and movement to erosional (dissipative) stages. A movement to any beach stage must be in the direction of the arrows (adapted from Short 1980, and Wright and Short 1983).

there is a danger of oversimplification and later work by the authors of the model points out the necessity to appreciate the significance of such factors as sediment size, more complex hydrodynamic processes and offshore factors, but these complexities should not be allowed to detract from the stimulus provided by testing this model against a wide range of beach-type coastal environments.

The second example of a general model, that was developed after years of painstaking research into separate components and was then synthesised into a whole, is the decaying delta-barrier island cycle as developed by Boyd and Penland. The production of this model (Fig. 2.4.3) was derived from a series of studies of individual coastal barrier beach and island landforms associated with the evolution of the Mississippi Delta and adjacent coastlines. In many ways it is an example of a space–time interaction, in that different temporal stages of the cycle can be seen along the present deltaic coastal plain of south Louisiana. The concept of the total delta plain needs initial explanation and is in its own right an example of a general coastal model, developed from the Mississippi but now recognised in all major world deltas.

The essence of both models is the process of delta-switching. This is illustrated by Figure 2.4.4 which shows the sequence of active lobes of the Mississippi Delta. In general, for hydraulic reasons the locus of active deposition with associated levées, sand bars, crevasse sedimentation and other active growth features switches to another location within the wider deltaic area. Once abandoned, the old lobe, denied the lifeblood of sediment and subject to regional subsidence, decays and land surfaces gradually convert to marsh, backswamp and ultimately sucumb to marine penetration. As the active delta decays, sand bodies which are normally derived from the small percentage of sand brought down by distributaries, are modified by wave action to form beaches and sandspits. With further subsidence these sand bodies, most of which have low beach and dune ridges, become detached and form barrier islands. As sinking continues, these long, narrow islands are increasingly at a distance from the coastline, although they still indicate the former position of the active delta front. Finally, wave action, especially during extreme events such as tropical storms, reduces the barriers to shoal areas which are ultimately dissipated. The sequence becomes a cycle when the shoal area is reoccupied as the active lobe of delta front sedimentation (during a later episode of delta switching) returns to its former position. tion.

Like the beach-stage model, the delta front model also encapsulates and explains associated features of delta margin morphology such as washover surfaces, chenier ridges, barrier inlets and dune forms. Unlike the beach-stage model, the delta front model has not been tested against other major delta systems, but there would seem to be no reason to assume that it would not be of general value as long as certain conditions are satisfied, i.e. the process of delta lobe switching, subsidence and a sufficiency of sand in the deltaic depositional environment and relative tectonic and climatic stability.

THE COAST IN THREE DIMENSIONS: THE TIDAL FACTOR

A major compounding factor in coastal geomorphology is the effect of tides, not only in the sense of the rhythmic vertical shift of energy impact, but also in those areas where tidal flow is restricted and therefore significant for erosion and deposition. Numerous papers, largely deriving from the barrier island inlets of the Atlantic coast of the USA have provided substantial insights into the relative importance of ebb and flood flows. The terminology of flood and ebb deltas and the relationship of such forms to tidal range and inlet geometry has now spread outwards from the original study areas, and in so doing has provided a central core to a continuum of coastal forms associated with tidal flows; a continuum with open estuaries at one extreme and deltas at the other. These general relationships are summarised in Figure 2.4.5 which has global applications.

The essential process explaining the various inter and subtidal forms is the separation of flow at different stages of the tidal cycle as a result of different relative water elevations. The addition of fresh water discharge into the estuary, lagoon or back bay increases complexity as water is ponded back with rising tidal water levels and released with increasing velocity as the tidal

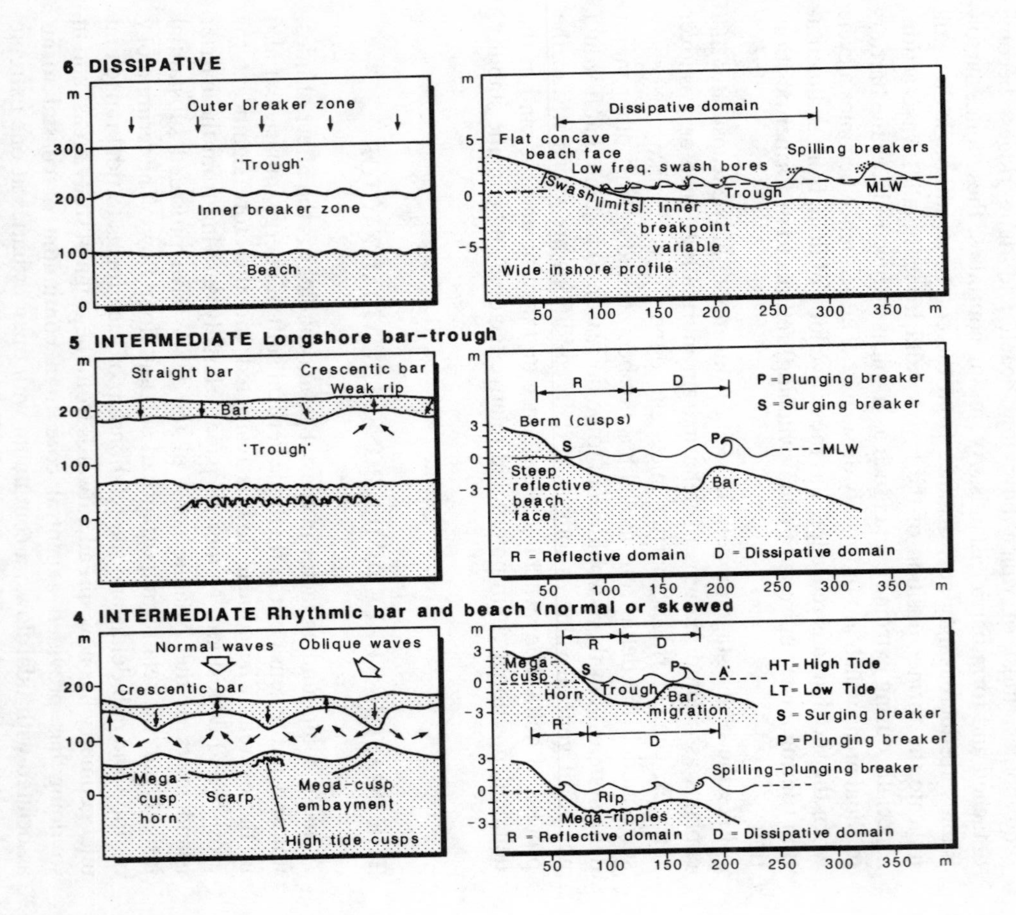
6 DISSIPATIVE
Outer breaker zone
'Trough'
Inner breaker zone
Beach
Dissipative domain
Flat concave beach face
Low freq. swash bores
Spilling breakers
Swash limits
Inner breakpoint variable
Trough
MLW
Wide inshore profile
5 INTERMEDIATE Longshore bar–trough
Straight bar
Crescentic bar
Weak rip
Bar
'Trough'
P = Plunging breaker
S = Surging breaker
Berm (cusps)
Steep reflective beach face
MLW
Bar
R = Reflective domain
D = Dissipative domain
4 INTERMEDIATE Rhythmic bar and beach (normal or skewed
Normal waves
Oblique waves
Crescentic bar
Mega-cusp horn
Scarp
Mega-cusp embayment
High tide cusps
HT = High Tide
LT = Low Tide
S = Surging breaker
P = Plunging breaker
Mega-cusp
Horn
Trough
Bar migration
Spilling-plunging breaker
Rip
Mega-ripples
R = Reflective domain
D = Dissipative domain

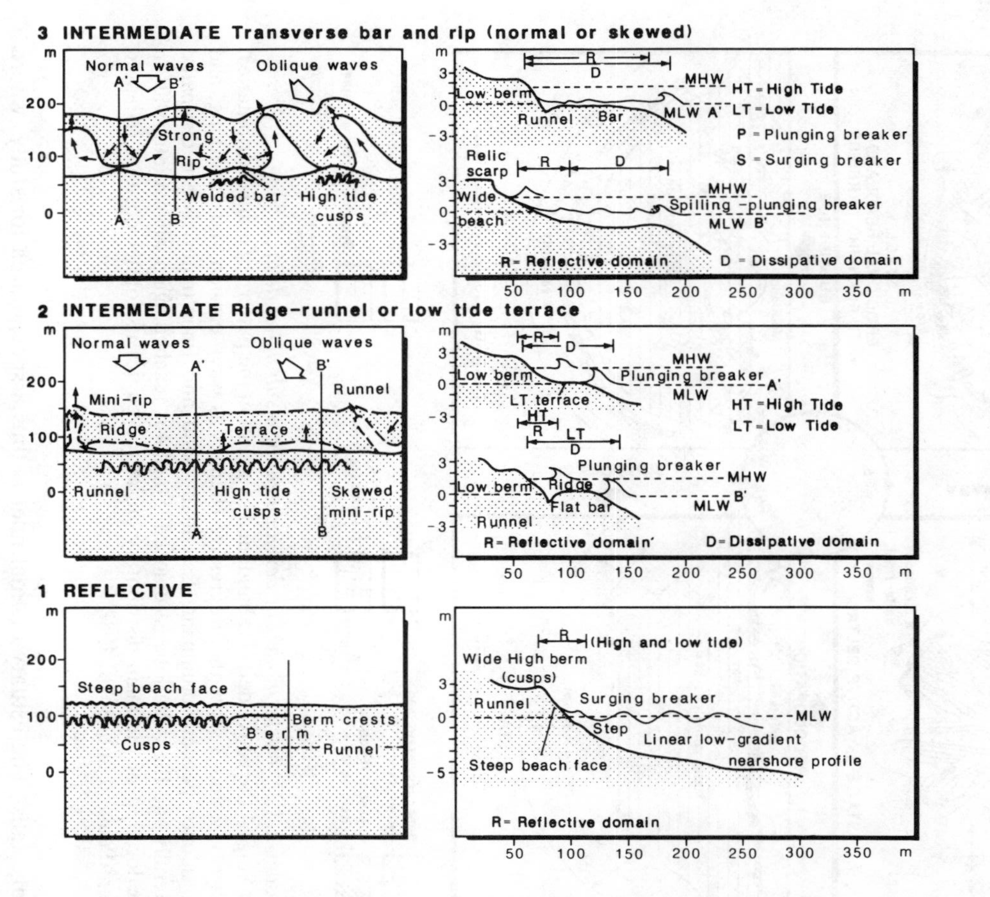

FIGURE 2.4.2 *The six major beach stages showing typical plan and profile details*
Beaches will pass through a cycle from Stage 1 to 6 but process can be checked or reversed if waves do not follow a normal rising and falling pattern – see Figure 2.4.1 (adapted from Short 1980, and Wright and Short 1983).

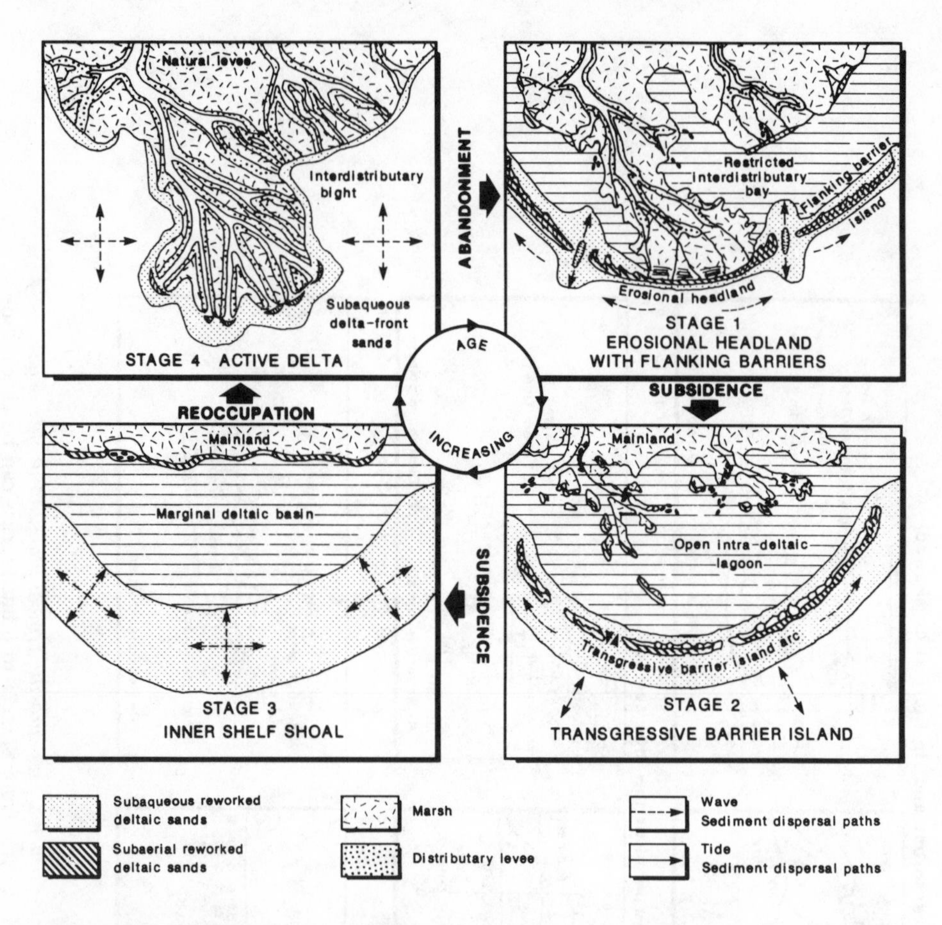

FIGURE 2.4.3 *Cyclic models of development of a sub aerial delta*
At Stage 4 the active delta is abandoned by a shift in the pattern of main distributaries. Stage 1, 2 and 3 show the progressive changes as wave erosion and reworking, and subsidence become dominant. Essentially sand and finer materials are dissipated laterally and the delta lobe is removed. Stage 4 is reached again when a later shift in the location of the main distributary outlet reoccupies its former position (adapted from Penland and Boyd, 1982).

stage falls. The estuary or tidal inlet is thus a special area for study, where fluvial and coastal geomorphology combine and where the overriding factor is the effect of tidal flows and elevations rather than waves. Other variables such as the ratio of fresh water discharge to the volume of tidal water entering the inlet or estuary will control mixing, and this in turn determines the nature of several other processes, including sedimentation. Numerous texts are available on the complex subject of tidal inlets, estuaries and similar coastal areas; areas that are of vital concern as a result of the vast number of people who choose to live there, and who depend on the waterway not only for

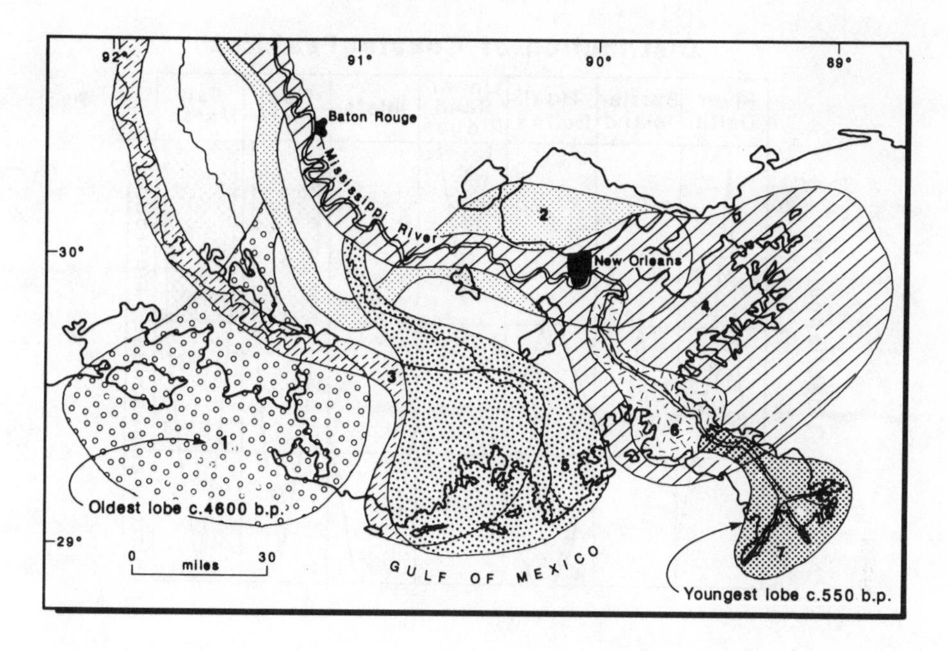

FIGURE 2.4.4 *The different locations of the Mississippi delta at various times in the past*
Delta switching is a common feature of most large deltas. Figure 2.4.4 can be compared with Figure 2.4.3 which shows the coastal geomorphological consequences of such switching (adapted from Kolb and Van Lopik, 1958).

industry and commerce but as a ready facility for the discharge of their effluents, wastes and the pollution products.

For coastal geomorphologists, the study of inlets is not only an area where an understanding of tidal phenomena is essential, it also demands an appreciation of three-dimensional space. Beaches, for example, are usually studied in two dimensions, either in planimetry or in profile, but rarely as a three-dimensional volume of material. There is a tendency to study maps or aerial photographs and produce outlines and surface patterns, thereby losing sight of the fact that beach changes are volumetric; a sandspit has height as well as length and width and so on. In inlet studies this two-dimensional view is untenable; the changing elevations of the surface of the sea relative to the inlet and the need to measure volumes of water moving in and out, enforce a three-dimensional view. Moreover, the study of ebb and flood channels which are, by definition, at different elevations and planimetric locations, emphasises the need to consider the vertical dimension. Thus, in the context of teaching coastal geomorphology, the study of inlets or estuaries, or preferably the continuum illustrated by Figure 2.4.5, seems to

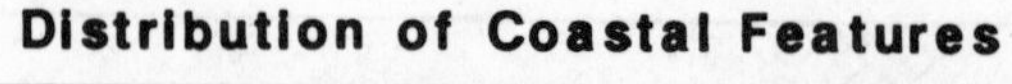

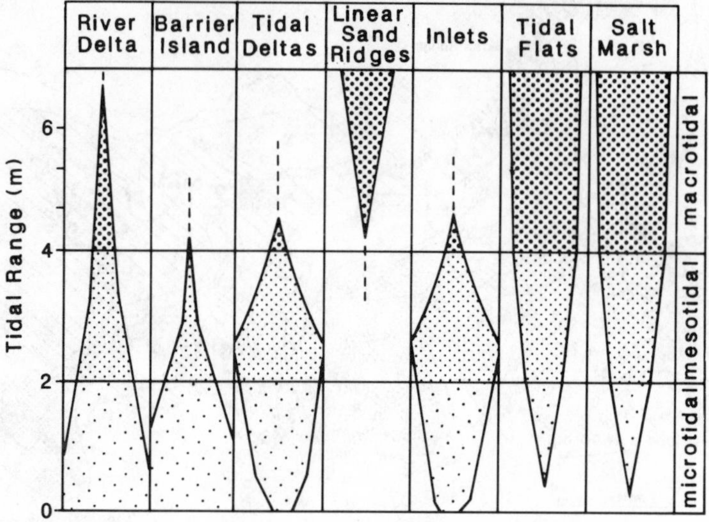

FIGURE 2.4.5 *Certain types of coastline are more common under specific tidal range conditions*
Coasts with small tidal ranges are dominated by wave energy; conversely high tidal ranges may produce strong tidal currents especially at coastal constrictions. Saltmarshes and tidal flats are also characteristic of coastlines with large tidal ranges (adapted from Hayes, 1975).

provide the best opportunity to introduce the importance of a three-dimensional view of forms and processes. It is also important to use these particular environments as case studies to illustrate the distinction between the tide as a direct factor in coastal geomorphology, and as an indirect mechanism whereby other processes are allowed to operate on different zones of the coast for different periods of time.

COASTAL CYCLES AND THE FREQUENCY OF EVENTS

A decade ago, few coastal geomorphologists in Britain and Western Europe would have had more than a superficial awareness of washover processes. Nevertheless, in microtidal, unconsolidated coastal environments, this process is at the root of much physiographic change. In general, with low tidal ranges, wave action and related processes are confined to a narrow vertical zone, but from time to time the sea surface is elevated by storm activity and wave driven forces impinge on upper beach and coastal edge surfaces. If these waves pass over or through the coastal edge then this is termed washover. The height and resistance of the coastal barrier is of critical importance, but with exceptional high-energy and high-water level conditions (e.g. during storms) the beach is driven on to and across the coastal barrier. The coastline may then retreat, dunes may be destroyed, vegetated surfaces are overwhelmed by sand and other materials, and the sea might break through to lower marsh, lacustrine or back-bay areas. The effects of such dramatic events are most evident in areas that are normally low energy coastlines. More recently, coastal geomorphologists have shifted their attention from the exceptional super-elevation event such as a hurricane, which may have a return period of the order of ten to thirty years, to lower magnitude but more frequent events in the same area; for example a return frequency of five or six times per annum. Thus the realisation that storm events are part of a spectrum and not a 'single exceptional occurrence' has become more widely appreciated.

A similar situation occurs in the North Sea basin where tidal surges are now recognised as being part of a series of events, with frequent low-magnitude changes superimposed on the predicted tidal cycles at one extreme and the rare flood disaster situation as occurred in 1953 at the other. Increasingly the coastal geomorphologist is using the terminology of coastal engineers, who design onshore and offshore structures to height and strength specifications according to statistical calculations of probable wave strength and tidal elevation conditions. The Dutch have always designed their sea dykes on the basis of such statistical extrapolations of tide and wave conditions, but it is only with the building of such constructions as the Thames tidal barrier, or media attention on the safety of expensive oil and gas production platforms in the North Sea, that the British public have been introduced to such phrases as 'designed to withstand the hundred-year event'. To the physical geographer, this concept is essentially the limiting end of a spectrum of energy levels. The question that remains unanswered relates not so much to the catastrophic hundred-year event as such, but to the problem of what is more important in explaining the genesis of any specific landform, repeated action over the years or the single major event?

PROBLEMS OF DATA ACQUISITION

Whilst great progress has been made, and models such as those described above facilitate and illuminate the study of particular coastlines by providing a cogent study framework, their construction highlights the need to overcome significant difficulties in the acquisition of basic data. The most serious deficiency relates to the calculation of the fundamental source of energy to drive the particular coastal system, namely waves. Instrumentation is available to measure wave period, height and frequency but it is expensive and difficult to deploy. In a section of his review of coastal landforms in 1983 entitled 'Coastal processes and responses: still in a backwash?', McLean comments on the lack of progress in the development of instrumentation for the crucial post-breaker zone. He also refers to a recently published text of geomorphological techniques which is unable to provide information on a coastal technique that does not predate 1975. Wave climate data must be available before such basic calculations as refraction or shoaling or set-up levels can be made. Without refraction calculations (which are relatively easily produced by any one of several computer programmes), longshore energy patterns cannot be produced. Without a knowledge of longshore energy, quantification of sediment transport alongshore is impossible other than by empirical or indirect methods. For some coastal areas reliable bathymetric contours are unavailable, as is information on the distribution of sea bed materials.

There can be few branches of geomorphology where the cost and difficulty of obtaining fundamental process information are so great. As a consequence of these difficulties, considerable use is made of hindcasting techniques using readily available meteorological statistics of strength, frequency and directions of winds with fetch vectors to calculate the incidence of waves at a coastline. With some notable exceptions, the reality of coastal geomorphological processes is characterised by a singular lack of direct measurement of its basic energy input, waves. The absence of appropriate wave measurements is particularly common in mid and high latitude regions where wind waves rather than swell waves are dominant, and the directions of wave incidence may be more variable.

THE REALITY AND PRACTICE OF COASTAL GEOMORPHOLOGY

The development of the beach cycle, delta cycle, tidal inlet models and the recording of baseline data that allows calculations to be made for the safe design of sea walls, harbours and offshore oil platforms require a vast input of time, expertise and money. In general, ideal locations are used as field laboratories where random or compounding variables are either eliminated or held constant, so that the system's linkages can be examined directly. In contrast, the individual coastal geomorphologist trying to explain or predict a given length of coastline feels a sense of despair, not only as a result of the lack of time, equipment or baseline data, but also as a consequence of the existence of local factors that do not seem to exist in the model or simulation. Many environmental scientists feel this deficiency and try to overcome it by beginning with the end product, the coastal landform, and using it as a summation of the creative processes and other factors. For example, the ecologist, studying vegetation faces the same problem of complexity and overcomes it in the same way, as is reflected in Goodall's contention that 'There is much to be said for the view that complexes of environmental factors determining plant distribution can be indicated and measured better indirectly through the plants themselves than by direct physical measures.'

All process studies are aimed at explaining a form, whether a beach gradient, a pattern of ripple marks, or the spacing of nearshore sand bars. Thus the first stage of enquiry is a careful study of the form of the coastline and its individual components. The skill that is necessary is to read the landforms and see them as indicators for use in comparing reality with the textbook models and other forms of simulation. The fit may be partial, and this is the spur to look for the reasons why there is a deviation from the model which was constructed under ideal conditions and environmental circumstances. Thus the traditional geographical skills of observation and recording leading to classification in the widest sense, once seen as sterile ends in themselves, are again of importance in the process of explaining and predicting coastal landforms.

Nevertheless, before this 'morphological approach' can be used, one condition must be satisfied; a sufficient number of detailed, thorough, process-based research projects must have taken place and, equally important, have been summarised and recorded in the scientific literature. Over the last decade most unconsolidated coastal forms have been the subject of detailed investigation. The models exist, the quantitative relationships have been established, and the proliferation of excellent textbooks (usually not comprehensive coastal texts, but at the level of individual groups of forms, e.g. beaches, saltmarshes, dunes, etc.) has ensured that these pertinent results are readily available. The next stage is under way, where subsets are being studied either on the basis of a fundamental property (e.g. differences between the response of shingle and sand beaches in the same process environment) or on the basis of clear climatic differences (e.g. the evolution of a sand beach in polar compared with tropical conditions).

A review of recent applications of coastal geomorphology reveals a growing demand for the morphological approach, notably in the area of environmental impact assessment. The reality of these requests is that time and money are not available on the scale that is required to set up process measurements, even assuming that the necessary field equipment and instrumentation are available. The developer, the coastal engineer, the planner and the writers of environmental impact assessments and others must work to relatively short time-scales, as does the teacher on the beach or on the clifftop, for the teacher is the epitome of the applied geomorphologist. In all these wide-ranging situations, the coastal geomorphologist is being asked to make predictions and to answer basic questions relating to coastal evolution. The pragmatic responses can only be based on careful morphological analysis aided by a collation of whatever process-related information is available and, on occasion, historical evidence of change. In theory, more reliable answers might be derived from a carefully designed series of tracer experiments or a series of monitoring measurements, or by placing a number of wave recorders in the surf zone, but these and similar process measurements are usually impractical. Nevertheless, tracer experiments, monitoring measurements and wave recording devices have been deployed elsewhere and the results so gained have been analysed to produce a wide range of viable models and systems. To use these models it is necessary to recognise and classify the coastal landforms, either in whole or in part, and on this basis to select the appropriate process–form model. The time dimension or the stage of development within this model must then be identified. Thereafter the systems linkages can be analysed. In this way theory and reality come together to produce explanation or prediction as required.

FURTHER READING

Three recent textbooks provide general accounts of the main areas of coastal geomorphology. There are differences in emphasis, but all contain discussions of essential processes and landform types:

Davies J. L. (1980) *Geographical Variation in Coastal Development*, 2nd edn (London: Longman).

Pethick J. (1984) *An Introduction to Coastal Geomorphology* (London: Edward Arnold).

Bird E. C. F. (1984) *Coasts* (Oxford: Basil Blackwell).

An older more substantial text with the value of a range of case studies from areas that are unfamiliar to British readers is:

Zenkovich V. P. (1967) *Processes of Coastal Development* (Edinburgh: Oliver Boyd).

The best textbook on general coastal geomorphology is:

Davies R. (ed.) (1978) *Coastal Sedimentary Environments* (Berlin: Springer-Verlag).

This comprises a series of chapters by expert authors on specific themes in coastal geomorphology: for example, deltas, coastal dunes and sedimentation.

Other books at a more advanced level which can be recommended strongly are:

Komar P. 1976) *Beach Processes and Sedimentation* (New Jersey: Prentice-Hall).

Schwarz M. L. (ed.) (1972) *Spits and Bars* (New York: Dowden, Hutchinson & Ross).

Schwarz M. L. (ed.) (1973) *Barrier Islands* (New York: Dowden, Hutchinson & Ross).

Schwarz M. L. (ed.) (1982) *Beaches and Coastal Environments*,

Schwarz M. L. and Bird E. C. F. (eds) (1985) *The World's Coastlines* (New York: Van Nostrand Reinhold).

The latter is a comprehensive text describing 135 regional coastlines in one book.

2.5
Lilies and Peacocks: a View of Biogeography

I. G. Simmons
University of Durham

I wonder whether it is possible to agree with many of the judgements of John Ruskin (1819–1900), for he once wrote that lilies and peacocks were useless, something which no ecologically minded geographer could ever accept. Yet he did say in 1862 that 'There is no wealth but life' and perhaps thereby provided a motto for modern geographers, for it is a phrase which conjoins a judgement based on human values with the products of many millennia of organic evolution. It is around such a theme that this view of biogeography is constructed.

CONTEXT AND DEFINITION

Biogeography is a word which has attracted more than one meaning and is used divergently by different groups of researchers and teachers. That it concerns the how, where, when and why of the presence on Earth of some 10 million species of living organisms is held in common, but thereafter we can recognise two main sets of users of the term. The first comprises those biologists who are interested in explaining the distribution of the various taxonomic units of plants and animals, and are generally concerned with changes over evolutionary time rather than the much shorter ecological time, which coincides roughly with the Holocene. The second set comprises those geographers who have an interest in the distribution and relationships of living organisms, and who have adopted the term 'biogeography'. They are generally (though by no means totally) interested in ecological time, and in the meso and micro-scale interactive systems usually studied by ecologists. Increasingly they are concerned with the connections between the products of the evolution by natural selection of both organisms and the systems of which they are components and the effects of human societies upon these species, populations and ecosystems. This review discusses the 'ecological time' group, because in pragmatic terms this is what most of the geographers involved have chosen to investigate and communicate.

That such studies are in keeping with the key themes of the present volume need not be doubted. At its simplest, 'Horizons' might imply a boundary of limit, to which no self-respecting scholar would confess, but the other concepts find a resonance in the biogeography of today's geographers. Process, for example, is highly appropriate to describe the systems-oriented model of ecosystems which looks at the flows of energy and matter, at changes through time, and at the impact of human societies. Such a concept as the ecosystem, or indeed the distribution of a plant or animal, can and should be studied at a variety of scales from the very small to the global, and the ideas of dynamic changes (whether in evolutionary or ecological time) and equilibria are the very language of systems description and analysis. The same is true of types of activity like environmental management, which is the deliberate transformation of nature; and environmental impact, which is more likely to be accidental, except in those countries which have statutory Environmental Impact Statements, where at least part of the connotation is legal.

It is possible to assert that biogeography is not necessarily a part of physical geography any more. The distribution and number of species has been changed by human agency (lilies, peacocks, smallpox and syphilis provide examples), the genetics of species have been altered by domestication (our border terrier is not very like a wolf to look at, but the lemons and pigs most of us come across are in fact domesticated strains), and most ecosystems have been subject to deliberate or accidental transformation. If this point of view is accepted then we have the core of the genuinely geographical bigeography proposed by Pierre Dansereau: a subfield based on the concepts that humans create new genotypes and new ecosystems.

PROCESS IN BIOGEOGRAPHY

A role for the ecosystem

Many of the fundamental concepts of biogeography can be defined in terms of processes. Individuals live, reproduce and die; populations rise, fall or become steady; energy flows and is lost to space as heat; and chemical elements are cycled from compartment to compartment of the world systems. Ecosystems change their constituent components through time, through self-generated feedback (autogenic succession) or through the forcing function of 'outside' events such as large-scale climatic change (allogenic succession). Thus even the word 'equilibrium' has usually to be qualified as 'dynamic equilibrium' since the unchanging situation that it implies is in general foreign to the stuff of biogeography.

Getting to grips with the description, explanation and manipulation of all this movement is difficult. In the last thirty years, the ecosystem concept formalised by E. P. Odum in 1971 as 'Any unit that includes all of the organisms in a given area interacting with the physical environment so that a flow of energy leads to . . . exchange of materials between living and non-living parts within the system . . . is an ecosystem', has been seminal. In particular, ecologists have been able to discuss ecosystems in terms of energy flow and nutrient cycling. The organic matter accumulation of the mature community over time can be measured and this rate of production of organic matter is an important aspect of ecosystems on the world scale. The increase in organic matter or total energy content over a given time appears as an increase in the 'living weight' of plants, i.e. an increase in biomass. The rate of accumulation of biomass (i.e. weight of living matter/unit area/unit time) is called net primary productivity (NPP).

Energetics, cycles and dynamics

In terms of energy at the global scale, about 520×10^{22} joules (J) of energy strike the top of the Earth's atmosphere every year, of which about 100×10^{22} J reaches the Earth's surface and is of a suitable wavelength for photosynthesis. Of this, some 40 per cent is reflected back into space by deserts, snow, ice and oceans, leaving 60×10^{22} J as the 'pool' for photosynthesis. Of this, a large quantity is respired and the remainder appears as NPP. The total biomass produced annually by green plants is estimated to contain about 170×10^{19} J (much of it from phytoplankton in the oceans), and so the average utilisation of the energy present in light of the right wavelength by the flora of the globe is about 0.2 per cent. Less than 1 per cent of that 0.2 per cent is consumed by man as food: maximisation of the interception of energy is clearly a concept of humans rather than an outcome of organic evolution.

A brief look at nutrients reveals that a characteristic feature of their pathways of nutrients is circulation between living organisms and non-living (abiotic) 'pools' of various scales. These pathways are generally cyclic: they involve the use and reuse of the atoms of the nutrients and, since they also include a non-living phase and involve both soil and atmosphere, they are usually called biogeochemical cycles. At a local scale, one principal result of an ecosystem's evolution seems to be the creation of 'tight' or 'closed' nutrient pathways which recycle essential nutrients within the system. Any losses are small and are balanced by inputs from the non-living parts of the ecosystems. The two major pathways for recycling nutrients are a return of animal and plant detritus by way of animal excretion and microbial decomposition. All these processes need energy for their metabolism, and this is one way which the energy content of an ecosystem is 'used'. An ecosystem which is to be relatively stable through time must, therefore, have adequate storage of both energy and nutrients to guard against periods when their supply is cut off.

To both of these basic aspects of energetics and biogeochemical cycles it is usually necessary to add population dynamics, for the energy and the matter flow concerned through individual organisms of a distinct species, each with its own behavioural characteristics. Ecological energetics by itself is, therefore, open to the charge made by one ecologist that it was like grinding up cows to make hamburgers: you couldn't be sure that a monkey hadn't slipped in somewhere!

Synthesis

The idea of the ecosystem is capable of application at any scale, from the underside of a stone in a stream to the whole planet. As with most systems, complexity increases with size and with this growth they are likely to exhibit emergent properties. These, put very simple, are characteristics of a whole system which cannot be predicted from our knowledge of the behaviour of the individual parts. Thus, the properties of hydrogen and of oxygen do not prepare us for the behaviour of water, and no amount of knowledge of soils, plants, animals and water can predict for us accurately the changes following certain forms of human impact. At a global scale, the wholeness of inorganic and organic fractions of this planet is exemplified by Lovelock's *Gaia* hypothesis. This suggests that the composition of the atmosphere is a consequence of life, not a given thing to which the biota has adapted. In policy terms, a significant implication of this concept is that the living systems of the earth survived the Pleistocene glaciations, when much of the temperate zone of the northern hemisphere was lifeless. Thus the regulators of a global organic–inorganic stability might likely be the organisms of the continental shelves and of the lowland equatorial forests, both with high NPP.

THE HISTORICITY OF MANIPULATIONS

No geographer will fall into the trap of regarding man–environment relations as ahistoric, but it is worth reminding ourselves that human-induced manipulations of nature have a long history, starting with hunter-gatherer societies of the Pleistocene and post-glacial periods. Such a viewpoint puts process in perspective.

Simple societies

Hunter-gatherers account for 90 per cent of the time since human evolution, and they have occupied most environments, from the ice-margins of the Neanderthals and near recent Inuit to the semi-deserts of the Saharan Neolithic and near recent Bushmen. Conventional wisdom long ignored the fact that hunter-gatherers might transform their environments, but newer studies have revealed some examples of their role as manipulators of ecosystems. At the end of the Pleistocene, for example, there occurred an apparently sudden and certainly irreversible demise of about 200 genera of large warm-blooded animals, a phenomenon which has been labelled 'Pleistocene overkill'. The phenomenon is perhaps most closely observable in North America: here, two-thirds of the mammalian megafauna disappeared at the same time as the first entry of man into the New World, but the hypothesis that the two events are causally related is still controversial.

As a tool for environmental manipulations fire has been of great utility to hunter-gatherers, with many instances to be found in the Antipodes. For example, studies in the Cape York peninsula of Australia have identified the usefulness of large stands of cycad trees (*Cycas media*) which, in yielding some 131kcal/m^2/yr of edible kernels, equal some cultivated crops. It is suggested that these stands are a consequence of firing the area repeatedly, thus encouraging the fire-resistant cycad at the expense of other trees. The cycad *Macrozania*, if fired, produces all its seeds simultaneously and at seven to eight times the unfired quantity.

If we turn to the impact of agriculture, we can see that its imprint has been strong since its apparent evolution in the ninth to sixth millennia BC in south-west and south-east Asia. Together with meso-America, these two regions saw the emergence of the first domesticated plants and animals, in which breeding of biota for human ends replaced that of natural selection. Animal populations, the vegetation, the form of slopes and valleys and the soil cover of the land units have all been altered, many of them in apparently irreversible ways.

Although not necessarily a focus of early domestications, the ancient Maya civilisation which flourished *c.*AD 250–1000 provides a good example of the impact of a preindustrial, agriculture-based culture. The Maya inhabitated what is now parts of Guatemala, Mexico (Yucatan), Belize and the fringe of Honduras, mostly in the lowlands but with some extension of occupied area into the upland areas that interrupt the plains. The Classic Maya were underpinned economically by a set of subsistence techniques. Of these, the most important spatially was probably shifting agriculture of maize, with sweet potato and manioc as other important domesticated crops. Other large-scale alteration of the land surface included the creation of raised fields in areas of swamp and river floodplain. Canals were dug and the mud dredged up formed a rich soil, while the canal harboured fish and turtles; each year the act of keeping free the canals added more rich silt to the fields. Therefore the Maya transformed much of the landscape they inhabited, though presumably patches of virgin savanna, grassland and forest remained in the uplands, and there were areas of open water and swamp whose ecology remained undiverted by human activity.

As another example of the environmental consequences of a type of preindustrial agriculture, we may look in more detail at the growing of wet rice, or *padi* which refers to the growth of domesticated rice in conditions which require it to be submerged beneath 100–150mm of slowly moving water for three-quarters of its growing period. The characteristic nutrient poverty of tropical soils is circumvented through the nutrients brought by water from the higher volcanic slopes on to the terrace; through the fixation of nitrogen by the blue-green algae which proliferate in the warm water; and via the release of nutrients by the decay of stubble and of other organic matter used to fertilise the fields. Given that the flood padi may also yield animal protein in the form of fish or crustacea, then its virtues as a food-producing system are convincing. All this comes about because a steep hillside clad with forest has had its ecosystem changed by human activity to something resembling an aquarium.

Impacts of industrialisation

Important though these periods were (and indeed still are in the case of preindustrial agriculture), their magnitude looks lower when placed alongside the effects of human communities with access to technology powered by fossil fuels such as coal, oil and natural gas. To flows of direct solar energy, wind and water, and biomass energy, can now be added that of photosynthetic fixation of energy many millennia ago. While on a global scale, these flows are still small besides the natural flows of the planet (Table 2.5.1), their regional and local concentration means that other ecosystems can be obliterated completely. Cities, for example, cause local and regional changes in biota and ecosystems (Table 2.5.2).

One major impact of industrialisation has been upon the patterns of the introduction and extinction of species. Knowledge of the history of extinctions is in general less reliable than that of introductions, until very recent

Table 2.5.1 *Global mean energy flows for various natural and human-induced processes*

Process or event	*Energy flow cal* m^{-2} day^{-1}
Solar energy to earth	7000
Weather	100
Primary production by plants	7.8
Hurricanes	4.0
Tides	1.54
Animal respiration	0.65
Cities	0.45
Fossil fuel combustion	0.11
War (non-nuclear)	0.05
Floods	0.04
Earthquakes	0.001
Volcanoes	0.0005

Source J. F. Alexander (1979) 'A global systems ecology model', in R. A. Fazzolare and C. B. Smith (eds) *Changing Energy Use Futures*, Vol. 2 (New York, Oxford: Pergamon), pp. 443–56.

times. The record rises markedly after AD 1800; before then, habitat alteration and the indirect effects of maritime traffic had begun to add to the species finished off by natural means, but the nineteenth century saw a rapid increase in all the causes to which species extinction has been attributed. About one-hundreth of the warm-blooded fauna has vanished since AD 1600, and in terms of all animals, the extinction rate of species and subspecies appears to be about one per year, compared with one every ten years from 1600–1950 and perhaps one every 1000 years during the great dying of the dinosaurs. Nor can the extinction of plants be forgotten. Habitat changes due to climatic change and human activity have ousted populations of plants at many scales. As might be expected, island floras have been especially vulnerable to introduced diversivores and herbivores like the pig, goat and sheep. Extinctions on continents tend to leave some specimens of the plant in another part of its range so that the diversity of their flora is not so easily thinned out as that of isolated islands. Nevertheless, one estimate suggests that the world higher plant flora comprises 240 000 species and that of the 85 000 species in the temperate zones, 4500 are threatened of which some 450 are endangered. The equivalent data for the tropics (150 000 spp of flowering plants in total) are virtually impossible to obtain, but there are forecasts that 50 000 tropical species may become extinct by the end of the century. Such is the interdependence in ecosystems that a disappearing plant can take with it ten to thirty dependent species such as insects, higher animals and even other plants.

One feature of this century has been the interaction of technology with science, and one consequence has been the growth of our ability to intervene directly at the molecular level of genetics. Genetic engineering is still at an early stage and the potentialities though enormous are not yet fully clear. It may, for example, become possible to breed plants with 100 per cent resistance to a set of diseases, or to give all crop plants the legumes' ability to capture nitrogen from the atmosphere. It is possible that a material economy based totally upon organic materials could be developed. We can also be sure that the military applications of this development are under active consideration.

Military implications

Global military expenditure in recent years has been equivalent to the total GNP of the sixty-five countries of Africa and Latin America, and the potential for environmental alteration has been nowhere more apparent than in the second Indo-China war. In southern Vietnam, 587kg/ha of munitions were delivered; 8.1×10^6ha (equivalent to the area of Connecticut or 66 per cent of the area of Wales) was hit, producing 350×10^6 craters, unevenly spread. The craters often breached hardpans which had held up perched water tables and thus made soils more arid. The flying metal killed wildlife and made trees difficult to use in sawmills as well as opening the bark to

TABLE 2.5.2 *A selection of 'downstream' effects of urban–industrial areas upon biota*

System	*Immediate 'downstream' area*	*Biological consequences*	*Long distance*
Atmosphere	Airflow resources normal characteristics	Unknown	CO_2+ particulate loading have global consequences
	particulate loadings increase: higher precipitation downwind.	Small	
	Particles fall-out downwind	Some plants intolerant of deposition on leaves	
	Photochemical smog drifts downwind	Some plants killed, others slower growth	
	NO_x, SO_x fall-out as acid precipitation	Death and dieback in some plants species, e.g. coniferous trees	
Ground and Surface water	Contamination of rivers and lakes: decreases with dilution	Severe consequences for biota: only a few organisms tolerant of severe contamination	Persistent substances may build up in food webs at some distance
	Flood hazard increase engineering of river, e.g. channel modification and cut-off	Faster flow in channels not tolerated by some plant spp; cut-offs form new habitat	Rates of delta formation and state of floodplain habitats may alter
	Water may be led off for aquifer recharge	Lower flows may affect e.g. fish populations	
	Water purified and reused	– as above –	

	Water-table lowered by pumping	land sinking may provide new habitats for wetland species	
	Rapid runoff from built-up areas	Flashy rivers inimical to some species of plant and animals	Areas flooded may increase: some plants affected, also perhaps migrant animal spp.
Landforms	Silt production increases during construction and is rapidly transported downstream	Silty rivers have different biota: high loads may choke gills of fish; prevent photosynthesis	Offshore and seabed biota affected–corals especially, for example.
	Solid wastes necessitate finding fill sites like quarries, gravel pits. Tip contaminants	Wildlife habitats near city obliterated; contaminants may enter rivers (*v.s.*)	Loss of environmental quality
	Transport and communication systems necessitate earth movement: cuttings, tunnels, embankments.	New habitats created for subdual plants; landscaping may encourage animals, e.g. deer near motorways	
Food supply	Intensification of agro-ecosystems	Reduction of biotic and landscape diversity, increase in abundance of pest species	Reduction of genetic variety

fungal rot. Herbicidal chemicals were used on a large scale for the first time in war with 86 per cent of the missions against forest and mangrove and 14 per cent against crops. These tactics caused deforestation over large areas, followed by nutrient dumping, the eutrophication of fresh waters and by the invasion of pioneer grasses. The forest is slow to recolonise for, in some areas, the lack of evaporation and transpiration has led to high water tables which inhibit tree regeneration. At present rates, climax forest may take 500 years to come to maturity. Defoliated mangroves appear to be showing no recolonisation and so coast erosion has accelerated since there is now no protection from storms.

The potential impact of nuclear weapons upon plants, animals and ecosystems is still more immense. Models of potential global atmospheric, climatic and biological consequences of a 5000MT nuclear exchange have been developed: these are necessarily imprecise but suggest very serious effects for the whole planet. In a war which involved both urban and military targets, the thousands of explosions would project tremendous quantities of soot and dust into the northern hemisphere's atmosphere. Within a week or two the pall of smoke and dust might coalesce into a single cloud surrounding most of the northern hemisphere, especially the mid-latitude belt of the USA, Canada, Europe, USSR, China and Japan. This cloud would absorb most incoming solar radiation because of its soot component; dust alone would allow more light through by reflection and scatter. Unless the soot were quickly scavenged out, most of the sunlight would be excluded and temperatures at the earth's surface would fall rapidly a week or so after the exchange.

In continental interiors, below zero temperatures would result, whatever the time of year. Such temperatures would return towards normal only very slowly. The sea would buffer coastal areas and islands from the most acute drops of temperature but the difference in temperature between ocean and continent would subject these marginal zones to months of violent weather. This is the so-called 'nuclear winter', though the period of cooling is probably too short to initiate another episode of glaciation. Nevertheless, it is possible that there would be rapid transport of aerosols to the southern hemisphere, at a level such that the cold and the darkness would envelop the entire planet. Destruction of the ozone layer of the atmosphere would mean that clearing skies would bring enhanced UV-B radiation. So plant and animal species able to cleanse their populations quickly of unfavourable mutations might soon dominate: weedy plants, insects and rodents are the most likely candidates. The biogeography of the apocalypse is no more attractive than any other aspects of the event.

CONSEQUENCES AND CONCLUSIONS

The starting point that biogeography can be about human-innovated genotypes and ecosystems now leads us to consideration of the relationship between the natural and the manipulated environments. In ecological terms, E. P. Odum summed up the ecosystems of nature and the ecosystems of man rather neatly in 1969. Those of nature are based on the characteristics of the different stages, which occur during succession in the development of ecological systems towards the self-maintaining equilibrium and is sometimes called the 'climax' or 'mature' condition. The end point of successions, the mature stage, is especially notable for the high degree of internal mutual dependence, for its tight nutrient cycles, and for the high degree of stability which is maintained, although piecemeal renewal is essential. In spatial terms, the trend is towards a mosaic of ecosystems moving towards maturity but with patches of early stages for example where new land is created, where catastrophic events have occurred, and where the death of large organisms such as senescent trees has taken place.

We can refine Odum's classification by designating four main types of ecosystem:

1. Early succession systems with low biotic diversity, linear food chains and a high net biological production.
2. Mature systems which generally show the opposite characteristics from type (1). Notably, a lot of biomass is supported by the energy flow, relative to type (1), and the food chains are predominantly web-like, with the detritus stage very important.
3. A mixture of (1) and (2); for example at times of general environmental change or where natural fires bring about juxtaposition of the two types of system.
4. Inert systems, with little or no life: volcanoes are one example, ice-caps another.

Ecosystems manipulated by man can be described in the same terms as natural ecosystems. Thus early successional phases of natural ecosystems are equivalent to the simple systems of modern agriculture; the mixed systems of type 3 would apply to areas of mixed forest and farmland; the mature systems are the same in both, often being deliberately protected as parks or reserves; and the inert systems have their equivalents in cities (which, like volcanoes, give off SO_2 and particulate matter) and some types of derelict land. Using the type and quantity of energy flow as a criterion, it is possible to combine both natural and man-manipulated ecological systems in one classification (Table 2.5.3). Here the source of the energy becomes a differentiating factor. Some systems are entirely solar powered and may use only the incoming solar radiation which they fix as chemical energy. Others may receive a natural subsidy of organic matter brought in by natural processes; estuaries are an obvious example of this. Yet others are subsidized by man-procured supplies of fossil fuels (i.e. stored photosynthesis). Some systems are augmented by fossil fuels but remain fixers of solar energy (e.g. mechanised agriculture, short-cycle forestry); others, like the city, are entirely powered by fossil (and

TABLE 2.5.3 *Ecosystems classified according to source and level of energy*

Ecosystem type	*Kilocalories per sq meter per year* *Range*	*(Kcal/m²/yr)* *Average estimated*
1. Unsubsidised natural solar-powered ecosystems: e.g. open oceans, upland forests. Man's role: hunter-gatherer, shifting cultivation	1 000–10 000	200
2. Naturally subsidised solar-powered ecosystem. e.g. tidal estuary, lowland forests, coral reef. Natural processes aid solar energy input: e.g. tides, waves bring in organic matter or do recycling of nutrients so most energy from sun goes into production of organic matter. These are the most productive natural ecosystems on the earth. Man's role: fisherman, hunter-gatherer.	10 000–50 000	20 000
3. Man-subsidised solar-powered ecosystems. Food and fibre producing ecosystems subsidised by human energy as in simple farming systems or by fossil fuel energy as in advanced mechanised farming systems: e.g. Green Revolution crops are bred to use not only solar energy but fossil energy as fertilisers, pesticides and often pumped water. Applies to some forms of aquaculture also.	10 000–50 000	20 000

4. Fuel-powered urban–industrial systems. Fuel has replaced the sun as the most important source of immediate energy. These are the wealth-generating systems of the economy and also the generators of environmental contamination: in cities, suburbs and industrial areas. They are parasitic upon types 1–3 for life support (e.g. oxygen supply) and for food; possibly fuel also though this more likely comes from under the ground except in LDCs where wood is still an important domestic fuel.	1000 000–3 000 000	200 000

Source Odum (1975).

in some cases, nuclear) fuels. This sequence also denotes an increase in absolute throughput of energy per unit area.

The major inference we can make from the compartment model and from Table 2.5.3 is that human strategy runs counter to that of nature. Whereas nature moves towards mature, self-maintaining systems, man moves towards one type of successional phase at the expense of the mature and mixed systems, and increasingly requires subsidiary energy to keep up productivity. He also increases the area of inert systems. In turn, many kinds of wastes reduce the diversity of organisms present in all types of system. A further implication is that the matrix of ecosystem types produced by the imposition of human strategies on those of nature may in future be subject to unpredictable and uncontrollable fluctuations, with the consequence that they will not be stable enough as a habitat for large numbers of the human species.

Sustainable futures

Once again, we are reminded of energy availability and flow as a key to many aspects of the behaviour, not only of natural systems but of human ones as well. In particular, it brings out the key notion of sustainability. If productive ecosystems such as agriculture, forestry and fisheries are dependent upon fossil fuels to maintain economic rates of output, then it is clear that such rates are not in the long term sustainable. To find such sources of energy we must look either to nuclear energy derived from atomic fusion or to the solar energy captured by devices such as photovoltaic collectors or living plants ('biomass energy'). The techniques of powering not merely a few marginal 'alternative' communities but whole societies accustomed to an urban–industrial lifestyle by such means are poorly developed, and nuclear fusion is scarcely feasible yet except in the form of military devices yielding uncontrolled explosions, but there must lie the future unless the whole human species is to abate its per capita energy demand by a large quantity.

It may be the case that today's world will produce such a sustainability while maintaining its 'normal' trajectories. But a large number of scholars seem to think that some reorientation of human behaviour is essential if sustainability is to be achieved. In a much discussed paper of 1967, the historian Lynn White suggested that the Judaeo-Christian tradition had given sanction to the profligate use and transformation of the earth's resources. What was now needed, he said, was not the concept of the dominion of man, but a return to something like a spiritual equality of all the inhabitants of the earth. Thus he held up St Francis of Assisi as a patron saint for ecologists, representing those who think that all forms of life should have the same rights before the law, so that man and nature co-evolve on a more or less equal basis. To some, intuitive communication between Gaia and her denizens is possible, and we are thereby able to change our consciousness of the Earth and hence redirect our attitudes towards, let us say, lower rates of resource use, lower rates of population growth and less environmental manipulation.

For those who find that this model is too strongly flavoured with mysticism and prefer something with ideals but with more apparent common sense, the microbiologist and humanist scholar René Dubos paraded the statue of St Benedict of Nursia, founder of the great monastic order. Their acts, which he found symbolic, were those of stewardship of the land: of good husbandry, of draining marshes and reclaiming fen and sour moor, and of erecting buildings that were not only functional but had a thrifty beauty as well. Extending the idea, Dubos quoted examples of productive and sustained agriculture in Europe and the USA, and the landscape gardens of eighteenth-century England, as instances of how man was indeed different from, and superior to, nature in an untamed condition. Yet his creations were not so drastic as to be unsustainable.

This anthropocentric idea has gained many adherents and apologists among biological scientists, philosophers and theologians alike: if the Franciscan model might be summed up in the word 'unity', then the Benedictine version is 'stewardship'. Thus, the achievement of biogeography in the last two decades has been first, scientific consolidation which has interfaced with the best work of ecologists and other biological scientists; secondly, the provision of an awareness that the living world is a precious resource of great variety and potentially infinite renewability; and finally, the realisation that linking the two there is the world of values, ethics and philosophy which in this book are properly the province of Chapter 4.4. At the same time, there emerges a focus on the management of the environment: both an applied science, and a consciousness about what we ought (and ought not) to do with the other living organisms of this planet, so that we may have peacocks and piglets as well as lemons and lilies. To Ruskin's comment about wealth and life, we might add his apothegem of 1859 and apply it to biogeography (and indeed perhaps geography as a whole): 'Fine art is that in which the hand, the head and the heart of man go together.'

FURTHER READING

The idea of a biogeography defined in such a way that it does not belong solely to physical geography is introduced by the first two references, whilst the awesome all-inclusive Gaia concept is most accessible in the third:

Simmons I. G. (1979) *Biogeography: Natural and Cultural* (London: Edward Arnold).
Simmons I. G. (1983) *Biogeographical Processes* (London: Allen & Unwin).
Lovelock J. E. (1979) *Gaia: A New Look at Life on Earth* (Oxford: Oxford University Press).

Introductions, naturalisations and extinctions are discussed in:

Leven C. (1977) *The Naturalized Animals of the British Isles* (London: Hutchinson).

Myers N. (1979) *The Sinking Ark* (Oxford: Pergamon Press).
Myers N. (1983) *A Wealth of Wild Species: Storehouse for Human Welfare* (Boulder, Col.: Westview Press).

For an ecological view of urban areas see:

Boyden R. *et al*. (1981) *The Ecology of a City and Its People: The case of Hong Kong* (Canberra: ANU Press).
Douglas I. (1983) *The Urban Environment* (London: Edward Arnold).

Immediate impacts of nuclear explosions and the 'nuclear winter' hypothesis are introduced respectively in:

Westing, A. H. (1980) *Warfare in a Fragile World: Military Impact upon the Human Environment* (London: Taylor & Francis (for SIPRI).
Harwell M. A. (1984) *Nuclear Winter: The Human and Environmental Consequences of Nuclear War* (New York: Springer-Verlag).

Three other thought-provoking sources are:

Barbour I. G. (ed.) (1973) *Western Man and Environmental Ethics* (Reading, Mass.: Addison-Wesley). (Includes a reprint of Lynn White's controversial 1967 paper on 'The historic roots of our ecological crisis'.)
Dubos R. (1973) *A God Within* (London: Angus & Robertson).
Dubos R. (1980) *The Wooing of Earth* (London: Athlone Press).

2.6
Soil Processes and their Significance

Steven T. Trudgill
University of Sheffield

Soils are a resource used by man and a component of environmental systems. As such, they repay careful evaluation of their quality and spatial distribution, and merit study of their role in the flow and transfer of chemical elements, water and energy in environmental systems. The awareness of resources issues and the role of soils in environmental linkages has increased in the last twenty years, and has tended to supercede some of the old soils teaching which was concerned solely with soil formation and differentiation. This is not to say that the study of podsols and gleys does not still form an important part of the curriculum, but rather suggests that the relationships between soils and agriculture and between soil study and biology have become increasingly important. As with many other studies, the focus on function and process has replaced the simple study of form; nevertheless, in soils, the relationships between form and process do remain reasonably clear, with soil horizons acting as the visible evidence of the action of soil processes.

The relationship between form and process is illustrated by the link between organic soils and cycling (Figure 2.6.1). In rapidly cycling systems, there tends to be an absence of organic matter on the surface, but a breakdown and incorporation in the *A* horizon of the soil. In cooler, wetter, less fertile conditions, organic matter tends to accumulate, yielding organic acids which lead to the chelation and mobilisation of iron and the development of podsolic horizons. Thus, visible horizon form can be taught as evidence of process. However, process is not always easy to assess and new ideas are constantly coming forward. It is perhaps, for this reason that the Soil Survey of England and Wales has tended to focus upon visible characteristics of soil profiles, rather than on inferences about origin, when devising soil classifications. Here, the interests of the educationalist and the soil surveyor have not always coincided. Soils teaching becomes more interesting if the discussion can focus on processes, genesis and interrelationships: soil nomenclature becomes less ambiguous and less contentious if it is based on visible characteristics, rather than on inferred process. In addition, technicali-

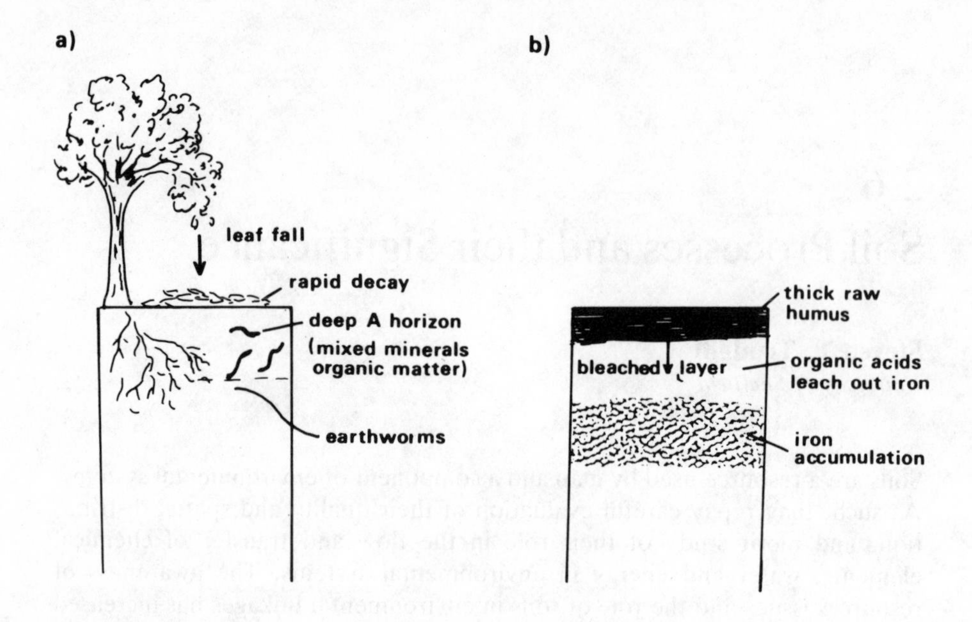

FIGURE 2.6.1 *Soil horizons and cycling*
(a) Rapid cycling: incorporation of organic matter by earthworms and a large A horizon.
(b) Less fertile soil, lower amount of cycling, little organic matter decay, organic matter chelates out a bleached, leached horizon, depositing chelated iron below.

ties in nomenclature have made soil types more difficult to teach in an interesting way to the less motivated students.

Thus there remains a diversity of approaches to soils teaching: for example, those who cling to the podsol and gley approach, those who deal with the new nomenclature, those who deal with soils in an applied context and those who focus upon linkages within ecosystems. A further dimension to soil study is that, professionally, many soil scientists in research institutes started life as physicists, chemists, biologists or in other disciplines. In school, most soils teaching occurs under the umbrella of geography or biology, or to a certain extent under geology or chemistry. In addition, much research goes on under the aegis of agricultural oganisations. There is perhaps a distinction beetween soil science in agriculture, soils as taught in geography and soils with a more biological angle. Sometimes these interests coincide and sometimes they diverge: it is perhaps best that none should neglect the others. This chapter therefore considers both the traditional aspects of soil formation and the importance of soil variability. Emphasis is given to the functional relationships with other parts of the ecosystem (water, landforms and plants), and to soil use and wider environmental impacts.

SOIL FORMATION AND DEVELOPMENT

Models are, in many ways, provisional mechanisms for dealing with real-world complexities. Traditionally, soil formation models have dealt with simple assumptions such as the relationships of soil types with the soil-forming factors, especially climate and parent material. Like many working models, these have become enshrined as monolithic entities to be learnt and reiterated, but as with all such enshrined models it should be remembered that they are representations of reality and are not reality itself. Real soil is complicated by a number of other factors, especially adjustment to climatic change – a topic which has received increasing attention in recent years and which qualifies the simplistic assumed association between soil type, parent material and climate. In particular, the formation of deeply weathered tropical soils is often associated with their longevity, as much as with their current climate; the characteristics of many UK soils are related to Pleistocene history as much as they are to their location in the temperate climatic zone. Many of the agricultural soil resources in Britain are those which were deposited or at least modified substantially during the Pleistocene glaciations, whilst in southern Britain, away from the glacial limits, Tertiary age landscapes were also modified by Pleistocene processes. In more tropical zones, the soils have not suffered wholesale removal and relocation by glaciation, but they have certainly suffered climatic modification. Here, many surfaces have been stable, and have undergone weathering since at least the Tertiary and in some cases back into the Cretaceous. Many Australian and African surfaces, for example, are thought to be of this order of antiquity.

True relationships between current climate, soils and parent materials can only be evaluated if a state of equilibrium exists between soils and climate; that is, if the soil visible today is completely adjusted to present climate. Unfortunately, the application of this important concept is difficult as it is problematical to arrive at indices of soil characteristics which are indicative of equilibrium conditions. However, it is probable that many of the soils in Britain are not yet fully adjusted to post-glacial conditions but represent soils which were becoming more leached in the Holocene, especially after the Atlantic period of increased wetness. They have also become increasingly modified by management, particularly in terms of tree clearances which tend to reduce the amount of nutrient cycling and allows leaching to dominate in the soil development process, encouraging podsolisation (Fig. 2.6.2).

A useful conceptual framework for the teaching of soil formation is the basic thesis that all soils have processes in common but that they are differentiated by the different combinations and relative magnitudes of the processes. Thus, in all soils there is a greater or lesser upward and downward movement of water, with soluble chemical elements; a greater or lesser mixing of organic matter, and so on. A fundamental process distinction, however, is that between lateritic and podsolic weathering (Fig. 2.6.3). Under

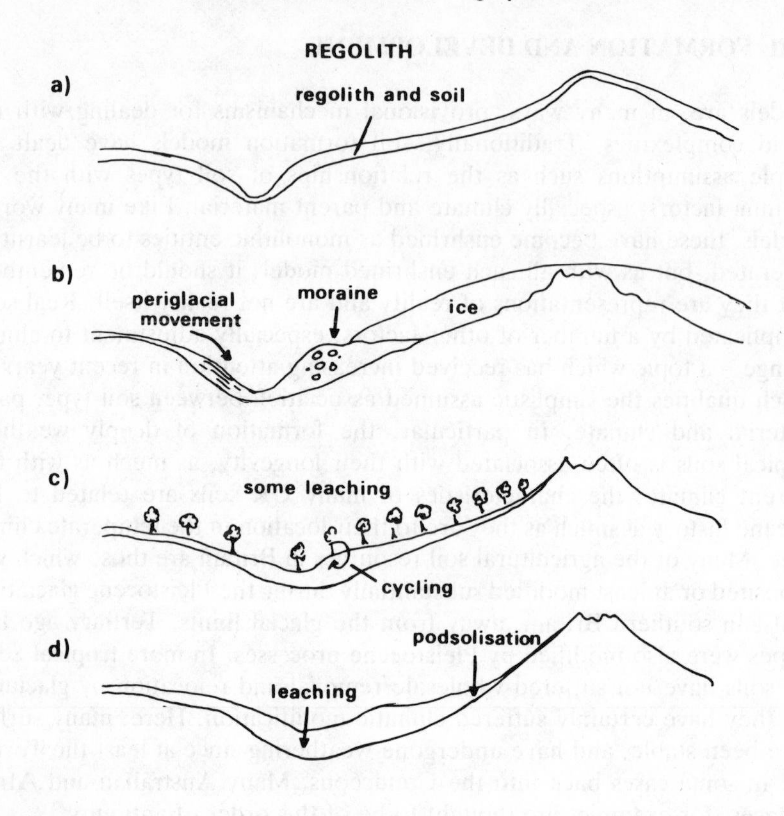

FIGURE 2.6.2 *Glacial and post-glacial soil changes*
(a) Preglacial
(b) Glacial
(c) Post-glacial
(d) Clearances by man

podsolic weathering, organic acids from accumulations of organic matter mobilise iron and aluminium, substances not highly soluble in water uncharged with organic acids. The mobilisation is achieved largely by chelation. This accounts for the presence of leached horizons under dark organic matter horizons. The iron is reprecipitated in relation to some combination of pH change, decay of the organic substances or saturation of the organic acids with metal cations. Under tropical conditions, the suggestion is that organic matter tends to decay more completely, with a corresponding lack of chelatory organic acids and the lack of a mobilisation of iron and aluminium. Silica is thus relatively more mobile, giving rise to lateritic soils where iron is left as a residue towards the surface of the soil. The widespread application of this idea has yet to be fully tested but has received some support on chemical grounds.

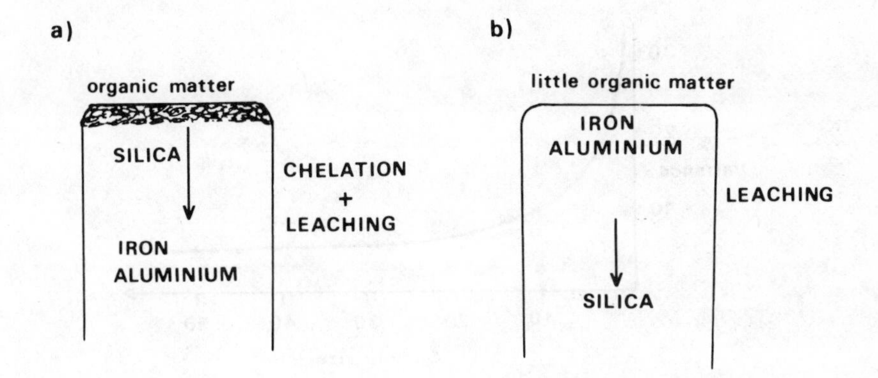

FIGURE 2.6.3 *Suggested schema of podsolic and lateritic weathering*
(a) podsolic
(b) lateritic

SPATIAL VARIABILITY

Soils are spatially heterogeneous, even when developed on similar parent materials under similar conditions. The challenge to soil science has been one of characterising spatial units of soil in a way that is faithful to the range of variation but yet presents spatial units which are manageable both in terms of cartographic representation and in agricultural practice. A Soil Survey mapped Soil Series has only a 60 per cent chance, on average, of representing the soil actually present on the ground. Thus, for many management purposes such as forestry or agriculture, a more detailed knowledge of soils is also required for the evaluation of planting strategy. General purpose maps remain, however, a fundamental way of evaluating soils and soil resources and of understanding the relationships between soil distributions and soil-forming factors. The focus continues to be on those aspects of soil which cannot be easily altered, such as texture and site, rather than on those which can be readily altered, such as fertility. In recent years, increasing attention has been paid to soil structure, a property easily altered in some soils.

Silt soils are especially prone to compaction if ploughed when too wet, that is at moisture content above field capacity. The evaluation of soil resources thus has to take into account not only the more permanent aspects of soil properties but also those factors which are sensitive to management, such as structure, and which can have a marked effect on crop productivity. Such factors, however, can vary markedly spatially and many farmers have experience of patches or corners of fields which suffer from compaction, making management on a per-field basis difficult, let alone on a mapping unit basis. Attempts at identifying spatial variability include taking progressively increasing number of samples in a fixed area until the variance decreases to an

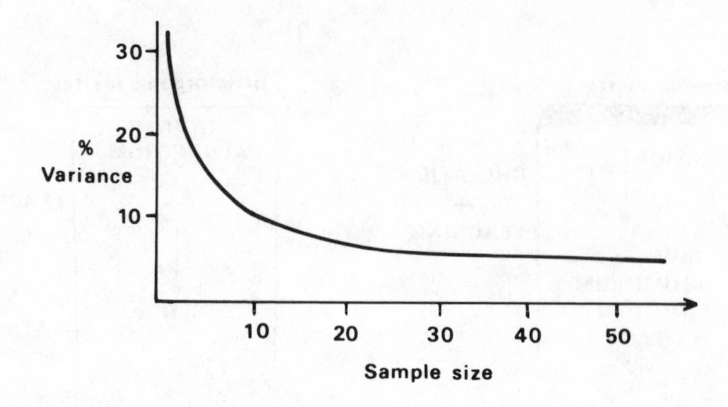

FIGURE 2.6.4 *An illustration of sample size and percentage variance in data from a sample*

acceptable level (Fig. 2.6.4). Other attempts have been made to review the periodicity of spatial variability, involving a more sophisticated range of techniques.

SOIL SYSTEMS

In the mid-1970s it was common to find geographers producing complex diagrams of interrelationships between variables by drawing labelled boxes with arrows, indicating directionality of influence or flow of matter – all this as if it constituted some kind of explanation for the system. Looking back, it is clear that such diagrams may have helped to organise the thoughts of the originator, but often took the reader so long to comprehend that they were unhelpful as a teaching mechanism. Certainly, it remains that systems thinking is vitally important rather than completely unhelpful. However, the wholesale rush to see everything in terms of boxes and arrows has now yielded to a more thoughtful consideration of linkages, and of the quantification of linkages rather than the simple representation of linkage as if, in itself, it proved something. People are researching in a system framework rather than just drawing systems diagrams. To a certain extent, this is old wine in new bottles and it is probably true that systems have appeared explicitly more in books aimed at students rather than in the pages of the *Journal of Soil Science*. Many researchers still set their objectives within a small portion of a system, rather than on the linkages of a large system, but soil has such a vital role in the regulation of the flow of nutrients, water and energy in ecosystems (Fig. 2.6.5), that such linked work is deserving of greater attention. One problem with this, though, is that science tends to be reductionist. It focuses on smaller and smaller portions of the real world, especially in the attempt to

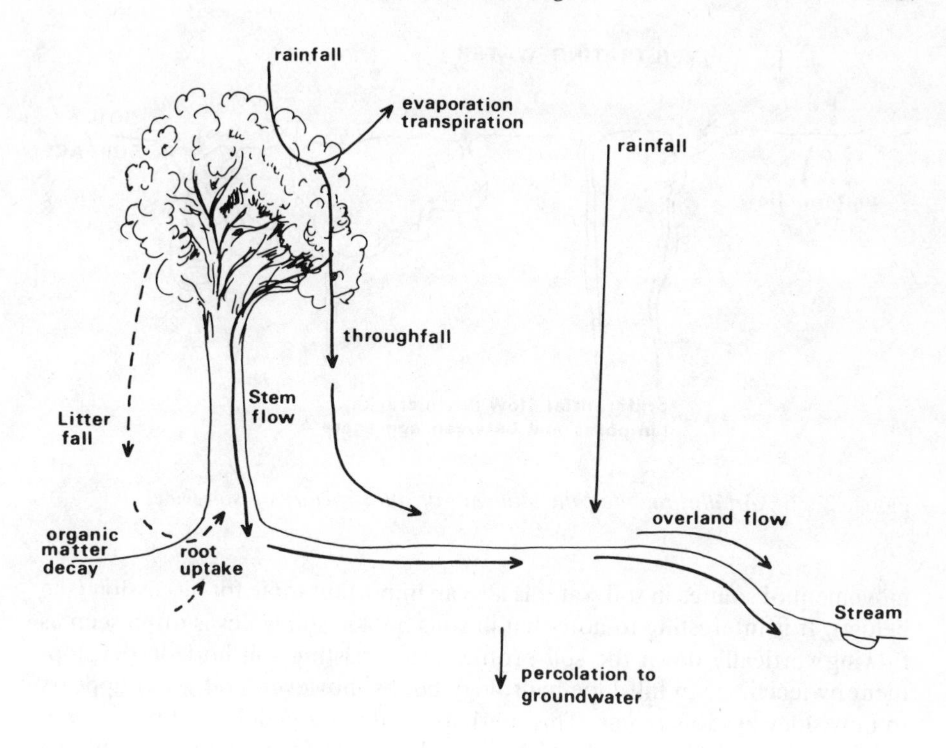

FIGURE 2.6.5 *Some components of nutrient, energy and water flow through soil in an ecosystem*

control for the effects of variables in an experimental method framework. Such an approach tends to advance the science better in terms of detailed mechanisms, but does not necessarily tell us more about whole systems. On the other hand, ecosystem analysis has often been at a level where it is difficult to see how the detailed mechanisms fit into the overall scheme; this is because such analyses tend to deal with input–output balances, say of nutrients, rather than with the greater detail of internal mechanisms. Nevertheless, systems studies have now achieved a sensible level of involvement in geography in general and also in soils.

SOME INTERDISCIPLINARY LINKS

Soils and hydrology

Systems thinking is implicit in many soil hydrological process studies, soil acting to regulate flow to bedrock aquifers and to streams, with studies of linkages and of quantities and directions of flow being important. The

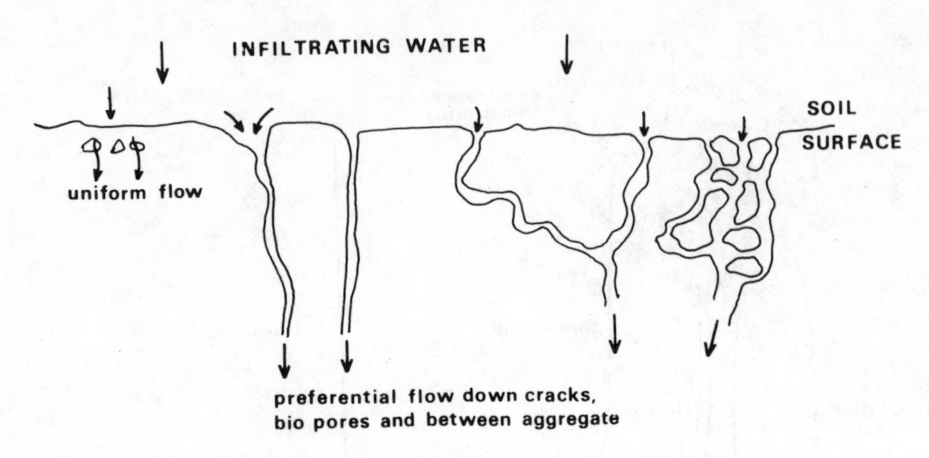

FIGURE 2.6.6 *An illustration of the influence of soil structure on soil water flow*

movement of solutes in soil water is also an important topic for discussion (see below). It is interesting to note that in soils books, soil water is often seen as flowing vertically down the soil profile, thus assisting soil horizon development by leaching. In hillslope hydrology books, however, soil water appears to flow sideways to streams. This, perhaps, reflects a pedological preoccupation with formation and a hydrogeomorphological focus upon contributing area processes.

The role of soil and soil water in geomorphology has been increasingly realised. Within this realisation has also been the evaluation of the role of soil structure. In well-structured soils,much of infiltrating soil water may bypass a proportion of the soil matrix, flowing down cracks and root channels (Fig. 2.6.6). This can become an important contribution to rapid flow in the initial stages of stream hydrographs, providing a rapid throughflow component and leading to discharge peaks similar to those which can be provided by overland flow. Such flow can also be important in the rapid surface to output transfer of surface-applied substances, such as nitrate fertiliser. It also complicates modelling procedures in terms of leaching predictions, in that chemical equilibrium conditions cannot always be assumed and because leaching is not wholly efficient if some of the soil matrix is being bypassed. Such bypassing flow is also termed 'preferential flow' and, because it may flow down larger pores (but not necessarily so), it can also be termed 'macropore flow'.

In general, the study of the relationship of soil water flow with soil structure is following the tendency for soil water models to take greater account of field conditions. Initially, and rightly so, soil physics dealt with uniform flow under control conditions. This enabled the basic relationships between water content, water tension and flow to be evaluated. However, the inability of theoretical models to predict flow in all field situations has instigated the elaboration of basic models so as to improve their scope of prediction. This follows a classic modelling procedure: evaluate the simple situation and then calibrate and revise for more complex ones.

Soils and geomorphology

Soils provide a process control in geomorphology, with soil acidity and water content exercising a major influence on regolith weathering. Soils also are developed on surfaces of contrasting ages, and thus geomorphological studies and soil studies can assist each other in the interpretation of soil age and landform age. Thus, the more deeply weathered, near-equilibrium soils can be found on the old land surfaces. Weakly developed or truncated soil profiles are to be found on younger and more sensitive surfaces.

In terms of process control, it is clear that if organically derived acids from soil surface horizons are neutralised within the soil profile, then bedrock weathering will be minimal. If, however, the soil profile is free of weatherable minerals, then bedrock weathering will be effected by such acids reaching the soil/bedrock interface. This consideration concerns the weathering of the bedrock but, in the former case where acids are neutralised in the soil profile, surface lowering will proceed because of the weathering of soil minerals themselves. Thus situations may be seen where the location of weathering and chemical erosion may vary between the soil profile and the soil/bedrock interface, but the net, overall effect, in terms of surface lowering, may be the same.

Soils vary spatially in many of their properties, as discussed above. It follows that there will be spatial variation in the rates of weathering in and beneath soils. This is a vital consideration in terms of the evolution of landforms. If geomorphology is to be concerned with the evolution of landforms, then landforms can be seen to arise in part by differential erosion; that is, some parts of the landscape are being eroded more than others. Such differential erosion may well be understood in terms of spatial variation in the soil characteristics which control weathering in and beneath soils. Thus soil acidity often increases upslope, suggesting that rates of erosion may increase upslope, leading to slope decline where the crest of the slope is lowered at a greater rate than the foot.

This should be balanced by hydrological considerations (Fig. 2.6.7). Water flow tends to increase downslope as water accumulates from upslope positions. Therefore, transport of weathering products can be facilitated downslope. However, water also tends to become progressively chemically saturated downslope as weathering products are moved downslope. It may well be that the location of the greatest solutional erosion is, then, somewhat downslope of the slope crest where acidity combines with a degree of water movement but before chemical saturation occurs. Such considerations are important in the understanding of the spatial variations in erosion rates, not simply downslope but across slopes and in relation to geological structures. In addition, chemical variation in superimposed material, such as a hetero-

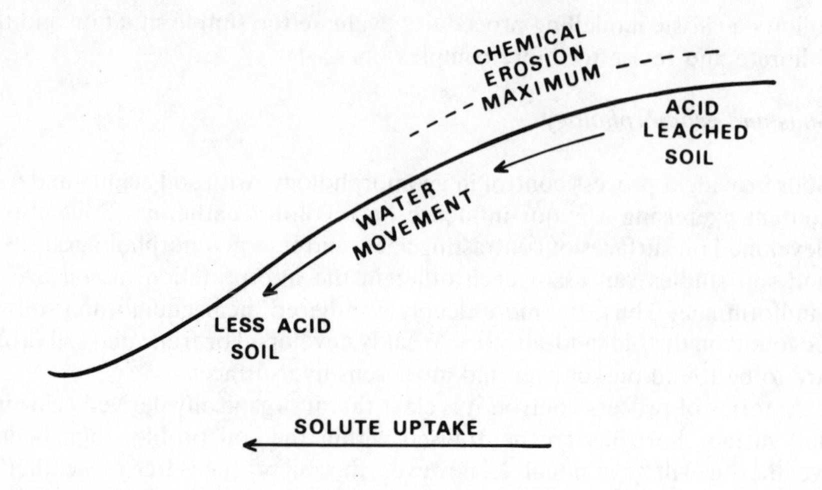

FIGURE 2.6.7 *Schema for the processes and distribution of chemical erosion on a hillslope*

geneous glacial drift, can have a marked effect on spatial variations in bedrock erosion rates and hence upon the progressive evolution of the landscape. Thus, in a system framework, ideas on soil spatial variability and hydrological processes can be combined to give a greater understanding about landform processes and forms.

Soils and biogeography

Soil is a factor in plant distribution as much as it is in landform distribution. In the past, some simplistic correlations have been made concerning soil–plant associations. Current thinking involves plant strategies and competition more. In addition, some established ideas, like that of succession have also been challenged. Many assumed successions are no more than zonations in response to environmental gradients. These are not stages in a changing succession and are liable to be present as only slowly changing zones, unless environmental conditions change. Thus, salt marsh zonation, for example, may in fact be relatively static and be present in relation to differing water cover and salinity, changing but slowly unless sea level changes. Many sand dune successions are also relatively stable and have already been through a succession of changes to arrive at their current form of zones – only to be revitalised into further succession if conditions change. Again, we have had the classic situation of initial models being modified, calibrated and revised in the light of current thinking. The important point is not to enshrine the model but to see it as just a current, working model, provisional upon further work. This is not to say that the original model is necessarily wrong, rather it is only a start and the priority is to specify the limits of its jurisdiction.

Soils are often taught in biogeography courses, although often separately from plants. Functional soil–plant linkage is relatively easily taught in terms of cycling of nutrients and energy, especially where soil organisms and soil organic matter are concerned. However, in terms of soil–plant relationships, soil is only one of many factors. The plant acts as an integrator of all the environmental influences upon it, including climatic as well as edaphic factors — but most importantly, most vegetation needs to be thought of in an historical dimension in terms of management by man. The vegetation of many upland areas is dominated by the influence of grazing and, since soils may be influenced by vegetation, especially the degree of nutrient cycling, soils are thus also influenced by man. Upland, podsolic soils under poorer grassland can often be interpreted as being so because of a lack of trees which would, if present, improve cycling and the flow of nutrients in the soil–plant system, reclaiming nutrients washed to the base of the soil profile by root uptake and returning them to the surface in leaf litter (see Fig. 2.6.1). The soil–plant relationship is thus a two-way affair, plants influencing soil in many ways and soil also being one factor in influencing plant distribution.

SOILS AND AGRICULTURE

Fertilisers

Soils and agriculture cover a wide range of topics but in recent years attention has been given to some specific aspects including the use of agricultural fertiliser and its losses to non-agricultural systems – perhaps illustrating the importance of systems thinking in terms of linkages between different components. In particular, there is current concern over the losses of fertiliser nitrate from soils to drainage waters. This represents a loss to the farmer and also an increase in nitrate levels in both groundwaters and surface waters (Fig. 2.6.8). In addition, there are two other main concerns about the presence of fertiliser nitrate in ground and surface waters. First, the nitrate increases the productivity of plants in the water systems, especially if combined with phosphate derived from soils by soil erosion or derived from domestic sources in sewage. Such enrichment with nutrients is termed eutrophication. In extreme cases, algal populations may respond to such enrichment to the extent that they can cover a water surface, blocking light to lower layers of the water body, leading to a loss of plant life at depth. Such ecological changes also have an effect on fish populations often giving rise to changes in species composition and fish size. Secondly, if the water is used for drinking, there are concerns for human health. High nitrate levels in drinking water have been associated with stomach cancer and also with the 'blue baby' syndrome: methaemoglobinaemia. Nitrate is more of a concern than phosphate, as it is more mobile in soil water. Current research has suggested that up to one-third of applied nitrate may be lost by soil leaching, especially under conditions of heavy rainfall. It appears that there should be closer co-operation between the soil physicist, the mathematical modeller and the field worker for a better understanding of this important topic. It is certain that farmers do not willingly waste fertiliser, and the relationship of fertiliser application time to subsequent weather conditions is an important consideration in understanding leaching losses.

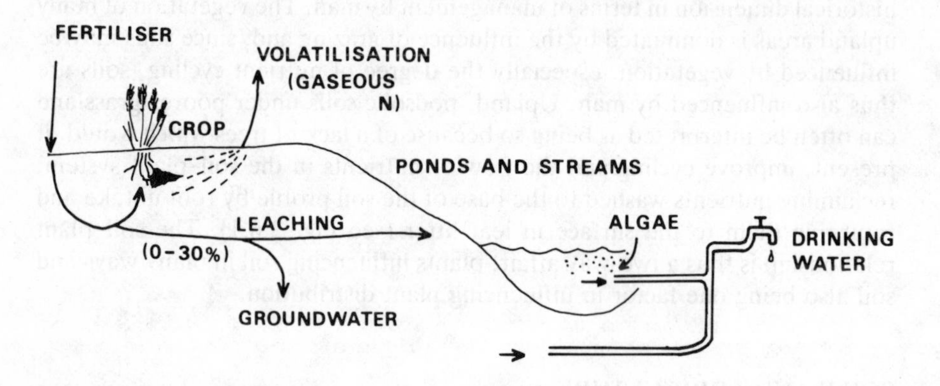

FIGURE 2.6.8 *Fate of applied nitrate fertiliser*

Soil erosion

With any environmental problem it is of interest to establish whether the problem cannot be tackled and solved because the solution is not technically or physically known, or whether it is because a solution is known but the implementation of the solution is limited because of social, economic or political considerations. Increasingly, it would seem that soil erosion is in the latter class as many technical solutions to the problem exist. These include careful evaluations of risk areas in terms of slope, vegetative cover, grain size of soils and the erosivity of rain water and runoff water. Also included are high technology solutions involving major engineering works and, often more effective, low technology solutions which require not only low installation outlay but also little effort in maintenance. The latter include simple terraces, brushwood holding fences and gabions (stone-filled wire cages) and, probably most effective, good land use management. This includes non-cultivation of the steeper and most vulnerble slopes, land use management going hand in hand with land capability evaluation (Figure 2.6.9). Socially sensible plans are often those which involve low technology, easily maintained structures installed at little cost and with little political upheaval.

Nevertheless, some of the soil erosion and conservation literature continues not only to neglect the social dimension, it often consists of detailed expositions of the erosion processes rather than suggestions concerning their

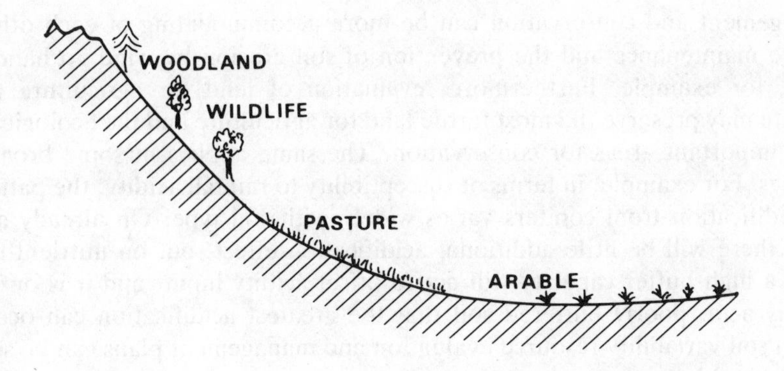

FIGURE 2.6.9 *Land use scheme for sloping sites, avoiding soil erosion by not cultivating the steeper slopes*

control. While an evaluation of erosion processes is a prerequisite for the implementation of conservation measures, it seems that the steps are not taken to link evaluations of erosion processes with conservation measures in a social framework.

SOIL MANAGEMENT AND THE ENVIRONMENT

The relationship between scientific study and effective action on environmental problems is an interesting one which has wide scope in the general context of soils, agriculture and the environment. In general, it can be suggested that management policies are based on implicit or explicit models of how the environment is thought to work. That is to say, that any management action will have an expected outcome. With the increase in systems thinking, the repercussions of management actions have become more widely seen, not only in terms of the intended target but also in terms of the wider environment. Such an approach is seen in the case of nitrate fertiliser and runoff water quality, but it is also apparent in the use of pesticides and herbicides. More recently, hedge removal has also become an issue in terms of wider conservation implications. In addition, the wide-ranging possible effects of acid rain on soils have become well publicised in recent years.

Clearly, in some cases there are conflicts between the use of soils for agriculture and other uses. Drainage is a good example: a soil is either drained or it is not – there is little room for compromise between agriculture and conservation. One can also question the need for agricultural intensification, especially in a European context, where it may be a case of adding yet more food to an overproductive EEC mountain rather than increasing exports to less developed countries to alleviate food shortage. In other contexts, soil management and conservation can be more accommodating of each other. Hedge maintenance and the prevention of soil erosion by wind go hand in hand, for example. Furthermore, evaluation of land for agriculture and wildlife may preserve the most fertile land for agriculture and the ecologically most important areas for conservation. The same applies in some broader settings. For example, in terms of susceptibility to rainfall acidity, the pattern of acidification from conifers varies widely with soil type. On already acid soils, there will be little additional acidification effect, but on nutrient-rich soils, a high buffer capacity will act to offset acidity input, and it is on the weakly acid, poorly buffered soil that the greatest acidification can occur. Again soil variability resource evaluation and management plans can be seen as closely related.

CONCLUDING PERSPECTIVE: HORIZONS IN SOIL STUDY

Such wide considerations have become more and more the province of soil study, and involve not only links with other parts of the physical system but also those with economics, sociology and politics. It is right, therefore, that soil study should continue within geography courses: geography is uniquely placed to tackle the issues involving physical, biological, chemical, economic, social and political considerations. But it should not be forgotten that such a broad evaluative skill as geographers have rests also in many ways on the specialist skills of the contributary disciplines. Soil study is very much the meeting of the disciplines and, as such, provides a good training within geography as much as it provides a crucial subject within biology, chemistry, physics, agriculture and engineering. This means that soil study has come a long way from 'Describe and interpret the profiles of a podzol and a gley soil'. It remains a foundation of the subject, but the links between soil properties like texture, structure and water-holding properties are fundamental to soil management and the evaluation of soil resources, and are the links between soils and other parts of the environment.

Soils geography thus appears to be a reasonably healthy subject, and should continue to be so if attention is paid to basic processes, developments in soil survey, soil properties, soil management for agriculture and linkages with other environmental components. No longer should the horizons of the soil scientists be limited by the study of the soil pit: that is essential, but a wider view in an applied context is also important. The priority here is to optimise soil productivity whilst minimising wider environmental degradation. In as far as this aim depends upon mastering not only the internal links between form and process but also the external relationship with other disciplines, it represents a fitting end to a review of processes in physical geography and an appropriate bridge to an examination of integrated fields of study.

FURTHER READING

Perhaps the best general starting point is the concept of the system itself as applied to soil. Relevant examples and generalisations will be found in the first three suggestions (of which the first is the best introduction), whilst the last book takes a more specific look at soils:

White I. D., Mottershead D. N. and Harrison S. J. (1984) *Environmental Systems: An Introductory Text* (London: Allen & Unwin).
Bennett R. J. and Chorley R. J. (1978) *Environmental Systems: Philosophy, Analysis and Control* (London: Methuen).
Huggett R. J. (1980) *Systems Analysis in Geography* (Oxford: Clarendon Press).
Trudgill S. T. (1977) *Soil and Vegetation Systems* (London: Oxford University Press).

Still at a general level, interesting perspectives which include the soil dimension (origins, relationships and changes) are provided by:

Ollier C. D. (1984) *Weathering* (London: Longman).
Gerrard A. J. (1981) *Soils and Landforms* (London: Allen & Unwin).
Grime P. (1979) *Plant Strategies and Vegetation Processes* (Chichester: John Wiley).
Goudie A. S. (1983) *Environmental Change*, 2nd edn. (Oxford: Clarendon Press).

The human dimension is specifically considered in:

M.A.F.F. (1970) *Modern Farming and the Soil* (London: HMSO).
Gasser J. K. R. (1980) 'Impact of man in catchments (i) Agriculture', in Gower A. M. (ed.) *Water Quality in Catchment Ecosystems* (Chichester: John Wiley).

Finally, three papers are suggested which have methodological and pedagogical implications:

Trudgill S. T. (1983) 'Soil geography: spatial techniques and geomorphic relationships', *Progress in Physical Geography*, vol. 7, pp. 345–60.
Nortcliffe S. (1984) 'Spatial analysis of soil', *Progress in Physical Geography*, vol. 8, pp. 261–9.
Trudgill S. T. (1977) 'Environmental Science/Studies: depth and breadth in the curriculum', *Area*, vol. 9, pp. 266–9.

PART II
EXPLANATION AND CONTROL

Section 3
Environmental Impact and Change: A Geographical View

3.1
Stable and Unstable Environments – the Example of the Temperate Zone

John Lewin
University College of Wales, Aberystwyth

It is the physical geography of the temperate zone that is likely to seem the most familiar to most readers of this book. In this zone members of literate and scientific societies, reading in English, have both their everyday experiences to go on, and the records of some centuries of analysing their environments to help them. Assuming a comfortable familiarity can, however, lead to some quite startling misunderstandings and difficulties in novel or changing situations. In the last twenty years we have begun to appreciate much more fully how temperate environments are themselves changing, especially because of the extension works and side effects of those very urban societies that have themselves fostered our modern understanding of 'environment'. In the longer term, it is now seen that in mid-latitudes climatic changes have been more dramatic and prevalent than previously thought, involving conditions sometimes very far from temperate. Recent studies have also produced new knowledge at an accelerated rate, sometimes involving radical new perspectives which resulted both from theoretical advances and from new research methods using the latest technology. It can be appreciated, too, that restricted levels of information have also been imposed through limitations of travel and linguistic capability, a noteworthy disability from which many 'western' geographers have suffered with respect to eastern Asia. Altogether and for most of us, the last few decades have shown that the temperate environment is truly a stranger place than we knew, with the sobering after-thought that our improved understanding of this intricate but distinctive zone still does not of itself qualify us to pronounce on world environments in general.

INITIAL CONSIDERATIONS: TOWARDS ACTUALISTIC ENVIRONMENTAL MODELLING

A consciousness of the limits to present knowledge is perhaps as appropriate an attitude for this age as were the confident pronouncements of earlier ones. Interestingly enough, it is now just one hundred years since W. M. Davis first gave broad outline to his cycle of erosion. According to his biographers, this happened at a meeting of the American Association for the Advancement of Science in 1884. His model was very clearly a product of its own time and place, inspiringly conceived in the intellectual – and physical – climate of eastern North America and western Europe in the late nineteenth century. The applicability of his ideas outside that environment has been greatly criticised, for example by W. Penck with his experience of the tectonically unstable regions of Turkey and South America, and by L. C. King with his field knowledge of southern hemisphere plainlands. Twenty years ago, in the academic context of that time, R. J. Chorley re-evaluated what he then called W. M. Davis's 'geomorphic system', emphasising the beguiling attractiveness of Davis's clear-cut model yet at the same time criticising its limited application, the dialectic basis of its presentation and supporting evidence, and its stultifying influence. Indeed one is now distinctly reminded, on rereading Davis, of the sure style and authority of other nineteenth-century writers like Charles Darwin and Karl Marx.

Davis's model was billed as 'geographical', but it involved neither those interactions between man and his habitable environment nor the climatic and biogeographical elements essential to a comprehensive 'physical' geography. Such matters had been explored at the time, notably in G. P. Marsh's *Man and Nature* of 1864, and have been much more studied since. Geomorphology, climatology and biogeography have, for much of their history, been researched as essentially separate disciplines (linked to geology, physics and biology as much as geography), but there are recurrent attempts to explore links between them. Within geography, the development of systems thinking, as for example promoted and developed by R. J. Chorley, has been most important in this respect. In the 1960s, a 'new' quantitative physical geography attempted a leapfrog into science, but on the back of statistics and cybernetics rather than the traditional basic sciences. Functional relationships between environmental entities that might be regarded as intrinsically physical, biological or chemical were established by statistical techniques. But such potentially geographical studies have not proved dominant; instead research in subdisciplines has continued more or less independently. Thus some of the earlier valuable quantitative work by geographers was in climatology where data were available for analysis and where in mid-latitudes geographers were able to show how fluctuating and variable supposedly stable climates were. In other fields, it rapidly became clear just how scarce numerical observations actually were, even in temperate environments. Geomorphologists found that they neither knew quantitatively what the forms of hillslopes actually were, nor as a matter of precise observation what processes were occurring on them, nor at what rates these various processes were operating.

In practice, the stimulating and liberating effect of new theoretical approaches to physical geography, together with much better contacts with other sciences, has had the result of inducing a considerable volume of field measuring and monitoring. Consequently there developed a demand for field and laboratory equipment for a whole range of studies, some new and some traditional, that were seen to require a more rigorous survey, observational and analytical approach. It is significant that the 1960s and 1970s were, for many temperate countries, times of expanding higher education and research funding, and that as the activity rate of academic geography increased, there was an insurgence of new blood into the profession which was not only able to promote new ideas and greater scientific rigour, but which also had the great good fortune of access to facilities for such things as computing, analytical chemistry, photogrammetry, radiometric dating and pollen analysis.

Thus in recent decades, studies of temperate environments may be characterised as being highly active and productive – though perhaps also as compartmentalised and studying many small problems rather than few large ones. Some areas of study were enhanced along existing lines, others expanded dramatically, whilst others became neglected. Expansion occurred in those areas where geographers and their research sponsors saw themselves as making significant contributions, both on grounds of theory, technical feasibility, and practical benefit. River catchment studies proved particularly appropriate for development. Water flows and storages can be directly observed over short time-spans (a runoff event, a hydrological year) that are meaningful for analysis and using field equipment that it is practicable to install; analysis of field observations gives valuable information that other disciplines have not attempted on the same scale; and there are also, in the background, important practical considerations involving erosion, water resources, and flood and drought hazards. Above all, there appeared to be within hydrology those opportunities for systems modelling on a more or less formal basis which the prepared minds of research geographers were looking for, and which in their own environments they had the chance to take up.

In geomorphology, two sets of studies seem to have gained most attention. Serving to replace the inspired but often speculative generalisations of the older denudation chronology, a strong group of researchers has been studying the actualities of Quaternary environmental change, piecing together in impressive detail such things as the pattern of late Quaternary deglaciation and the course of vegetation or sea-level change during the present interglacial. Again newly available technology or expertise has proved important, for example in dating or in pollen analysis. A preoccupation with the past is justified in particular by many geomorphologists who follow J. Budel in believing that 95 per cent of the total relief of mid-latitudes consists of forms inherited from pre-Holocene conditions. In this sense, mid-latitude landforms are not temperate at all, but derive from prior cold or even pre-Quaternary quasi-tropical conditions. Other than in high mountains, the remaining 5 per cent involves coastal or alluvial environments, or that thin but vital skin of soils whose slow generation but potentially rapid destruction may be of such human concern. The second group of geomorphologists has concerned itself with just these sensitive environments, and has been inspired particularly by the approaches to field hydraulic processes introduced by L. B. Leopold, M. G. Wolman and J. P. Miller in *Fluvial Processes in Geomorphology* (1964). A vast amount of patient field monitoring has been undertaken, involving, for example, stream sediment transport and rates of soil movement on slopes. A general pattern to this is only slowly becoming apparent, and as time has passed researchers have also become more involved with process detail itself, as described in previous chapters, rather than with anything overtly geographical.

What has emerged, both in geomorphology and in other fields of physical geography, is indeed much more realism with regard to processes and a much greater amount of information on the rates and manner in which these processes operate. This has meant that physical geography has become difficult to comprehend overall, because its components range from studies parallel to those of palaeobotany on the one hand, to the physics of sediment transport on the other. As a result, it is certainly not easy to construct an integrated, regional 'temperate environment' appraisal – because few physical geographers have recently been thinking in that way on that scale, and because information is still just not available for many of the situations for which it is desirable.

SLOPES, RIVERS AND FLOODPLAINS IN TEMPERATE ENVIRONMENTS

By way of example, and because it is an area in which so many individual pieces of research have been completed, let us generalise about what has recently been discovered about sediment movement in temperate fluvial systems. This involves diffuse 'wash' processes and mass movements on hillslopes; streamhead processes like natural piping and gullying; and channel processes of bank erosion and bedform change. Four themes to recent research can be identified.

Rates of processes

In 1983, I. Saunders and A. Young summarised results from some 270 publications involving instrumented surveys on rates of surface processes (dominantly concerning slopes) comparing them with 140 reviewed previously. They noted that very little indeed was published about such processes before 1960. For the most part temperate regions come out as having the slowest process rates – though in this zone as in others such generalisations are greatly complicated by variable relief, tectonic activity and human activity. It is perhaps upon accelerated erosion rates that attention has become most clearly fixed (Fig. 3.1.1), and the understanding and prediction of soil loss is proving to be an important study field.

Channel change studies have been growing in number especially since 1970, both using direct field monitoring and the analysis of historical maps and air photographs. It is now appreciated that in many 'stable' environments, the recycling of alluvial sediments by shifting rivers is an important and complex activity. Figure 3.1.2A shows channel changes on a reach of the River Severn near Welshpool. Considerable reworking of the valley floor is indicated in the last 300 years, with the detail of change being complex and sometimes challenging to our preconceptions. Thus the development of individual meander loops may involve such phenomena as deposition on the outside of bends and the formation of asymmetrical and compound meander loop forms, as this example shows.

FIGURE 3.1.1 *Eroding soils on loess near Oxford, Mississippi, USA*

There have been some attempts to relate patterns of change to particular environments (for example, exploring the possibility that the hydraulic geometry relationships between channel widths, depths and velocities occur with regional differences), but for the most part forms are related to local hydraulic conditions or in a more general way to stream discharge magnitude

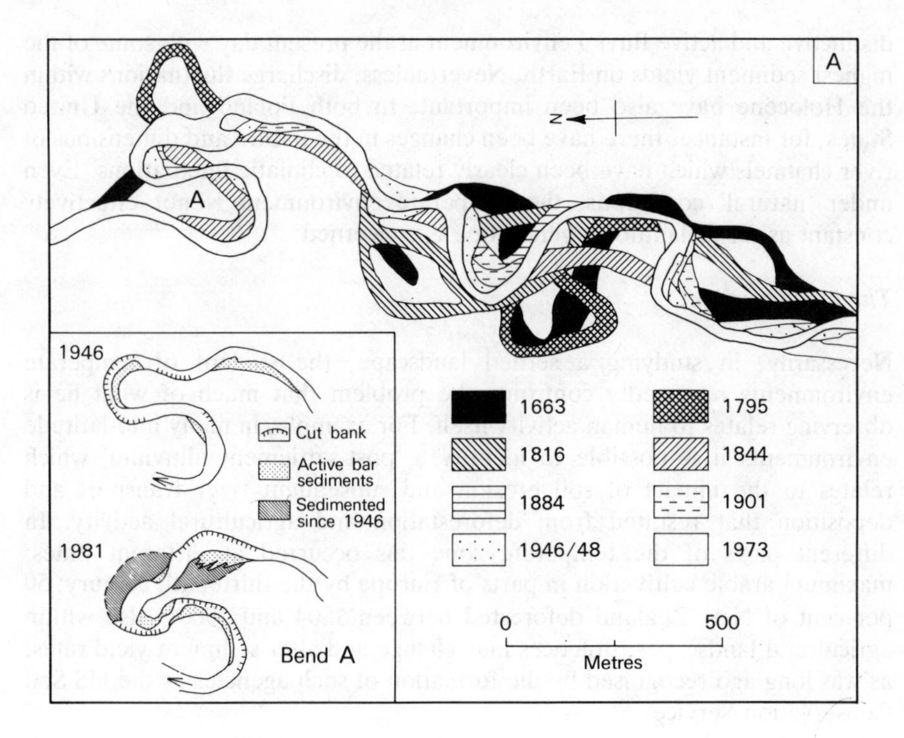

FIGURE 3.1.2A *Channel changes on a reach of the River Severn near Welshpool*

and frequency. What is emerging, is that many temperate environments (unsurprisingly) are less characterised by extremes than others, so that floodplain forms tend to build up progressively over years without the catastrophic effects of 'superfloods' as found in many semi-arid environments. This allows changes to be predicted, given a knowledge of past rates and an understanding of the way channel forms change over time.

Inheritance and change

It is appreciated that the material currently being transported or reworked by streams is notably inherited from prior cold conditions and from former conditions of much higher sediment transfer activity. Glacier sediments and outwash, and periglacial deposits, are important in providing material which present Holocene rivers are slowly but progressively reworking. The timing of prior glaciation (which varies spatially in the temperate zone) is important in understanding the nature of present day activity, as is the nature and disposition of preweathered material. In this respect, the vast easily eroded loess deposits of certain more continental temperate environments are highly significant: this makes, for example, parts of North China an extremely distinctive and active fluvial environment at the present day with some of the highest sediment yields on Earth. Nevertheless, discharge fluctuations within the Holocene have also been important. In both Poland and the United States, for instance, there have been changes in the nature and dimensions of river channels which have been clearly related to climatic fluctuations. Even under 'natural' conditions, the temperate environment is not effectively constant as far as landform generation is concerned.

The human impact

Necessarily, in studying a settled landscape, the student of temperate environments repeatedly confronts the problem that much of what he is observing relates to human activity itself. For example, in many mid-latitude environments it is possible to identify a 'post-settlement alluvium' which relates to the impact of soil erosion and subsequent river transport and deposition that resulted from deforestation and agricultural activity. In different parts of the temperate zone this occurred at different times: maximum arable cultivation in parts of Europe by the thirteenth century; 50 per cent of New Zealand deforested between 1804 and 1900. Also within agricultural landscapes, practices may change and alter sediment yield rates, as was long ago recognised by the formation of such agencies as the US Soil Conservation Service.

Rivers have also been grossly affected by direct control and urbanisation, the latter a physically complex and culturally varied process involving building construction, paving, storm water culverting, river channelisation and so on. Urban hydrology has itself become a specialised study field in which geographers have participated effectively. Geomorphological side effects of features like reservoirs installed for the provision of urban water supply or irrigation have also in recent years begun to be studied in considerable detail. Reservoirs may affect the whole river downstream by trapping sediment which is thus unavailable for further transport and deposition; the regulation of flow may also affect the frequency of events competent to transport sediments. This may lead to physical problems of downstream channel scour, reservoir capacity decrease, a decrease in channel size and a channel pattern adjusted to different sediment loads, and eventually to effects on floodplains as a whole.

Finally, and to a variable extent, fluvial processes have come to involve the dispersal of urban, agricultural, and industrial waste as well as natural materials, and here it is appropriate that the new-found process realism of physical geographers should be used to understand the pattern of waste dispersal and deposition. For example, mid-Wales was the site for intensive mining for lead and zinc in the nineteenth century; although mining is no longer active, metal pollution continues to affect floodplain and other environments because of the environmental recycling of mining wastes and by-products. Given an understanding of fluvial processes, however, it is possible to predict where floodplain soils are most highly polluted, sometimes with some quantitative precision. Thus metal concentration may be seen to depend on the volume of waste output in relation to natural sediment flow, the distance of mine waste input points up-valley, the timing of floodplain sedimentation in relation to that of mining activity, and the detailed sedimentary environment involved – especially the sediment sizes found there.

That such human activities may have effects on the geomorphology of river channels and floodplains, may be illustrated by a reach of the River Ystwyth (Fig. 3.1.2B). Here a meandering channel in 1800 was replaced by a braided and aggrading one in the late nineteenth century in association with mining activity. In this century, the channel has become incised again in older mining era sediments. It is important to appreciate that environmental changes may both take place within stable environmental systems, yet may also result from external changes arising through such things as climatic fluctuation or human impact, both being related through the dynamical system changes discussed by John Thornes in Chapter 1.2.

Systems variety and systems interaction

The final point concerns matters that are highly geographical: systems that are non-physical and non-fluvial interact with ones that are, whilst there is a

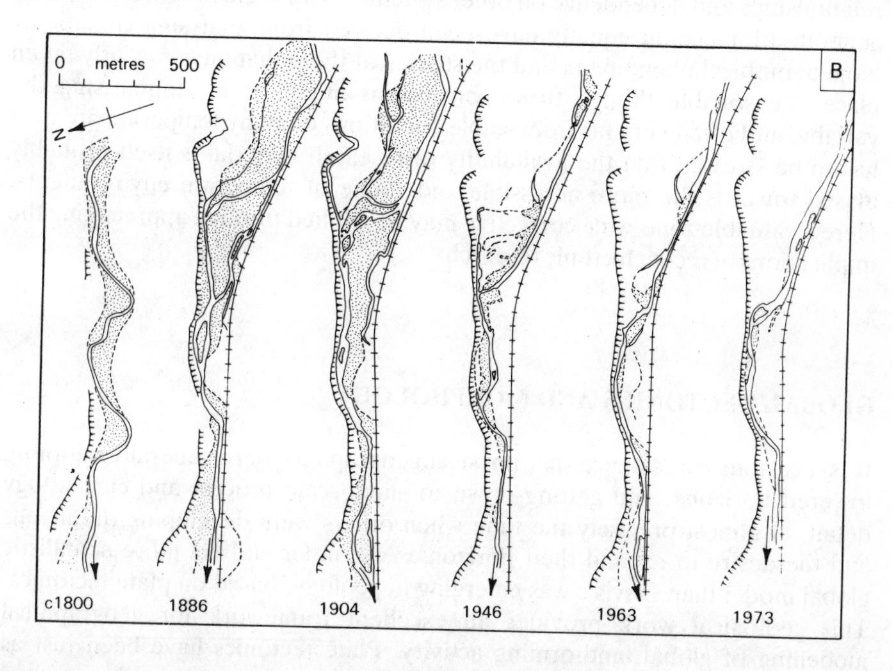

FIGURE 3.1.2B *Channel changes on a reach of the River Ystwyth in relation to mining activity*

spatial variety to fluvial systems that is particularly important. Concerning the first, we know that aeolian, chemical and biochemical processes are important in both natural denudation and in metal pollutant dispersal alike, such that even during single weather events materials may move variably in solid or solute forms. Spatial variety is found within river catchments at the small and large scale. For example, both sediment transport rate and channel pattern have usually been related to available stream power, defined as the product of stream slope, discharge and the unit weight of water. This may be maximised in the middle reaches of contemporary rivers in upland Britain and it is here that channel change rates are maximised and floodplains are dominated by lateral sediment accumulation forms like point bars. Lower down, despite increased discharges, gradients are lower and sediments finer and more cohesive. Lateral shift rates are minimal and sedimentation is dominantly overbank in type. On a continental scale, by contrast, sediment supply is often dominated by headwater mountain areas, such that the geomorphological functioning of the temperate zone is dependent upon those adjacent mountain environments which are discussed in a later chapter.

Thus, even by looking briefly at fluvial systems, we might characterise temperate environments as ones of high recent research endeavour, moderate rates of measured process activity concentrated in certain subenvironments, large effects of inheritance and of human activity itself, and strong interrelationships and dependence on other systems. With some modification, such generalisations could equally have been derived from analysing climatic or biogeographical phenomena and the studies of them that have recently taken place. Reasonable though these conclusions might be in emphasising the variable and often very far from stable conditions of many temperate areas, it has to be stressed that the availability of research material is itself probably biased towards the more accessible and stable of temperate environments. Here a valuable zone-wide corrective may be applied through appreciating the implications of recent tectonic research.

GLOBAL TECTONICS AND MORPHOLOGY

It is a curious coincidence that physical geographers were generally adopting lowered horizons, and getting down to small-scale process and chronology detail, at almost precisely the time when others were developing the means and the desire to expand their horizons. A grander and yet more actualistic global model than Davis's was emerging in geology, based on plate tectonics. This geological work provides an excellent framework for geographical modelling of global landforming activity. Plate tectonics have been just as dependent on a new technology of discovery and on a wealth of individual research effort as has been applied in physical geography, whilst studies have embraced both active and recent tectonic activity, and the stratigraphy of sediments that derive from the erosion of positive relief elements produced at various places and at different times in earth history.

Lessons must be learnt that are vitally important for physical geography. There is now available a growing body of evidence which tells us where and when mountain building actually occurred, together with the timing, volume and type of sediments that progressive (or at times, balancing) erosion of such upstanding or uplifting areas produced. We can now see, on a realistic basis of painstakingly accumulated data and not just in a generalised model, what has actually happened to landform development on a continental scale and in the longer term. This development is variable spatially, not so much by zonal climatic regions, as according to tectonic environment. There are three related conclusions that seem important for present purposes:

Theoretical separation of uplift and erosion in time, as was proposed by Davis, is now known to be unrealistic for the bulk of continental sediment transfer. The stratigraphic record shows quite conclusively that tectonic and erosional activity are maximised together in high energy environments, but that for many temperate environments this is not present or has long passed. High erosion rates have also characterised many arid environments in the geological record (aridity produced in the past, as today, by the configuration of continents with respect to atmospheric and oceanic circulation), so that the drier mountain-fringing parts of the temperate zone, as in the southern USSR, may be ones of significant landforming activity.

Persistent global variety of land forming activity, in terms of nature and rates, is an inevitable consequence. High-relief tectonically active mountain belts can sustain relief renewal for geologically extended periods of time. Here there may be a genuine dynamic equilibrium between rates of uplift and rates of sediment removal. There may also be distinctive assemblages of processes in high-relief environments that characteristically include catastrophic slope instability and debris flow-type activity. Such areas contrast at the other extreme with stable shield or craton areas, at times subjected to marine incursion and withdrawal, and where very old landform elements indeed may be found, sometimes disrupted by rifting. Intermediate are orogenic belts which may have ceased to be active and which are in process of reduction and incorporation into continental cratonic cores.

Continents, including temperate zones, have a variable geometry incorporating the three major relief settings outlined above in different ways. At the largest scale, actual denudation systems may incorporate orogenic areas (commonly sediment-supply areas situated asymmetrically on continental blocks, as in the South American Andes), cratonic sediment-transfer zones in which persistent low-relief continental conditions have prevailed, and marginal sedimentary basins where eroded material has accumulated in inland, deltaic and marine environments.

Fascinatingly enough, the work of W. M. Davis and his two major 'evolutionary' critics W. Penck and L. C. King, can be related to the three morphogenetic continental environments suggested above – even though each appeared to propose evolutionary models of worldwide applicability. Thus Davis's cycle best fits orogenic belts in the course of abandonment. Penck's relation of morphology to rate of uplift may be assessed in tectonically active zones, whilst King's receding scarplands and extensive surfaces can be expected in rifted and flexuring cratonic environments. Plate tectonics has rendered their supposedly global models regional according to tectonic zone.

The significance of such conclusions in giving variety to the temperate zone that might otherwise be inadequately appreciated from the margins of the north Atlantic may be reinforced by considering the geomorphology of Japan, for which the results of many studies are now very fortunately available in English translation, and comparing it with Britain. Parts of tectonically unstable Japan, with its frequent earthquakes and sixty historically active volcanoes, have undergone uplift of more than 1500m or subsidence of over 1000m in the last 1–2 million years. High relief and high precipitation (averaging 1–2000mm) and precipitation intensity (in the southwest a daily precipitation of 100mm occurs once a year on average) lead to mass movement and debris flows, and a landscape of deeply dissected mountains and fringing depositional plains with alluvial fans and braided rivers. By contrast, Britain has been tectonically almost stable (though certain movements are beginning to be appreciated), has high relief and precipitation only in parts of the north and west (and there not approaching the levels achieved in Japan), and has a landscape subjected to much lower erosion rates at present dominated by localised soil loss and the reworking of some alluvial areas by meandering rivers.

ON THE HORIZON

In the near future we should see study of internationally better and less well-known temperate environments developing in several ways. First, the process realism of recent years is likely to continue in fields ranging from urban climatology to natural stream sediment transport. Such studies in managed, man-modified environments ought to continue to be both extremely interesting and of practical value. Secondly, both through the proliferation of these studies and through larger-scale and more extensive monitoring techniques (notably derived from remote sensing) it ought to become increasingly possible to make statements that are relevant to the variety and functioning of the whole of the temperate zone, and to be less restricted to a limited spectrum of environments. Here the crucial importance of other environments discussed in the next few chapters has to be appreciated, and both knowledge for its own sake, economic development, and human welfare all strongly depend on not allowing our horizons to be restricted to certain mid-latitude environments alone.

Finally, we may see approaching a more satisfactory synthesis between the information provided by small-scale geographical work of recent decades, the broader approaches as presented by plate tectonics, and acceptable theoretical concepts. For example, an energy-balance approach to broad-scale landform development is possible in which erosional energy (quantitatively defined, somewhat like stream power, in terms of the product of regional gradient and unit runoff) is related to materials resistance (including effects of variable hardrock lithology, inherited preweathered material, and vegetation cover), and known sediment yield rates can be related to relief change rates such that landform developments over time can be modelled and checked against the stratigraphic record of yielded sediment. If the last two decades have seen the impact of new theoretical approaches and the refining of field monitoring and analytical techniques, together with the accumulation of a wealth of empirical information, perhaps the next may lead to their greater synthesis and predictive use.

FURTHER READING

A wealth of techniques and examples appropriate to a geographical analysis of systems, presented in a rigorous but not easily assimiliated style, is fully covered by:

Bennett R. J. and Chorley R. J. (1978) *Environmental Systems: Philosophy, Analysis and Control* (London: Methuen).

The underlying notion of climatically controlled regional geomorphologies which emphasises the inherited nature of most mid-latitude landforms is introduced through the life work of an eminent European geomorphologist:

Budel J. (1982) *Climatic Geomorphology* (Princeton, NJ: Princeton University Press).

The first of the following texts examines the technological, biological and earth science aspects of the city (the environment in which half the world's people will live by the end of the century), whilst the second provides an up-to-date documentation of the effects of people on environment (in the tradition of G. P. Marsh's *Man and Nature*):

Douglas I. (1983) *The Urban Environment* (London: Edward Arnold).
Goudie A. (1981) *The Human Impact: Man's Role in Environmental Change* (Oxford: Blackwell).

A more specialised volume which deals with climatic and other environmental changes as well as hydrological matters, and which also contains chapters in which the contemporary concerns of relating sediment transport and morphology to discharge are quantitatively studied, is:

Gregory K. J. (ed.) (1983) *Background to Palaeohydrology* (Chichester: John Wiley).

Finally, an up-to-date summary of much detailed work, allowing the nature of the contemporary British river environment to be assessed, is provided by:

Lewin J. (ed.) (1981) *British Rivers* (London: Allen & Unwin).

New Physical Frontiers: Instability and Change

3.2
The Polar and Glacial World

David Sugden
University of Edinburgh

Some 60 million years ago Antarctica was welded on to Australia and linked to South America by a mountain chain. Greenland was just beginning to separate from North America and the Eurasian basin of the Arctic ocean was beginning to form. Temperatures were much higher. Temperate forests ringed the Arctic and Antarctic coastlands, while vegetation thrived at the South Pole where today the mean annual temperature is −50°C. Ice-sheets were non-existent and valley glaciers were confined to the highest mountains: sea-ice was unknown and permafrost was absent except possibly in high mountains. The world was devoid of pingos and tundra polygons.

Today the polar world is dramatically different. The Antarctic is now a continent supporting an ice-sheet whose broad convex summit is over 3000m high and inundates all but a tiny fraction of the land (Fig. 3.2.1). Mean winter temperatures are below −70°C. The ice-covered continent forms the centre of the atmospheric circulation of the southern hemisphere with westerlies sweeping uninterruptedly over subantarctic seas. Downslope drainage winds cooled by the ice radiate out and are diverted by the Earth's rotation to form coastal easterlies. The winds power ocean currents and affect the seasonal extent of sea ice, which in winter effectively doubles the size of the continent. For most of the last 2–3 million years large ice-sheets also covered much of the Arctic and extended into northern mid-latitudes (Fig. 3.2.2).

The evolution of today's polar environment and its present nature results from the interplay of plate tectonics, atmospheric and oceanic circulation and their relationship to ice-sheet build-up and decay. Each of these is conveniently viewed as a process–form system. Each operates on a different scale in space and time. Thus plate tectonics affects the distribution of land and sea over the whole globe on time-scales of tens of millions of years, while ice-sheets may build up and decay over continental areas in tens of thousands of years; ocean circulation may change in a matter of 1000 years. An understanding of the physical geography of the polar world thus involves study of the behaviour of these various systems, the complex way they interact with one another, and their sensitivity to change – and acts as an

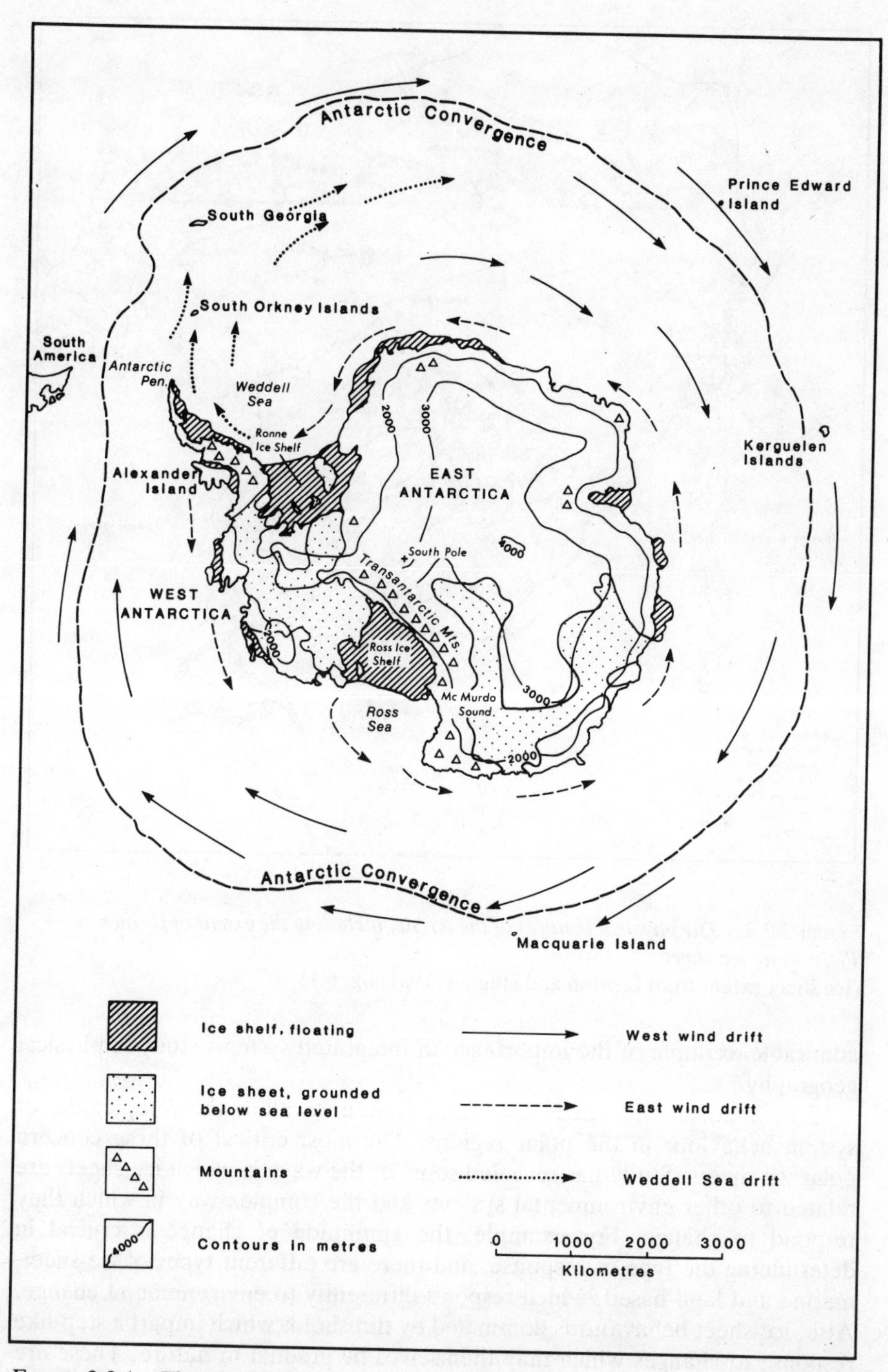

FIGURE 3.2.1 *The physical features of the Antarctic*

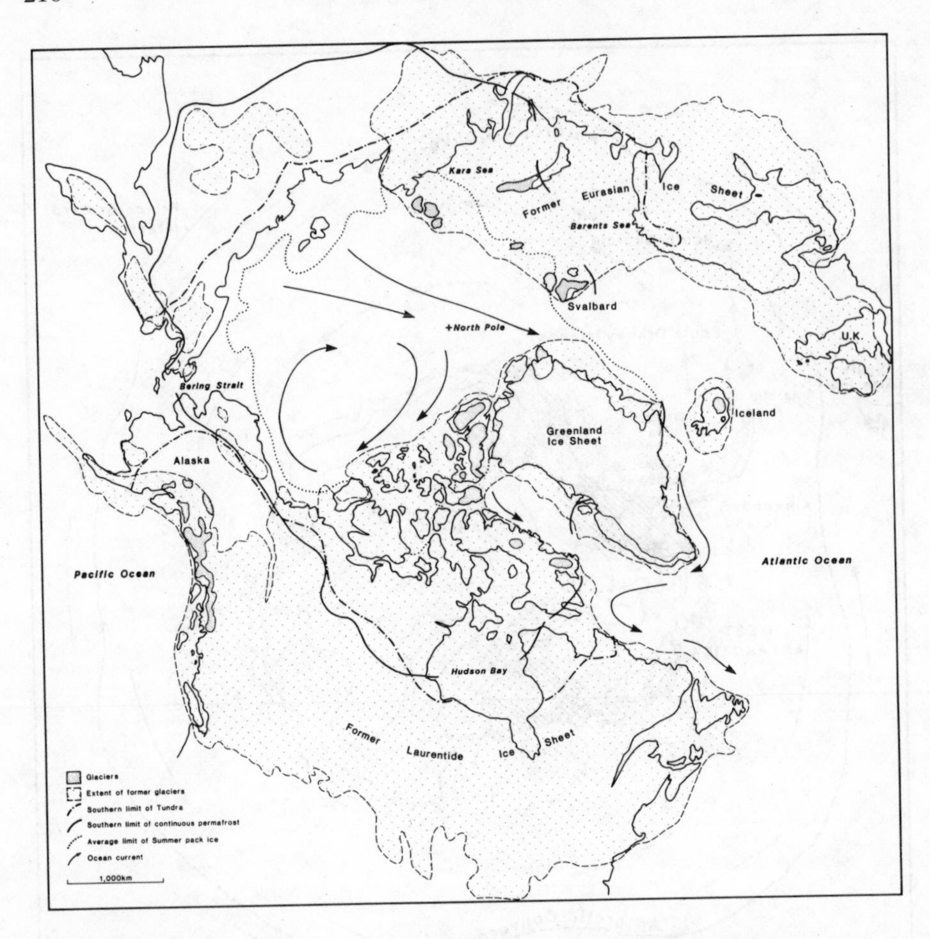

FIGURE 3.2.2 *The physical features of the Arctic, including the extent of former Pleistocene ice sheets*
(Ice sheet extent from Denton and Hughes, 1981, fig. 8.1)

admirable example of the importance of integrated systems study in physical geography.

Many important advances have come from unravelling the complexities of system behaviour in the polar regions. The most critical of these concern polar ice sheets. Study has revealed some of the ways in which ice-sheets are related to other environmental systems and the complex way in which they respond to change. For example, the amplitude of change is crucial in determining the type of response, and there are different types of ice-sheet, marine and land-based, which respond differently to environmental change. Also, ice-sheet behaviour is dominated by thresholds which impart a step-like response to changes which may themselves be gradual in nature. These are the themes of the present chapter.

INITIATION OF THE GREENLAND AND ANTARCTIC ICE-SHEETS

As so often occurs with initial discoveries, the first outline of the initiation of polar ice-sheets (based on deep sea sediment cores) provided a shaft of insight which tied together many previously puzzling observations. A variety of evidence was used, namely the amount of ice-rafted quartz grains, the composition of the microfauna which reflects sea temperatures and, most important, the stable oxygen isotope ratios ($^{18}O/^{16}O$) of calcium carbonate tests of microfauna. The ratio at any one time reflects the temperature of the water and the amount of ice on the land. The latter dominates oxygen isotope ratios since it is the isotopically ligher ^{16}O which builds up preferentially in ice-sheets, causing the sea water to be enriched in the heavier ^{18}O.

Plotted over time, the oxygen isotope ratios in the deep sea sediments show a progressive increase in $\delta^{18}O$ throughout the Tertiary indicating progressive cooling. Superimposed on this trend are a number of steps related to key stages in the growth of polar ice-sheets (Fig. 3.2.3). The first glaciers calving into the sea were present in Antarctica 26 million years ago, as shown by the first unambiguous evidence of ice-rafted debris. Presumably glaciers were present earlier in high mountains, perhaps as long as 38 million years ago. A

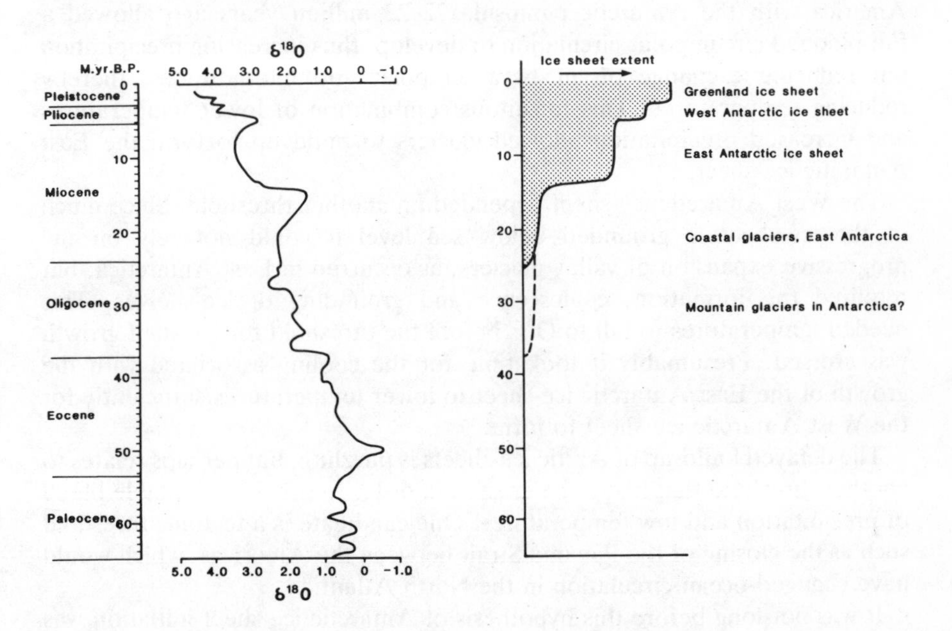

FIGURE 3.2.3 *Compression of the ^{18}O records of benthic foraminifera of Tertiary age in low-latitude deep-sea cores and the inferred stepped build-up of glaciers and ice sheets*
(^{18}O records from Savin *et al.* 1981, fig. 1, p. 424)

second major change took place approximately 14 million years ago and reflects the build-up of the East Antarctic ice-sheet. This was marked by a general ocean cooling, a sharp expansion of sea-ice and a change in the isotopic composition of the ocean as light oxygen isotopes were abstracted from the sea and stored in the ice-sheet. A third important change occurred 5–6 million years ago when the West Antarctic ice-sheet built up over island archipelagos and intervening seas, and a fourth step reflects the expansion of ice-sheet glaciation to the Arctic. Although ice-rafted debris shows there were tide-water glaciers 4–6 million years ago, the first full ice-sheet glaciation did not begin until 3.2 million years ago.

It is likely that these distinct steps in the build-up of polar ice-sheets relate to distinct thresholds in system response. The key initial threshold may have been the development of a circumantarctic ocean and atmospheric circulation in response to changes in the distribution of land and sea. First, the breakaway of Australia opened a sizeable ocean around Antarctica by 38 million years ago and was responsible for the initial cooling of ocean water and initiation of mountain glaciers in Antarctica. The morphology of one such mountain group detected beneath the East Antarctic ice-sheet by radio-echo sounding shows glaciated valleys which may relate to this initial glaciation (Fig. 3.2.4). Secondly, the breaching of the mountain chain linking South America with the Antarctic peninsula 22–23 million years ago allowed a full-blooded circumpolar circulation to develop, thus increasing precipitation but reducing exchange of air between polar and mid-latitudes, thereby reducing temperatures. This fortuitous combination of lower temperatures and increased precipitation allowed glaciers to build up to form the East Antarctic ice-sheet.

The West Antarctic ice-sheet depended on another threshold. Since much of the ice-sheet is grounded below sea level it could not rely on the progressive expansion of valley glaciers, as occurred in East Antarctica, but required the formation, coalescence and grounding of ice-shelves. This needed temperatures to fall to O°C before the threshold for ice-shelf growth was crossed. Presumably it took time for the cooling associated with the growth of the East Antarctic ice-sheet to lower temperatures sufficiently for the West Antarctic ice-sheet to form.

The delayed build-up of Arctic ice-sheets is puzzling, but perhaps relates to another threshold that had to be crossed in order to provide a suitable blend of precipitation and low temperatures. One candidate is a tectonic threshold such as the closing of the Panama Strait between the Americas, which would have changed ocean circulation in the North Atlantic.

It was not long before this hypothesis of Antarctic ice-sheet initiation was challenged by independent evidence. The problem is unresolved and illustrates the complexity of the argument and the difficulties of interpretation. The key evidence comes from the potassium-argon (K-Ar) dating of volcanic rocks which were erupted subglacially. These have been discovered at several sites in Antarctica on nunataks (Fig. 3.2.5). The rocks at Mount Early and

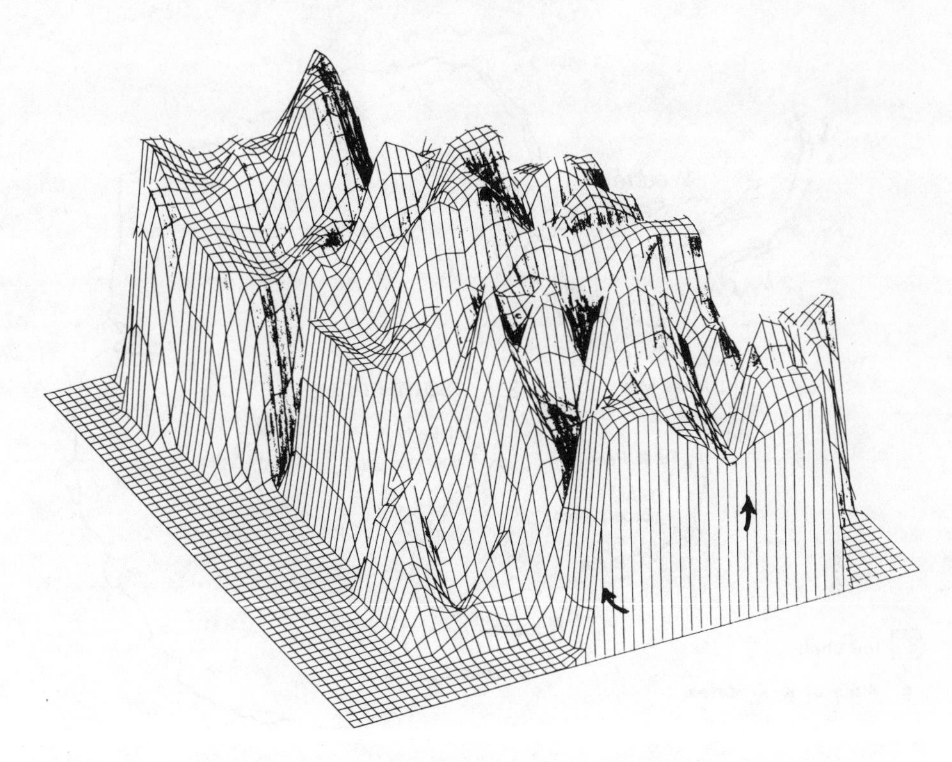

FIGURE 3.2.4 *Reconstruction of the alpine scenery of the Gamburtsev Mountains in East Antarctica revealed by radio-echo sounding. Arrows show troughs.*
(Reproduced by courtesy of David Perkins)

Sheridan Bluff were apparently erupted 15–18 million years ago while those in West Antarctica formed some 14–27 million years ago. These dates imply that the ice-sheets were in existence long before the dates suggested by the marine evidence. Perhaps the volcanoes erupted beneath small local glaciers or ice-caps. Perhaps the stepped growth of Antarctic ice took place earlier than suspected with, for example, the East Antarctic ice-sheet building up 38 million years ago. Or perhaps the interpretation of either set or both sets of evidence is incorrect.

The implications of the build-up of polar ice-sheets are profound. By 3.2 million years ago the present pattern of polar physical geography was broadly established. The growth had itself accentuated the initial cooling started by changes in atmospheric and ocean circulation. The significance of the ice-sheets in affecting polar temperatures can be illustrated by the observation that, were the Antarctic ice-sheet not present, the temperature at the South Pole would be almost 30°C warmer simply because of the lower altitude. If one also takes into account the high albedo of the white ice-sheet surface, then this figure should be increased still further.

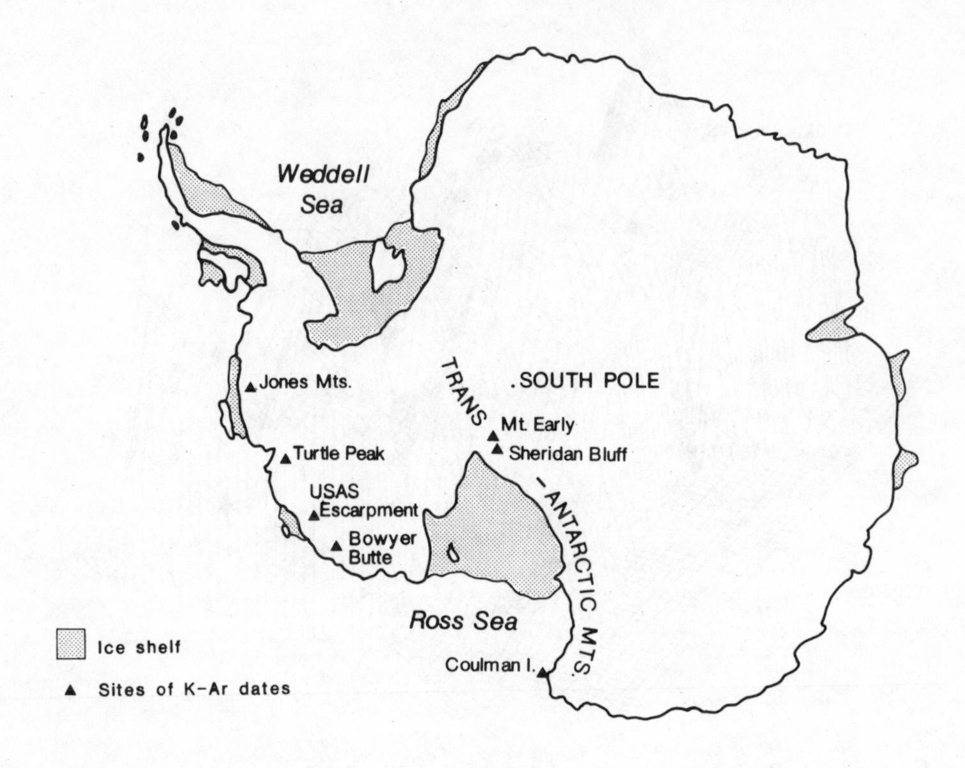

FIGURE 3.2.5 *The location of volcanic rocks in Antarctica whose ages suggest ice was present earlier than 14–27 million years ago*

OSCILLATIONS OF POLAR ICE-SHEETS

Since their inception, polar ice-sheets have been waxing and waning with a periodicity of about 100 000 years. This periodicity seems convincingly linked with changes in solar radiation received in northern hemisphere mid-latitudes in response to a cyclical variation of around 100 000 years caused by the eccentricity of the Earth's orbit, which modulates shorter cycles related to variations in the tilt of the Earth's axis (41 000 years) and changes in precession (affecting the Earth's distance from the sun in any one season) with cycles of 23 000 years and 19 000 years. Spectral analysis of oxygen isotope variations in deep-sea cores suggests precisely these periodicities over the last half million years. The saw-tooth shape of these cyclical variations in stable oxygen isotopes gives some indication of the pattern of ice-sheet build-up and decay. Typically it takes 50 000–100 000 years for the ice-sheets to build up, often with sharp and minor fluctuations, then abruptly deglaciation occurs and is followed by some 10 000 years of interglacial conditions.

At present, the link with solar radiation variations seems clear. However, there is little understanding of the mechanisms by which ice-sheets respond to external change.

Before discussing these mechanisms, it is first necessary to outline the scales of the oscillations in polar regions. During glacial maxima the Antarctic ice-sheet extended to the edge of the continental shelf. Evidence of this expansion is revealed by submarine glacial troughs, ice scouring and lodgement till on the offshore shelf. The biggest change was in West Antarctica where grounded ice extended over the Ross and Weddell Seas. Altogether there may have been as much as 80 per cent more ice on the continent.

In the Arctic major ice-sheets covered much of northern North America, north-western Eurasia and all of Greenland (Fig. 3.2.2). In addition, extensive ice-caps were centred over mountains. Sea ice or an ice-shelf filled the Arctic ocean and extended into the Atlantic and Pacific. Unlike the Antarctic, much of the Arctic lowlands remained free of ice and was therefore exposed to uninterrupted periglacial conditions for millions of years. With a sea level some 100m lower than that of today, extensive tundra existed in Alaska, north-eastern Siberia, the Bering Straits and the offshore platform of both coasts.

Interglacial conditions are likely to be represented by conditions today. The North American and Eurasian ice-sheets disappeared on several occasions, and indeed such periods are largely responsible for the oxygen isotope fluctuations seen in the deep-sea cores. In view of the great size of these two ice-sheets, their repeated build-up and decay represents one of the more remarkable illustrations of an extreme response to a modest insolation change. The Antarctic and Greenland ice-sheets shrank but then stabilised. Although there have been suggestions that both disappeared in past interglacials, there is no evidence of this. If the present existence of the Greenland and Antarctic ice-sheets is typical of interglacials, then it is important to establish the mechanisms involved. In particular, why did the Greenland ice-sheet survive when its larger, higher neighbour in North America disappeared?

A coherent understanding of ice-sheet response to change is still elusive. Perhaps the fundamental difference is between land-based and marine-based ice-sheets. Land-based ice-sheets formed the main part of the East Antarctic, Greenland, Eurasian and North American ice-sheets. Such ice-sheets may build up to thicknesses of 3–4km and their gradients and profiles are largely influenced by the strength of glacier ice. Flow is dominated by mechanisms of internal deformation in response to the mass balance of accumulation and ablation. In areas with a high input of snow they flow fast; with a low input of snow they flow slowly. On land, the ice-sheet terminates where the rate of ablation is able to melt the amount of ice being carried towards the margin. In snow they flow slowly. On land, the ice-sheet terminates where the rate of ablation is able to melt the amount of ice being carried towards the margin. In areas of low snowfall this point is reached in areas with relatively cold summers. Where snowfall is high, the ice may extend into relatively warm mid-latitudes in order to achieve sufficiently high summer melt rates to achieve balance.

The critical factor concerning the fluctuations of land-based ice-sheets is that they respond to changes in mass balance and thus are directly influenced by climate. There is a consensus that, especially in mid-latitudes where snowfall is adequate, the temperature and length of summer are critical in that they control ablation. A succession of cold summers will produce less melting and an immediate advance of the ice margin, while a succession of warm summers will lead to immediate retreat of the ice margin. Changes in snowfall will influence the amount of ice that has to be melted by ablation and will obviously also affect the position of the ice margin, albeit after a lag of some years as the effects of any change in the higher accumulation area take time to be transmitted to the margin. However, in most areas such variations are small in comparison to the dominant effect of summer temperature in controlling ice extent.

There is an exception in high polar locations such as north-east Greenland and the McMurdo dry valleys of East Antarctica where, in spite of very cold summers which favour minimum ablation, there is little snow. In such situations the critical factor limiting glacier extent may be the amount of snowfall received by a glacier (an increase in snowfall leading to an advance). Such precipitation effects may completely dominate ablation effects. Of particular interest is the fact that at high latitudes variations in precipitation and temperature may affect glaciers in contrary ways. Since it is the low temperatures which cause the low precipitation, warming can lead to an increase in precipitation. Thus in the case of glaciers where snowfall is the limiting factor, an advance may occur at a time of warming. This produces the fluctuations which are out of phase with those in more maritime or temperate latitudes.

Marine-based ice-sheets are grounded below sea level over much of their extent. The best existing example is the West Antarctic ice-sheet which incorporates islands archipelagoes and intervening bedrock lows extending over 1600m below sea level. Examples of marine ice-sheets in the past include the Hudson Bay and Atlantic fringe of the Laurentide ice-sheet. Marine ice-sheets exist because the weight of ice is sufficient to cause them to ground below sea level. In this sense they are analagous to a grounded iceberg, and their survival and extent is highly sensitive to ice thickness and sea level, in that thinning induced by mass balance changes or a rise in sea level will increase flotation. The grounding line which marks the point of ice-sheet flotation reflects such changes, and there are reasons for believing that retreat would, through positive feedback, accelerate thinning and precipitate collapse of an ice-sheet within a matter of centuries. It is important to stress that, unlike land-based ice-sheets, marine-based ice-sheets are very sensitive to sea level change.

A long-term goal is to understand how different ice-sheet types respond to any changes in solar radiation. There are numerous important unknowns; for example, which insolation changes are important and in which latitude or hemisphere they must occur? However, at this stage it is interesting to look at the potential of an approach trying to link some of the key complexities. One such attempt has been made by George Denton and Terry Hughes in a model resting on the assumptions that insolation variations affecting the southern margins of the northern hemisphere ice-sheets are the basic cause of the sheet fluctuations, and that the effect is transmitted to other ice-sheets via sea-level changes to produce globally synchronous ice-ages.

They suggested that full glaciations result from sharp summer cooling between 50° and 75°N which triggers growth of ice-caps in Canada and Scandinavia and the spread of sea ice in Arctic seas. As the ice-sheets build up and world sea level falls, marine-based ice-sheets grow in the Antarctic and parts of the Arctic. Growth of the ice-sheets further depresses temperatures and aids continued growth. Deglaciation occurs when high insolation increases melting along the margins of the land-based northern hemisphere ice-sheets. Apart from directly causing ice retreat, the resulting sea-level rise destroys marine-based ice-sheets. The process is self-accentuating, with decreasing ice-sheet size leading to higher temperatures and sea level. In combination, these remove the two biggest northern hemisphere ice-sheets. In the cases of the Greenland and East Antarctic ice-sheets, deglaciation stops when another threshold is reached and the sensitive marine-based fringe gives way to a stable land-based ice-sheet. The West Antarctic ice-sheet is an anomaly, and could be close to its threshold of survival today.

The dating and timing of ice-sheet growth and decay can be used to support this hypothesis. Figure 3.2.6 shows the solar radiation variations at two latitudes: the 41 000 year tilt cycle is dominant at 75°N and the 23 000 year precession cycle at 50°N. It is the latter that dominates melting of the southern margins of the mid-latitude ice-sheets, but the 41 000 year cycle reinforces or counteracts the effect. When both are in phase, ice growth or decay is enhanced in both southern and northern sectors, and produces major ice-sheet fluctuations. When they are out of phase, the effect is muted. The 100 000 year periodicity of ice ages is mainly a reaction to these shorter-term radiation changes. Only when the ice-sheet has grown to its maximum extent in mid-latitudes is it sufficiently sensitive to increased radiation to melt sufficiently to trigger the feedback sea-level processes necessary to cause full-scale deglaciation.

Denton and Hughes' model must be regarded as provisional. One reason is that it assumed that the critical factor affecting ice-sheet behaviour is insolation in northern mid-latitudes. If the critical areas are elsewhere then there will be different relationships between elements of the system. In this context it has been argued that during the last and present interglacials the subantarctic ocean temperatures warmed up several thousand years before those elsewhere in the world and also before the volume changes as indicated by ocean $\delta^{18}0$ values. If this is substantiated, then it is difficult to see how the northern hemisphere ice-sheets could initiate changes in the Antarctic. Indeed it seems

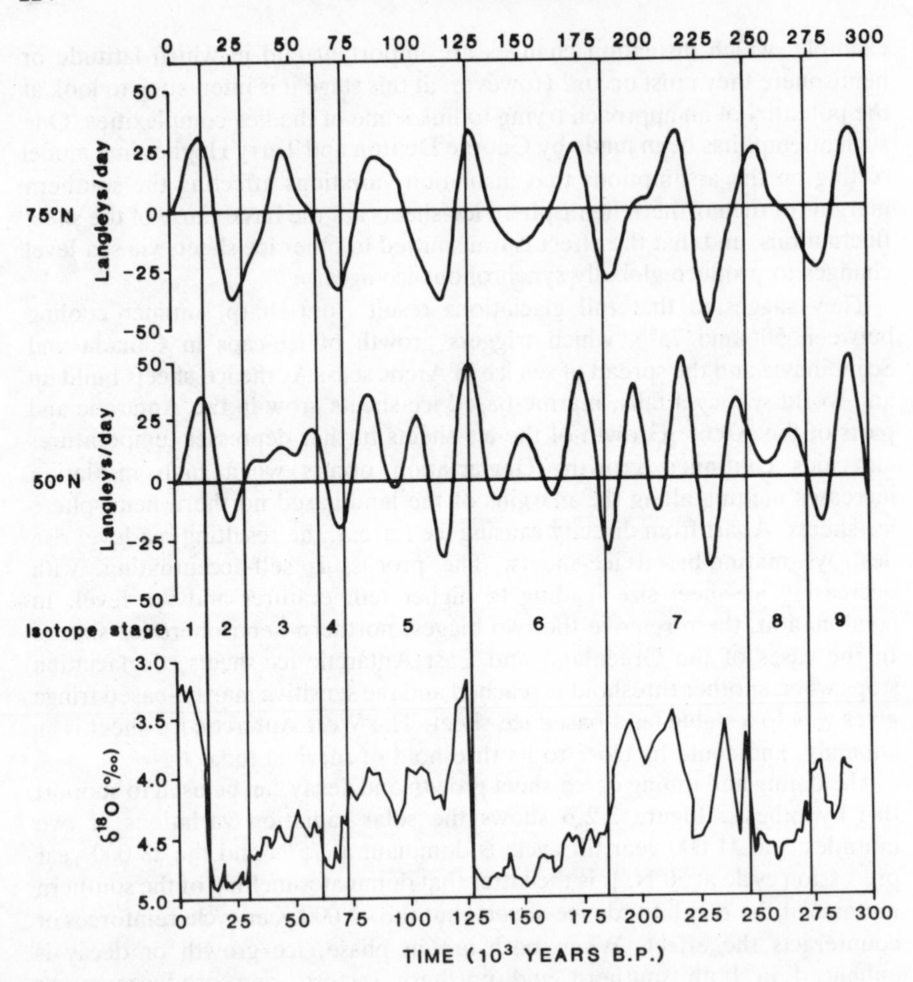

FIGURE 3.2.6 *Comparison of the northern hemisphere half-year insolation record at 75°N and 50°N and the oxygen isotope record from a deep-sea core (V 19–29)*
The oxygen isotope record is widely assumed to be a good approximation of the amount of ice on the land. (After Denton and Hughes, 1983, fig. 2, p. 130)

more likely that the response in northern mid-latitudes lags behind that elsewhere, perhaps because of the time taken for ice-sheets to build up and decay.

Another reason for regarding the model as provisional is that there are numerous feedback effects, some of which amplify and some of which suppress the original change. Until these can be clarified, even the dominant direction of change cannot be established. Some idea of the complexity is illustrated in Figure 3.2.7. Whereas the main positive feedbacks accentuating initial warming have been mentioned above, there are negative feedbacks which delay the response and may dominate regional glacier response. One

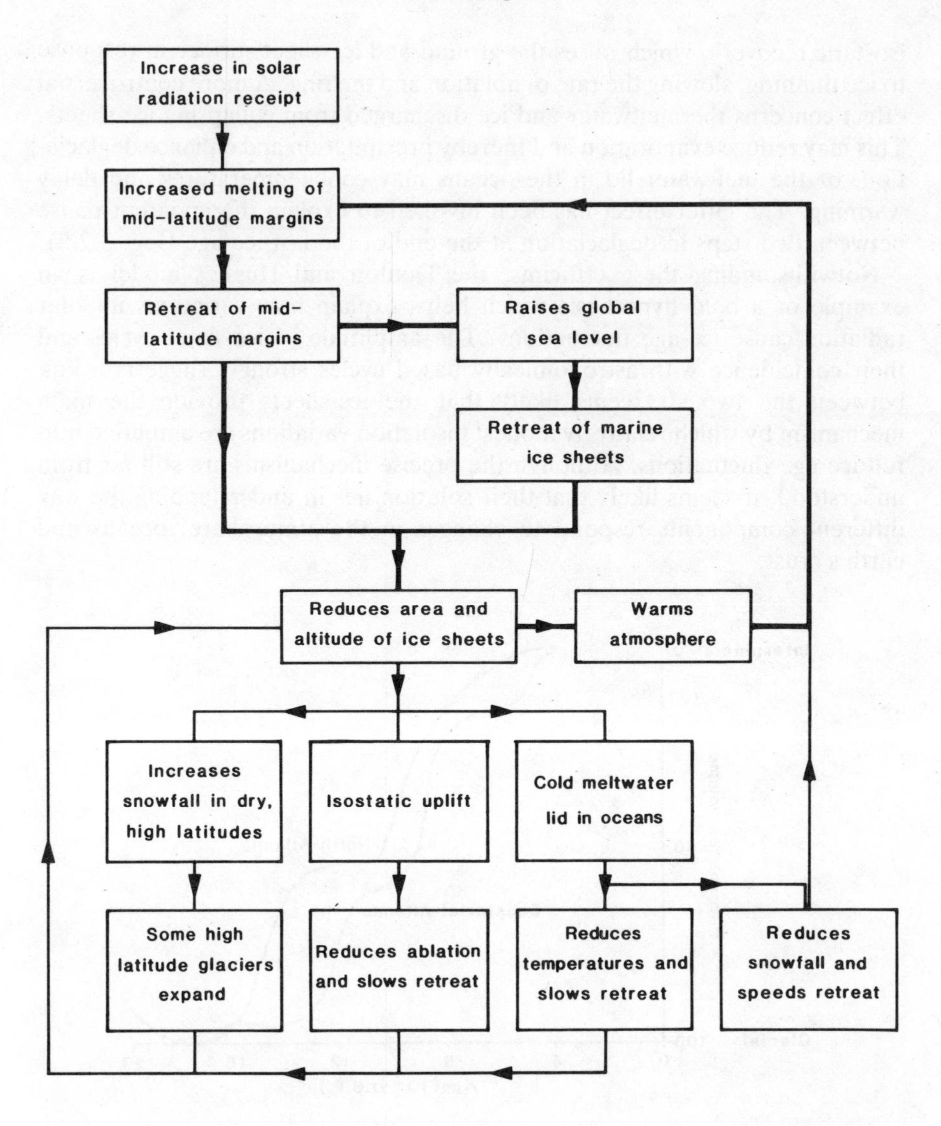

FIGURE 3.2.7 *Flow diagram illustrating some feedback effects associated with the response of ice-sheets to an increase in solar radiation received in northern hemisphere mid-latitudes*
Heavy lines indicate the main positive feedback loop which accentuates the initial change. Other lines indicate subsidiary effects, both negative feedback (which counter the initial change) and positive feedback.

possible effect of the decline of mid-latitude ice-domes is to increase precipitation at high latitudes. Such a situation has probably affected north-eastern Greenland where the ice appears to have attained its maximum extent only after the thinning of the Laurentide ice-sheet reduced the latter's blocking effect on easterly tracking depressions. Another effect relates to isostatic recovery, which raises the ground and ice-sheet surface in response to ice thinning, slowing the rate of ablation and melting. A more controversial effect concerns the meltwater and ice discharged from collapsing ice-sheets. This may reduce evaporation and thereby precipitation and enhance deglaciation, or the meltwater lid in the oceans may cool temperatures and delay warming. The latter effect has been invoked to explain the apparent pause between two steps in deglaciation at the end of the last ice age (Fig. 3.2.8).

Notwithstanding these criticims, the Denton and Hughes model is an example of a bold hypothesis which helps explain how variations in solar radiation cause ice-age fluctuations. The amplitude of ice-sheet cycles and their coincidence with astronomically based cycles strongly suggests a link between the two. It seems likely that the ice-sheets provide the main mechanism by which relatively modest insolation variations are amplified into full ice age fluctuations. Although the precise mechanisms are still far from understood, it seems likely that their solution lies in understanding the way different components respond to changes in the atmosphere, oceans and earth's crust.

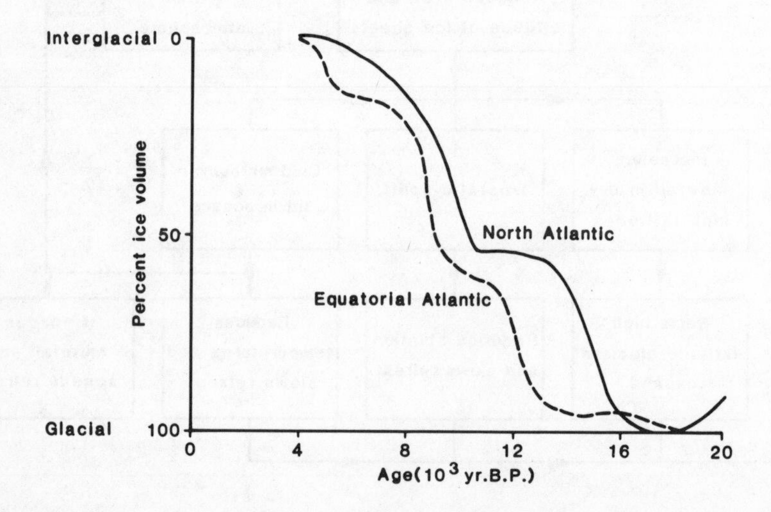

FIGURE 3.2.8 *A schematic curve showing the stepped nature of the last deglaciation, based on the interpretation of deep sea cores*
(After Ruddiman and Duplessy, 1985, fig. 1, p. 3)

SOME METHODOLOGICAL ISSUES

In the late 1950s very little was known about ice-sheets, and glacial studies focused almost wholly on Alpine valley glaciers. Rapid progress in the last thirty years was stimulated above all by the International Geophysical Year of 1957 which opened up studies of the Antarctic ice-sheet. The manner in which this progress occurred has many lessons for physical geographers, and it is worth highlighting some of the issues involved.

Scale

To most physical geographers trained in the 1950s, the scale of polar ice-sheets is daunting. Whereas traditional training often involved work in an areas of a few tens or hundreds of square kilometres, an ice-sheet demanded study at a subcontinental or continental scale. In place of detailed fieldwork, a variety of devices had to be accepted to increase areal coverage at the expense of detailed resolution. At first air photographs were the key, but subsequently satellite imagery allowed the features of ice-sheets to be identified and landscapes or formerly ice-covered areas to be classified. In addition, radio-echo sounding traverses of the Antarctic ice-sheet resolved details of large areas of the subglacial Antarctic landscape. However, even here the individual flight lines are often separated by tens of kilometres; a resolution level leaving many geomorphological questions unanswered. Theoretical work has leaned heavily on modelling and here it is fortunate that ice-sheets, through the predictability of ice behaviour, prove amenable to modelling at a large scale.

The time-scale too has required adjustments. In the 1950s, many thought the Antarctic ice-sheet formed in the Pleistocene. Oceanographers soon showed that this was a massive underestimate, and now those studying the evolution of the ice-sheet need to master a variety of new dating and sedimentological techniques.

Hypotheses

Another characteristic of progress is the way glaciologists and geomorphologists have employed hypotheses. Ideas of surges of the Antarctic ice-sheet, imminent collapse of the West Antarctic ice-sheet, synchronous and/or asynchronous ice-age fluctuations all abound in the literature. The cynically minded reader will say that these are merely wild ideas which purport to be serious hypotheses simply because there is so little field evidence to constrain them. On the other hand, it might be suggested that they are just the kinds of bold hypotheses that are necessary to stimulate the critical tests that mark the progress of science. Certainly, the Denton–Hughes model of ice-sheet fluctuations focuses attention on particular tests – for example, whether or not Antarctic glacial fluctuations precede those in the northern hemisphere by a few thousand years.

Assumptions

When applying a systems approach, the complexity of the possible links often seems overwhelming. A key step, well illustrated by polar ice-sheets, is the employment of assumptions which form the focus of further research. One key assumption that is helpful in modelling ice-sheets is that they are in steady state. For example, if morphology, input and output of mass and energy remain constant, then it is possible to model the flow, age and temperature regime of the ice-sheet. This is a vital way of dating ice-cores, interpreting landforms, reconstructing past climates from glacier extent, and so on. The problem, of course, is that few ice-sheets are in steady state. The history of their build-up and decay implies continual change such that they rarely, if ever, achieved steady state.

Another important assumption is that one can use modern analogues to extrapolate into the past. It has often been assumed that the characteristics of the existing Antarctic and Greenland ice-sheets apply to former ice-sheets. Thus it is assumed that the Laurentide ice-sheet had one major ice-dome comparable to that of the similar sized East Antarctic ice-sheet. Again details of former lapse rates, thermal regime, erosional and depositional processes are often assumed to be similar to those observed in association with modern ice-sheets.

However, the Antarctic and Greenland ice-sheets may be different from former mid-latitude ice-sheets in several crucial respects. The fact that they are still present, while mid-latitude ice-sheets are not, is convincing evidence that they are different in at least one respect! Mid-latitude ice-sheets may have been thinner than the Greenland and Antarctic ice-sheets, partly because they never had time to reach equilibrium, and partly because much of their bed included soft deformable sediments. If so, there may have been several minor domes. Precisely this argument applies in the case of the Laurentide ice-sheet where there are conflicting views as to whether it was single-domed or multidomed.

A further type of assumption is the surrogate measure. This is used where the variable of interest is unmeasurable, but an alternative feature can be measured instead. A classic example is the way $\delta^{18}O$ in marine sediments indicates ice volume on the land. This assumption lies at the heart of the recognition of glacial cycles on the basis of deep-sea cores. However, not only does it downplay the effect of ocean water temperatures on the $\delta^{18}O$ ratio, but it assumes that the same proportion of light isotopes are stored in the ice at all stages of a glaciation. This is clearly not true, because the $\delta^{18}O$ of snow accumulating in the centre of a cold high ice-sheet is different from that of snow falling on a warmer low ice-sheet, such as would have existed at the onset of glaciation. Further, the last ice to melt from a remnant of a large ice-sheet is likely to have a lower proportion of light isotopes than the first ice to melt. Together these effects will cause the changes in the $\delta^{18}O$ content of the oceans to lag behind changes in the ice volume on land. Perhaps the isotopic evidence that the Antarctic leads the northern hemisphere is an artifact of this effect.

Ruling hypotheses

Sometimes bold hypotheses escape critical testing and long outlive their usefulness. Indeed, they can spread into other disciplines and become a fixed point of reference difficult to dislodge. One such ruling hypothesis concerns the West Antarctic ice-sheet, particularly the view that it collapsed during the last interglacial. The idea was postulated on the basis of water-worn glacial deposits in the Transantarctic Mountains, in a place where it is far too cold for water to exist today. These deposits could perhaps reflect warmer interglacial conditions if the West Antarctic ice-sheet disappeared and maritime conditions affected the Transantarctic Mountains. One result would have been to raise sea level by 5–6m. Subsequently, former sea levels around the world at 5–6m were attributed to West Antarctic ice-sheet collapse. In turn, these were later used as evidence of the instability of the West Antarctic ice-sheet. The absence of the West Antarctic ice-sheet during the last interglacial is still frequently referred to, especially by those seeking evidence that global conditions were warmer than those of today. In the meantime, the initial evidence has been reinterpreted and the deposit is now known to be considerably older than the last interglacial. And so the idea lingers on through inertia and has escaped critical testing, even though it is now devoid of its original and only piece of supporting evidence.

Interdisciplinary approaches

Many disciplines have contributed to the understanding of polar ice-sheets. The value of an interdisciplinary approach is evident in the increasing understanding of the way polar ice-sheets behave. One example is the co-operation between theoretical modellers and field scientists. Twenty years ago there was a clear division between these two groups. Glaciologists were producing models which they found difficult to test because of lack of field evidence. At the same time field scientists, geologists and geographers among them, were collecting data at a small scale throughout the world but with little interest in process or modern theory. One of the real successes of the International Glaciological Society has been to forge increasingly effective links between the two groups so that understanding is progressing as a joint enterprise.

PRIORITIES AND LOGISTICS

Polar ice-sheets are key features of global physical geography because of their size and oscillating behaviour. They control sea level; influence the temperature and circulation of the atmosphere; create water masses and affect ocean circulation; fluctuate sharply in size in response to cyclical changes in solar radiation receipt, amplifying the signal to such an extent that all parts of the globe are affected. These changes occur on time-scales that are sufficiently short to be of direct human interest. In particular, they have implications for the long-term exploitation of ecological resources both on land and sea. There is also a danger that atmospheric pollution might change the world's atmosphere and unwittingly cause the ice-sheets to cross a threshold and amplify the initial change. More positively, we may wish to modify natural changes to our own benefit, for example by burning fossil fuels to stave off the next ice age. Set against the obvious importance of studying polar ice-sheets, progress achieved in the last thirty years is impressive and has involved many nations, especially those with land in the Arctic or with a claim to land in the Antarctic. The UK claims a large sector of Antarctica and has a long tradition of polar research as is recognised by the existence of the Scott Polar Research Institute, the British Antarctic Survey, and the headquarters of the International Glaciological Society. However, perhaps it is useful to ask whether research in British Antarctic Territory is as effective as possible.

One can note two major constraints which restrict UK Antarctic research. The first is the difficulty of access. Research is restricted in both space and time, since the essential ship support favours work in coastal locations in the summer. The reliance on ships also involves a large time penalty which makes it difficult for those scientists who do not work full time for an Antarctic organisation to participate in Antarctic research. The second major constraint concerns half the readers of this chapter who are unable to work in British Antarctic Territory because they are women. Unlike many other Antarctic enterprises which do employ women, a female scientist in the UK is effectively barred from working with the British Antarctic Survey, at least if she wishes to use the bases and associated logistics.

CONCLUSIONS

This review has focused on polar ice-sheets, whose inception and subsequent oscillations are the key to the creation and stability of today's polar environment. Oscillations of the ice-sheets which have characterised the Pleistocene Ice Age are the result of the amplification of orbitally induced cycles of solar radiation receipt. The response, which is complex, involves distinct ice-sheet components whose size and location determines their response to different limiting variables. Most progress in understanding ice-sheets has occurred in the last thirty years, based on the large-scale approach, the bold use of hypotheses, explicit use of assumptions and an interdisciplinary outlook. Both the intellectual and practical importance of the subject, and the freshness of its approach, fully justify its recognition as one of the highlights of physical geography over the past two decades.

FURTHER READING

An excellent introduction to the field is provided by an exciting and enthralling book which is very popular with students:

Imbrie J. and Imbrie K. P. (1979) *Ice Ages* (London: Macmillan).

Three more detailed sources from the research literature follow. The first two offer splendidly bold and general hypotheses, whilst the third gives a clear visual summary of the radio-echo sounding work of the Scott Polar Institute (Cambridge).

Denton G. H. and Hughes T. J. (1983) 'Milankovitch theory of Ice Ages: hypothesis of ice sheet linkage between regional insolation and global climate', *Quaternary Research*, vol. 20, pp. 125–44.

Mercer J. H. (1978) 'West antarctic ice sheet and CO_2 greenhouse effect: a threat of disaster', *Nature*, no. 271, pp. 321–5.

Drewry D. J. (1982) 'Antarctica revealed', *New Scientist*, 22 July, pp. 246–51.

Other specific topics in the chapter may be difficult to follow up because the sources are either specialised or very recent. Two papers in the following edited volume make it worth obtaining through an inter-library loan scheme. The first summarises marine evidence concerning the inception of the Antarctic ice-sheet. The second is a fine example of the way marine cores are interpreted, and one which argues that subantarctic fluctuations lead those in the northern hemisphere:

Kennett J. P. (1978) 'Cainozoic evolution of circumantarctic palaeoceanography', pp. 41–56.

Hays J. D. (1978) 'A review of the Late Quaternary climate history of Antarctic seas', pp. 57–71.

Both in: Van Zinderen Bakker, E. M. (ed.) (1978) *Antarctic Glacial History and World Palaeoenvironments* (Rotterdam: Balkema).

3.3
The Mountain Lands

Jack D. Ives
University of Colorado, Boulder

Western society has become increasingly fascinated with mountains since the second half of the nineteenth century, yet this fascination with some of the world's most dramatic landscapes has changed focus significantly over the last hundred years. The age of exploration and reconnaissance mapping is over, and with it the heroic age of mountaineering. However, the widespread realisation that the mountain lands form the physical, and often the political, framework of an ever-shrinking world, ensures that interest in them is not likely to diminish. The mountain lands have been pivotal to much research in the natural sciences. Most significantly, mountains are perceived as high energy landscapes in the sense that high altitude, steep slopes, water, freezing temperatures and gravity combine to ensure that landscape-forming processes operate much more rapidly than in low-energy environments (areas of subdued relief). This means that landscape developments occur relatively rapidly, and thus can be measured more accurately over a given period of time. This factor is offset, at least in part, by relative inaccessibility, ruggedness of terrain and inclement weather, which may limit time available for actual research in the field.

These general considerations have been somewhat displaced by a contemporary increase in concern about the apparent rapid rate of environmental deterioration that seems to be occurring as a result of a range of human interventions. Concern for the preservation of wilderness, the establishment of biosphere reserves, development of two-season tourism in mid-latitude mountains, and rapid growth of subsistence farming populations in low-latitude mountains, have brought mountain environmental issues towards the forefront of development and conservation politics. There is a growing perception that landscape instability, in the form of slope movements, deforestation, soil erosion and downstream sedimentation, changes in the hydrological cycle, collapse of mountain societies, and large-scale migrations of people, not to mention the ramifications of climatic change, are having an increasingly costly impact on the neighbouring valleys, plains, and lowlands. Yet such impacts, to the extent to which they can be proven, are primarily due to the imposition of commerce and exploitation on mountain regions by lowland societies.

Mountain development and conservation of mountain resources have become interrelated themes in an increasingly resource-starved world. Public opinion, however, is far outreaching our scholarly understanding of mountain landscape dynamics. We are alerted by the news media of the crisis facing Nepal, of catastrophe in the Himalaya, of the devastation of the east slope of the Andes, or the threatened future of the Alps. Cause and effect relationships are widely assumed, and since these environmental strategies have massive foreign aid implications and socio-economic impacts, it is timely to undertake a review of mountain geomorphology as one element of our understanding of mountain landscape dynamics.

The development of mountain geomorphology can be divided into three phases. During the first, prior to the 1950s, research efforts were largely descriptive, with an overwhelming concern for form and cyclic landscape evolution. While there were a number of early harbingers of the quantitative, process-oriented era, and a certain degree of overlap, the 1960 publication of Anders Rapp's seminal treatise on mountain slope development in Karkevagge, northern Sweden, provides the critical turning point. A second phase can be identified, characterised by quantification and numerous studies of mountain slope processes (reflecting the powerful impact of the IGU Commission on the Evolution of Slopes).

The third phase overlaps in time with the second, but is introduced to emphasise the growing need for careful and rational application of our increasing understanding of mountain dynamics to the solution of practical problems associated with the use and management of mountain resources. Geomorphic knowledge in mountain areas developed primarily from field description and measurement at 'natural' sites (often unvegetated), and through controlled laboratory experimentation. Yet geomorphologists recognised man as a major geomorphic agent several decades ago. Human activities range from those of the engineer, miner, dam-builder, logger, road-builder, soldier, to the subsistence farmer. It is necessary, therefore, to develop a better understanding of the man-mountain relationship, to include detailed study of the geomorphic implications of subsistence farming and uncontrolled population growth. Political, economic, sociological, and behavioural factors must also be taken into account. Of course, at this point it can be argued that the term geomorphology has been qualified beyond reasonable limits and must be replaced. This requires a holistic approach to mountain landscape dynamics, and the need to bridge the chasm between natural and human science becomes much more than a policy plank of Unesco's Programme on Man and the Biosphere (MAB) – it becomes a vital necessity. A fuller understanding of mountain dynamics, and its successful application to mountain management decision-making, may well prove essential for world social and economic stability, and should provide justification for a major development of mountain geomorphology. Nevertheless, this view in no way invalidates the need for more exacting research on the two main aspects of mountain geomorphology – continuous slow-acting processes and large-scale catastrophic events of such infrequency that it is difficult to incorporate them into standard process models.

DEFINITION OF THE MOUNTAIN ZONE

German geographers define mountains as having a relative relief in excess of 1000m, whilst geographers from eastern and central North America seem prepared to accept 300m – perhaps reflecting their different habitats. The Germanic *hochgebirge* (high mountains) and *mittelgebirge* (middle mountains) describing the Alps (mountains built during the most recent orogenesis) and the Hercynian massifs (uplifted and block-faulted, such as the Vosges and the Schwarzwald) make a useful, if limited, geographical formulation. Carl Troll, father of 'mountain geoecology', more rigorously defined hochgebirge as those high mountains that extend above the Pleistocene snowline, have been glaciated, and have several vegetation belts and a tree line. As landscape personality is determined by systems of landforms that recur throughout the high-mountain world, degree of glacial dissection must be considered in relationship to the degree of interfluve preservation. Thus two types of glaciated high mountains can be differentiated: Alpine and Rocky Mountain. In the former the higher sections are composed predominantly of precipitous slopes, and gently sloping, low energy enclaves are confined to the broad, alluvium-filled valley floors and the characteristic trough shoulders. The Rocky Mountain type displays wide interfluves, intermixed with arêtes and horns, which are subject to mass wasting processes and incur relatively low rates of denudation (Figs. 3.3.1 and 3.3.2).

The most dramatic scenery is most often associated with the hochgebirge, such as the Dents du Midi, the Lhotis-Nuptse ridge, or the Staunning Alps of East Greenland. This has attracted mountain geomorphological research to what is effectively a minor, if spectacular, part of the total area of the mountain lands. It has also led to the recognition that altitude and latitude combine as controls on the limits of this peculiar set of landforms in a high altitude–low latitude to low altitude–high altitude continuum – the arctic–alpine continuum of cold stressed environments. It also tends to underlie the special research concentration on mechanical weathering and mass movement, and the comparative neglect of the much vaster area of mountain lands below the upper timberline, within which people live.

PROGRESS IN MOUNTAIN GEOMORPHOLOGY 1960–85

Before 1950, mountain geomorphogy was primarily descriptive and hypothetical. The interest in high mountains tended to link glaciology and mountain

FIGURE 3.3.1 *The alpine type of high-mountain landscapes* (hochgebirge)
This view shows the Aletschhorn and the Oberaletsch glacier with its pronounced neoglacial moraines and trimline. Enclaves of 'low energy' slopes are confined to the trough shoulders and glaciated bottoms. The foreground low-energy enclave is Riederalp, site of an extensive modern tourist development, Canton Wallis, Switzerland.

geomorphogy and – on account of the broad relationships between altitude, low air temperatures, fluctuations across the freezing point and high latitudes – it also led to close affinities between arctic and alpine research, and periglacial and permafrost studies.

The early geographers and geologists had outlined the broad lineaments of mountains. Before the end of the nineteenth century the general accordance of summit levels (*gipfelflur*) had been identified as remnants of old erosion surfaces, though no hypothesis was entirely satisfactory because of the

FIGURE 3.3.2 *The Rocky Mountains type of high-mountain landscape* (hochgebirge), *Colorado Front Range, Indian Peaks Wilderness Area*
Note the combination of horns and arêtes with broad, high interfluves subject to periglacial mass wasting. The broad ridge on the left is Niwot Ridge, designated a MAB reserve in 1979. View west by south on to the continental divide.

difficulty of dating such 'old' surfaces and the lack of information on rates of tectonic uplift and denudation. Nevertheless, outstanding geomorphologists such as C. A. Cotton wrote systematic descriptions of high-mountain landscapes which reached the level of literary masterpieces.

The quest for exact measurements began to take form during the last decades of the descriptive era. Highly stimulating disputes on such topics as freeze–thaw cycles and the so-called meltwater variant of the bergschrund hypothesis of cirque erosion between Lewis and Johnson led to Battle's determination to measure processes, both at the bottom of bergschrunds and in the Cavendish Laboratory. Glaciology began to attract engineers, physicists, and mathematicians, and rapidly developed as a respectable science in the 1950s. Yet it still has not explained adequately the landforms so graphically described by Cotton. While Battle's work was an inspiring example of geomorphic method, he effectively invalidated the meltwater variant of the bergschrund hypothesis, and was obliged to fall back on the old descriptive approach in his final statement of a hypothesis for cirque formation.

Also in the 1950s, on a firm basis of Swedish glaciological and fluvial research, Anders Rapp instrumented many slope facets in Kärkevagge, northern Sweden. Rapp collected data on a wide range of slope processes operating in this arctic–maritime mountain environment between 1952 and 1960. His work fell naturally into two broad divisions:

qualitative – what types of geomorphic processes are active and what erosion and accumulation forms do they create? Are the processes and forms controlled by certain climatic or other factors?
quantitative – how frequently do the forms and processes occur and what is their ranking in terms of total denudation accomplished?

The factors of spatial distribution of specific forms and processes were examined, and their degree of continuity and frequency of occurrence (temporal variation) were studied. Rapp's inventories have become classic components of process, or climatic geomorphogy, and included: direct observations; examination of fresh debris upon the snow in late spring; examination of fresh debris upon the vegetation cover in summer; comparison of old and new photographs of the same localities; measurement of the yearly supply of pebble falls from the rock walls on to small carpets of sacking; collection of water samples from snowmelt rivulets and streams and determination of total content of salts and invisible microsediments. Eight years of painstaking work indicated that the following processes, ranked in order of importance, were dominant in Kärkevagge:

1. transport of salts in running water,
2. earthslides and mudflows,
3. dirty avalanches,
4. rockfalls,
5. solifluction,
6. talus creep.

Glacial erosion had been excluded, it was not possible to place slope wash on the list, and frost-bursting was not included since it is not a transporting process and is difficult to measure directly. The annual production of rock-waste by frost-bursting on rock walls, however, was calculated at 100–400 $t.km^{-2}$ of wall surface, which would probably give it first ranking over the above-listed processes. Rapp concluded that, without more data from other valleys with other types of slope, these six processes may have quite a different order.

Rapp's work set the stage for process studies in many parts of the world, but already in his conclusions were recognised the problems that still beset any full understanding of climatic geomorphogy. Despite the arduous collection of data over eight years, the results cannot be extrapolated with confidence to Kärkevagge's neighbouring valleys, let alone to Nepal or western Canada nor can they be extrapolated backward in time, nor used as a basis for prediction. In addition, despite Rapp's ranking of transport of salts in running water, most subsequent research has concentrated on mechanical weathering and transport of coarse debris.

Representativeness in space and time

This problem is best introduced through Rapp's own data. In the Narvik–Kärkevagge area 107mm precipitation was recorded in twenty-four hours, and 175mm in seventy-two hours (Riksgränsen, 5–7 October, 1959). The total precipitation from July to October 1959 inclusive was 794mm compared with the 1901–30 average of 308mm. The October rains were the heaviest since the Riksgränsen climatological station was established in 1904; the recurrence interval of such downpours may exceed 100 years. The ensuing material transfer by mudflows, debris flows, and mountain torrents represented by far the single largest geomorphic event during the entire period of observations. It can be argued that more geomorphic work can be accomplished during a single catastrophic event than all movement minutely measured for a decade. The effects of the hundred-year event also may be dwarfed by even more spectacular occurrences: for example, a very large landslide that occurred in the Langtang Himal, Nepal, 25 000 or more years ago. This landslide displaced approximately $10km^3$ of debris through a vertical distance of up to 2000 metres. Similarly, at Koefels in the Tyrolean Alps about 9000 years ago a landslide displaced $3km^3$ of debris through up to 1000 vertical metres.

Three other examples may be mentioned here. First, in 1970 a debris flow, resulting from an earthquake-induced avalanche near the summit of Huascaran (6768m) in the Peruvian Cordillera Blanca, covered $150km^2$ with debris, annihilated the town of Yungay, and killed 18 000 people. Second, in October 1968, rainfall varying between 60 and 120mm fell on the Darjeeling area, Himalaya, during a three-day period at the end of the summer monsoon when the ground was already saturated. Some 20 000 debris flows were released; the 50km road between Siliguri on the Plains and Darjeeling at 2200m was cut in ninety-two places; and approximately 20 000 were killed, injured or displaced. Third, in south-east Iceland 155mm rain in twenty-four hours on 29 July 1982 caused several dozen debris flows in the small area of Skaftafell National Park. I have photographed and observed this area qualitatively from 1952 to 1984, and estimate that (with the exception of the very high rates of glacial erosion, deposition and associated jökulhlaup and fluvio-glacial activity) this single geomorphic event accounted for more 'work' than all other processes combined over the last 100 years.

The regularity, or irregularity, of occurrence of the extraordinarily large events, such as the Koefels landslide, poses a serious problem to any attempt to rank geomorphic processes and deduce long-term denudation rates. There is a wide range of evidence supporting the notion that large landslide activity in the Alps peaked during the latter part of the last glaciation. Late glacial retreat of large valley glaciers exposed over-steepened rockwalls to mechanical stresses and collapse, leading to the majority of the spectacular landslides and rockfalls. On a lesser scale, the same period was also characterised by much more rapid construction of talus and alluvial cones than is occurring today.

Rates of activity

In the same context of variations in geomorphic activity through long periods of time, rates of mass wasting on Colorado Front Range interfluves have been shown to vary by an order of magnitude during the Holocene; current rates are lower than at any other time. We can therefore see that mountain Geomorphology faces some critical problems of variation in rates of erosion, not only by different processes, but also in space and in time.

On Niwot Ridge in the Colorado Front Range, current rates of horizontal movement in stone-banked and turf-banked terraces and lobes range from 4 to 43mm/yr, compared with intervening and more stable areas which indicate a rate of less than 1.0mm/yr. Movement, however, appears to be largely confined to the upper 50cm of the waste mantle. This represents an overall denudation rate of 0.01–0.14mm/yr for interfluves with an order of magnitude variation through the Holocene. Rapp estimated cliff-face retreat at 5m in 10 000 years and 30m in 30 000 years, compared with estimates of 7.6m in 10 000 years for the Colorado Rockies. Roughly comparable results have been calculated from measurements of talus accumulation over time in the Canadian Rockies and other areas.

The comparative neglect of chemical weathering in alpine research has been partly justified on the assumption that low temperatures and absence of free water during sub-zero periods inhibit chemical action; mountain streams, except glacier-fed ones, are usually clear. This argument is offset, however, by the realisation that low temperatures enhance carbon dioxide solubility, and this may be an especially important factor in limestone regions.

The challenge to conventional hypotheses

To a large extent, the 1960–85 period of intensive alpine process study constitutes a major advance in our understanding because several formerly cherished and intellectually satisfying hypotheses have been invalidated or seriously challenged. Although Battle had demonstrated in the 1950s that temperature fluctuations in bergschrunds are of neither sufficient magnitude nor frequency to support the meltwater variant hypothesis, no further significant work in this area was attempted for some time. School and undergraduate texts, however, perpetuated the assumption that nivation was a dominant high-mountain process. Partly because of the hypothesis of accelerated freeze–thaw activity around and beneath snowbanks, one mode of cirque development was assumed to embrace progressive enlargement of nivation hollows until snowbanks became small cirque glaciers and a further increase in erosion occurred as glacial processes took over.

This hypothesis of cirque formation was re-examined as a result of Thorn's impressive study of nivation in the Colorado Front Range. He demonstrated that freeze–thaw cycles do not increase in frequency beneath and around snowbanks, and that removal of sand, silt and clay-sized particles increases to magnitudes of between twenty and thirty times as a result of nivation, when compared with surrounding control slopes, though the specific processes are sheetwash and rivulet flow. Chemical weathering is also accelerated two to four times in a snowbank environment. This suggests that, at current rates, it would take 500 000 years to excavate a modest-sized nivation hollow (300 × 200m). Also, once the hollow becomes large enough for snow to accumulate sufficiently for its survival throughout the ablation season, the role of snow becomes protective. Thus the hypothesis that nivation hollow enlargement leads directly into cirque formation breaks down.

The argument that frequency and magnitude of freeze–thaw cycles, assumed to increase in arctic and alpine environments, accounted for the observed increase in quantities of frost-shattered rubble has also been invalidated. Freeze–thaw frequency in climatological screens actually decreased meridionally from southern to northern Canada, and only the annual freeze–thaw cycle penetrates more than a few centimetres below the ground surface in a high-arctic environment.

These hypothetical relicts of the early descriptive era of mountain geomorphology probably achieved their long-standing prominence because researchers attempted to explain the extensive frost-riven waste mantle of cold-climate environments in terms of contemporary processes. Not enough attention was paid to the probable great age of blockfields, for instance. The long debate concerning the development of blockfields, or mountain-top detritus, has involved discussion of the extent and thickness of the Late Cenozoic continental and mountain ice-sheets and the survival of vascular plants on nunataks throughout the ice ages. This long, intense dispute between geographers, geologists, and biologists merely highlights prevailing uncertainties concerning rates of erosion and their variation through time, as well as the difficulty of accurate dating of substrates.

In summary, twenty-five years of progressively more rigorous process studies has led to a number of vital conclusions. We have not gathered, nor been able to gather, adequately representative data; slope processes operating in mountain areas today are not primarily responsible for the mountain forms upon which they are operating; infrequent catastrophic events cannot be compared statistically with slow continuous slope processes; several major hypotheses have been invalidated, or found wanting, and are not yet replaced with alternative explanations.

These conclusion should not be regarded as pessimistic. Mountain geomorphology has advanced prodigiously since 1960. It is important to know what is happening on contemporary mountain slopes and to understand why and how a wide variety of processes are operating (leading into the applied aspect of natural hazards assessment). It is equally important to understand spatial and temporal lack of representativeness, and to realise that many high mountain areas are rising as rapidly as, or more rapidly than, the processes of denudation are wearing them down. In this regard, the recent Chinese work in Tibet and the Himalaya has demonstrated a very high rate of contemporary uplift. The current progress in plate tectonics and the advances in dating techniques (see Chapter 1.1) have provided an array of powerful tools for future research.

A model of alpine slope processes

The systems of processes which operate in alpine environments have been described by Caine as a cascade of sediment fluxes involving a variety of materials and controls. These fluxes have not received an equal amount of attention. They are also unevenly distributed, although it may be possible to define general patterns of relative significance, such as variation with elevation. Caine has proposed four systems, each with its own set of controls, responses, and levels of activity: the mountain glacial system; the coarse debris system; the fine clastic sediment system; and the geochemical system. These systems are distinct but they interact and material moves between them. The mountain glacial and coarse debris systems are most characteristic of high mountains for they tend to be restricted to areas of greatest elevation and strongest relief.

The glacial system The distribution of mountain glaciers is controlled by elevation and climate which influence mass balance and thus equilibrium line altitude and glaciation limit. This system is found at the highest altitudes where the movement of water in the solid phase is important for transporting debris derived from rockfall, as well as from erosion at the ice-rock interface. In presently glacierised mountains this system is probably the most efficient form of erosion, although great fluctuations in the extent of its dominance have occurred over the last 25 000 years.

The coarse debris system This system is also highly characteristic of high mountains. It involves the transfer of coarse detritus between rockwalls and talus and associated deposits at lower elevations, as well as large-scale failure. In maritime climates it may be tributary to the glacial system, with glaciers as the main transporting agent. In continental climates it may be considered as a closed system, with rock glaciers forming its down-valley extensions. Climatic changes, which induce glaciation, will open the system and permit intermittent removal of accumulated debris as glacial till. Most studies have concentrated on the talus and rockglacier components; in contrast their source areas, the rock walls, have received scant attention.

The fine sediment system This system may be regarded as open. Some material is derived from aeolian dustfall and much may be exported by flowing water to lower elevations. It has received much attention from researchers interested in mass wasting on alpine slopes. It responds most readily to environmental controls involving the hydrological cycle; to freezing and thawing of the ground; form and intensity of precipitation; and distribution of snow cover. There is a great variation in rate of activity between wet and dry sites; in wet sites rates of movement are much faster than those at lower elevations.

The geochemical system There has been relatively little work on geochemical exchanges in high mountains except in areas of alpine karst. As suggested above, the classical viewpoint is that chemical weathering is insignificant. However, recent studies support the contrary view of Rapp. While the total mass of solutes may remain slight, its importance probably derives from the efficiency of transport by running water through the system. It is also highly responsive to hydrologic controls.

Three case studies compared

Figure 3.3.3 summarises process studies from three different high-mountain catchments. They differ in area across three orders of magnitude, and also according to climate, structure, and lithology. While additional data could be added, they would not significantly influence the generalisations. Despite the disparities between the three sites, some important common features are evident which may be applicable to all high-mountain catchments. Two components in the geomorphic mass fluxes dominate in all three catchments. The coarse debris system accounts for between 10 and 60 per cent of all erosive power; contrasts between sites are inversely related to catchment size and relative relief – this is most probably influenced by the proportion of the total area of cliffs and taluses. The importance of solute transport almost equals this (from 12 to 50 per cent) and is related directly to catchment size. The record from the largest catchment implies that fluvial transport and deposition of clastic sediment is the most important. Although Rapp and Caine did not make comparable estimates, it is unlikely that fluvial processes are significant in their much smaller catchments.

Figure 3.3.3 and Table 3.3.1 also indicate that there are many studies devoted to talus shift, solifluction, soil creep, and other processes of slow mass wasting that are relatively unimportant – less than 15 per cent of all work done with relative importance decreasing as the size of catchment area increases. Finally, except the largest catchment, erosion and transport through the fluvial system, although difficult to evaluate, appear slight. This may indicate a lack of coupling between hillslope systems and the fluvial system in high mountains. It is important to note that this discussion is based upon average estimates and does not include the issue of temporal variability.

Mountain geomorphology, as emphasised above, has concentrated on high mountains. Within this context it must be understood that all geomorphic studies have demonstrated a difference in degree, if not in kind (excepting the glacial system) of erosion between mountain and lowland regions. This difference is consistent and measurements, however unrepresentative they may be, range from five times to one or two orders of magnitude, based upon river load estimates (i.e. the denudation rate), reservoir sedimentation, and geological data. This statement is made with the understanding that much of the coarse debris transfer relates to movement within a generally closed system – in other words it represents redistribution within the high mountains rather than absolute removal from them.

NATURAL HAZARDS AND MOUNTAIN HAZARDS MAPPING

Engineers have long been involved with slope instability. The remarkable concentration of geomorphic research on alpine slope processes, has led many researchers to extend their studies to a consideration of the implications of various forms of mass movement coming into contact with human lives and property. Thus there has occurred a significant involvement with the applied field of natural hazards research.

Winter sports developed rapidly in the European Alps after about 1950 and increased the number of people and amount of property at risk to snow avalanches. This led to systematic studies in Switzerland, where avalanche zoning plans became part of the legal process of land-use zoning. In Austria, a combination of the study of avalanche and 'wildbach' (mountain torrent – debris flow) allowed designation of a 'traffic light' colour scheme for cartographic representation of degree of risk. Risk, in this instance, is a function of magnitude (or impact pressure), frequency of occurrence, and interception with humans and/or property.

The next steps in the study of natural hazards in mountain regions were taken by the Geographical Institute of Berne University guided by Bruno Messerli. This involved depiction of all hazards in a particular area on a single map, and recognised the interrelationships among various types of slope movement and the impact of slope modification by human intervention. The Grindelwald geomorphic and combined hazards maps (scale 1:10,000) represent the single most important advance in this area of applied mountain geomorphology. Concurrently the Institute of Arctic and Alpine Research, under contracts to the author, developed a snow avalanche forecast model for the Colorado Rocky Mountains and, using remote sensing techniques and fieldwork, produced a rapid and inexpensive system of regional natural hazards mapping (scale 1:24,000) for land-use planning purposes. Collaboration between the Bernese and Coloradan teams produced a combined hazards

(a) The Upper Rhine Catchment

Sediment System	Morphologic Units	Sediment Fluxes
Glacial and Coarse Debris Systems	Glacierized Valleys; Steep Bedrock Slopes and Talus	Glacial Transport (62.8); Rockfall (554.2); Snow Avalanches (104.6); Mudflows (70.35); Talus Creep; Rock Glacier Flow (3.0)
Fine Sediment System	Waste Mantled Slopes	Solifluction and Soil Creep (53.5)
Fluvial and Geochemical Systems	Stream Channels; Valley Floors; Fans and Lakes	Fluvial Transport and Deposition (3386); Lake Sedimentation (10412); Solute Transport (2781)

(b) Kärkevagge, Scandinavia

Sediment System	Morphologic Units	Sediment Fluxes
Glacial and Coarse Debris Systems	Glacierized Valleys; Steep Bedrock Slopes and Talus	Glacial Transport; Trace; Rockfall (13.0); Snow Avalanches (14.6); Mudflows (64.25); Talus Creep (1.8)
Fine Sediment System	Waste Mantled Slopes	Solifluction and Soil Creep (3.53); Slope Wash (≈0.0)
Fluvial and Geochemical Systems	Stream Channels; Valley Floors; Fans and Lakes.	Fluvial Sediments (not studied); Solute Transport (91.0)

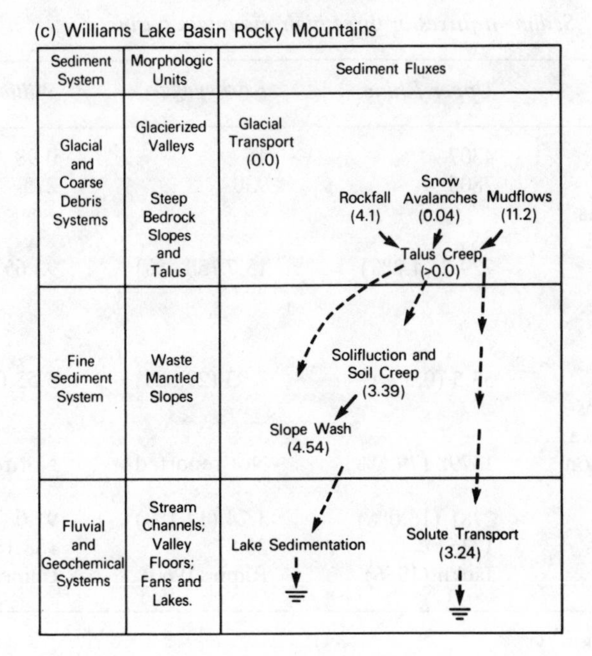

FIGURE 3.3.3 *Sediment fluxes in three high-mountain areas*
(a) The upper Rhine Basin (Jackli, 1956).
(b) Kärkevagge, northern Scandinavia (Rapp, 1960).
(c) Williams Lake Basin, Colorado (Caine, 1976).
All fluxes are based on vertical transport (10^6 J/km^2/yr).

map for a test area in the Colorado Front Range and the base data for a high-mountain environmental atlas.

The Swiss–Coloradan collaboration continued under the auspices of the IGU Commission of Mountain Geoecology with the establishment by United Nations University of a project on Highland–Lowland Interactive Systems. This supported an attempt to transfer the mid-latitude high-mountain experience to tropical and subtropical mountain areas of the 'Third World'. A wide range of studies is being undertaken in north Thailand, Nepal, southwest China, Ecuador, and Ethiopia. The focus is on slope stability, deforestation and soil erosion, subsistence farming and population pressures, and local farmers' perceptions of hazards. One outcome was the foundation of the International Mountain Society and publication of the quarterly journal: *Mountain Research and Development*.

The Nepal mountain hazards mapping project

This facet of the UNU project offers an excellent example of integrated applied research, involving preparation of prototype maps depicting natural

TABLE 3.3.1 *Sediment fluxes in three high mountain basins*

Catchment	*Upper Rhine*	*Kärkevagge*	*Williams Lake*
Area (km^2)	4307	15	0.98
Relief (m)	2800	930	275
Coarse debris and bedrock slopes	729.2 (4.2%)	15.7 (58.4%)	93.65 (49.8%)
Soil and fine sediment mantled slopes	53.5 (0.3%)	7.93 (29.5%)	3.53 (1.9%)
Channel transport and lake sedimentation	13798 (79.5%)	Not reported	Not reported
Solute flux (output)	2781 (16.0%)	3.24 (12.1%)	91.0 (48.3%)
Total	17362	26.87	188.18
Source	Jäckli (1956)	Rapp (1960)	Caine (1976)

Units: 10^6 J/km^2/yr

hazards in three areas of Nepal. In addition to the actual mapping (land use, geomorphology, vegetation, base map, and mountain hazards and slope stability), studies were made of extant geomorphic processes, farming practices, and hazard perception and response. The project is still incomplete and never acquired sufficient funding to permit as broad and as detailed approach as was deemed desirable. Nevertheless, it did attempt to combine both natural science and human science strategies as well as a training component for young Nepali co-workers.

The project also demonstrated how much the 'western' geomorphologists, who initiated the project, had to learn from their multilingual Nepali colleagues and from the local people. In addition to the contribution to mountain geomorphology, it has shown that complex mountain regions, especially where densely populated, are extremely dynamic. In many instances human or human-augmented slope processes may far outweigh 'natural' processes, but the human processes may also counteract the 'natural' ones. Thus the casual visitor to the Kakani–Kathmandu area, or even the agency 'expert', will perceive extensive deforestation, terrace farming on slopes up to 45°, and a large number of debris flows, and conclude that large-scale instability, or even catastrophic environmental collapse, is occurring. This view has entered the conservationist literature.

However, our studies in the Kakani–Kathmandu area led to the conclusion that the local farmers counter a significant proportion of the down-slope mass transfer by reterracing debris flows and repairing hillslopes in order to maintain agricultural productivity. They have an extensive understanding of the 'natural hazards' for which, because of the conspicuous human component, we substituted the neutral term 'mountain hazards', and over several generations have evolved sophisticated counter-strategies. While our data base is not adequate to determine the total mass fluxes of a representative area (this is an obvious and vitally needed next step), the research has led to a number of socio-economic and political conclusions. Rates of mountain environmental degradation are often overestimated and the indigenous reclamation is usually ignored; the local subsistence farmer, despite his need for assistance, is often a knowledgeable natural scientist yet he is frequently regarded as the ignorant despoiler of the environment. Land ownership and general socio-economic conditions more effectively dominate slope stability than either the assumed misuse by local people, or the 'natural' climo-geomorphic processes. Finally, foreign aid is frequently misdirected because problems identified as amenable to 'development' solutions often evolve according to current bureaucratic fads rather than through holistic analysis and involvement of the local people. These tentative conclusions should perhaps be regarded as hypotheses which need further testing. This is particularly important in view of the current tendency to ignore the need to develop a historical perspective. It is the lesson of the mountain zone that environmental trends are often simplistically assumed to be the result of population pressures of the last few decades.

CONCLUSIONS

This overview has attempted to demonstrate progress in mountain geomorphology over the last twenty-five years. While great strides have been made in our understanding of slope processes, the problems of extreme variability in space and time remain as a challenge for the future. The overview has also emphasised that research has concentrated heavily on a very small proportion of the total area of mountain lands, the more spectacular high-mountain belt dominated by cold-climate processes. While the knowledge gained from this concentration probably justifies the bias, in my estimation the major future challenge lies in the inhabited lower mountain belts where man may be the dominant geomorphic agent. Nevertheless, further progress in solving the spatial and temporal variability enigma will probably be made most efficiently in the high-mountain belt and, therefore, studies here should proceed hand in hand with human geography to give a holistic approach to mountain landscape dynamics that crosses the traditional divisions between physical and human geography. In fact, a fuller interdisciplinary approach becomes increasingly important for contemporary research in mountain lands.

It is equally apparent that inhabited mountain regions are highly complex fields for study. Thus, international collaborative efforts are vitally needed, including standardisation of objectives and methods, shared data banks, and identification of minimum data needs. In these ways research will be able to proceed much more efficiently (world data centres, after the pattern of the IGY Centres, would be a desirable organisational goal).

In these closing years of the twentieth century, a period of financial and economic stress and, particularly from our point of view, of increasing scarcity of research funds, this recommendation for an expanded effort, even the establishment of a world network of mountain research activities, may appear impractical. In this respect it is worth considering the fate of some highly expensive development projects in mountain regions. The engineer's estimate of the useful life of a large hydro-electric scheme is often significantly reduced because no attention was paid to the dynamics of the mountain catchment; uncontrolled deforestation, for instance, serves to fill the reservoir quickly with silt. Poorly designed roads lead to instability of extensive mountain slopes and serious down-stream damage. Resource mismanagement in mountain lands in general, often due to ignorance, or to the definition of short-term objectives, has the potential for devastating losses in the surrounding and densely populated lowlands and coastal areas. The problem is exacerbated by the extreme unreliability of much of the available data, and the tendency to generalise to the point of absurdity. Is it not vital, then, that mountain geomorphologists should accept the obligation, and grasp the opportunity, to prove the value of their work?

The argument for a future human-oriented mountain geomorphology can be proposed in another way. With the contemporary pressures and resource-use conflicts facing the mountain lands, much high-mountain research, and its handmaiden, arctic research, begins to appear effete, the perpetuation of an élitism that is a hold-over from the gentlemen adventurers of a previous era who were privately funded. Applied mountain geomorphology need in no way be perceived as lacking in intellectual challenge if this reputation is to be overcome and the potential of the subject is to be realised.

ACKNOWLEDGEMENTS

Support of The United Nations University has been responsible for most of my work in tropical and subtropical mountain areas. Much of this has been developed in close collaboration with Professor Bruno Messerli.

FURTHER READING

In order to reflect the range of sources available, the literature recommended below ranges from classic to modern, from specific to general, and from 'pure' to applied. The scene is set with three important papers, two typifying the substantial mastery that had been achieved by the beginning of the period reviewed in the chapter, and the last indicating progress in process study:

Lewis W. V. (ed.) (1960) *Norwegian Cirque Glaciers*, Royal Geographical Society Research Series, no. 4.
Rapp A. (1960) 'Recent development of mountain slopes in Kärkevagge and surroundings, northern Scandinavia', *Geografiska Annaler*, vol. 42, pp. 65–201.
Benedict J. B. (1970) 'Downslope soil movement in a Colorado alpine region: rates, processes and climatic significance', *Arctic and Alpine Research*, vol. 2, pp. 165–226.

A source which draws together an impressive interdisciplinary team of authors and illuminates both mountain and polar topics is:

Ives J. D. and Barry R. G. (eds) (1974) *Arctic and Alpine Environments* (London: Methuen).

This tradition is continued in two books which focus on specific mountain zones, but at the same time provide a strong general framework:

Ives J. D. (ed.) (1980) *Geoecology of the Colorado Front Range: A Study of Alpine and Subalpine Environments* (Boulder, Colorado: Westview Press).
Caine N. (1983) *The Mountains of Northeastern Tasmania: A Study of Alpine Geomorphology* (Rotterdam: Balkema).

Two wide-ranging surveys of the mountain zone viewed as an interface between people and their physical environment are:

Price L. W. (1981) *Mountains and Man: A study of Process and Environment*. (Berkeley: University of California Press).
Messerli B. and Ives J. D. (eds) (1984) *Mountain Ecosystems: Stability and Instability* (Boulder, Colorado: International Mountain Society).

3.4
Change and Instability in the Desert Environment

Andrew S. Goudie
University of Oxford

Deserts cover approximately one-third of the Earth's land surface, constitute one of the world's major ecosystems and provide a significant contribution to the global economy. They provide at least a fifth of the world's food supplies, over half of the world's production of precious and semi-precious minerals, and a substantial proportion of oil and natural gas reserves. They are also areas of very rapid technological and demographic change, with, for example, many instances of quickly developing urbanisation. Many of the great deserts are of considerable antiquity. For example, recent studies of aeolian sediments in dated ocean cores have shown that materials have been removed by deflation from arid surfaces in the vicinity of the present Sahara since the Cretaceous and Tertiary. However, during the course of these millions of years the deserts have undergone a series of major changes at a variety of scales. The purpose of this chapter is to explore the nature, causes and consequences of this variability, and to examine some of the ways in which it relates to human activities.

ASPECTS OF VARIABILITY

Rainfall variability

One of the most important characteristics of desert rainfall is that it is highly variable from year to year. This interannual variability (V) can be expressed by a single index:

$$V\% = \frac{\text{mean deviation from the average}}{\text{the average}} \times 100$$

European humid temperate stations may have a variability of less than 20 per cent whereas variability in the Sahara ranges from 80 to 150 per cent. This implies that although mean annual rainfall levels are so low there can still from time to time be individual storms of surprising size, so that maximum falls in twenty-four hours may exceed the long-term annual rainfall means. Such storms can cause severe damage as a result of flashflood generation, but it would be wrong to give the impression that all desert rainfall occurs as storms of such ferocity and intensity. As Table 3.4.1 shows, the average rainy day only produces 2.56–9.75mm of rain, and such figures are typical of many mid-latitude humid temperate stations. Nonetheless, extreme storms do occur (Table 3.4.2).

TABLE 3.4.1 *Rainfall per rainy day in arid areas*

Area	*Number of stations*	*Range of mean annual rainfall*	*Range of number of rainy days 0.1 mm*	*Average mm of rainfall per rainy day*
USSR	12	92–273	42–125	2.56
China	6	84–396	33–78	4.51
Argentina	11	51–542	6–155	5.41
N. Africa	18	1–286	6–57	3.82
W. Africa	20	17–689	2–67	9.75
Kalahari	10	147–592	19–68	9.55
Combined	77	1–689	1–155	6.19

TABLE 3.4.2 *Extremes of precipitation in arid areas*

Location	*Date*	*Mean annual precipitation mm*	*Storm precipitation*
Chicama (Peru)	1925	4	394mm
Aozou (Central Sahara)	May 1934	30	370mm/ 3 days
Swakopmund (South West Africa: Namibia)	1934	15	50mm
Lima (Peru)	1925	46	1524mm
Sharja (Trucial Coast)	1957	107	74mm/ 50 mins
Tamanrasset (Central Sahara)	Sept. 1950	27	44mm/ 3 hours
Bisra (Algeria)	Sept. 1969	148	210mm/ 2 days
El Djem (Tunisia)	Sept. 1969	275	319mm/ 3 days

The flashfloods that feature so strongly in the folklore of desert areas may, therefore, result from the occasional extreme event or from moderate precipitation events, falling on surfaces that are especially suited to flood generation. These include alluvial fans that have been rendered indurated and impermeable by calcium carbonate (calcrete), as near Las Vegas, Nevada, or fine-grained surfaces sealed by a clay crust. Statistical analysis of discharge records also demonstrates the extreme variability of river flows in arid lands. The coefficient of variation of annual flows (C_v) (the standard deviation of annual flows divided by the mean annual flow) is an important measure of hydrological variability. Values for selected arid and non-arid areas are shown in Table 3.4.3.

TABLE 3.4.3 *Hydrological variability of surface runoff in selected areas (from Data in McMahon, 1983)*

Area	C_v *of annual flows*
(a) *Arid areas*	
Australia	1.27
East Mediterranean	1.25
North America	0.65
South Africa	1.14
(b) *Non arid areas*	
Australia	0.67
North America	0.3
Europe	0.2
Asia	0.2

Current climatic fluctuations

The rainfall of desert regions is also known to undergo longer-term fluctuations that may last for a few decades or so. Such long drought periods may contribute to desertification and water shortage, but their very existence is the subject of controversy. Figure 3.4.1 shows annual rainfall curves over the last forty years. Certain stations (e.g. Jodhpur and Phoenix) show high variability from year to year but no consistent trend, whereas others (e.g. Alice Springs, Abéché and Agadez) seem to have experienced an apparent downward trend in rainfall. Changes have been especially dramatic in the Sahara and neighbouring sub-Saharan zones (Fig. 3.4.2a). Wet conditions were widespread in the 1950s, with decadal averages ranging from 15 per cent above normal in the southern areas to 35 to 40 per cent above normal along the Saharan fringe. Such favourable conditions came to an end in the 1960s and by 1968 the drought was beginning in earnest. Deficits for the six years 1968 to 1973 averaged 15 to 40 per cent below the mean. There was a slight lull in the mid 1970s, but large rainfall deficits returned between 1976 and 1983. Although the present Sahel drought is especially serious, other lengthy dry spells have occurred twice before in this century (i.e. around 1912 and in the early 1940s).

Such variability in precipitation is reflected in changes in river discharges and lake areas. Figure 3.4.2b shows how the discharge data for the Senegal river in West Africa broadly parallel the rainfall fluctuations, with three periods of specially low flow in this century. The low flows since 1968 have had serious consequences for irrigation schemes. Likewise, Lake Chad has shown a systematic decrease of its level. Between 1963 and 1973 its surface area had been divided by three and its volume by four. There has been considerable debate as to whether such hydrological and climatic fluctuations display any cyclical behaviour which might have predictive value. Cycles of various lengths have been postulated (e.g. 200 years and eleven years) but there is no convincing evidence yet of any useful and consistent periodicities.

Climatic changes

Over longer time-spans the world's arid zones have shown further substantial changes, for the events which led to the expansions and contractions of the great ice-sheets during the Pleistocene also led to major environmental changes in lower latitudes. One of the most impressive indications of such changes is the presence of relict sand dunes in currently humid areas. Such fossil dunes have now been found in the High Plains of the USA, the Llanos of South America, the Mato Grosso of Brazil, West and Central Africa, large parts of Australia and the Thar of India. For such dunes to form, rainfall levels were probably less than 100–300mm, yet present rainfall totals in areas of these fossil dunes can exceed 750–1400mm. The extent of this change has been described thus by Sarnthein:

> Today about 10 per cent of the land area between 30°N and 30°S is covered by active sand deserts . . . sand dunes and associated deserts were much more widespread 18,000 years ago than they are today. They characterised almost 50 per cent of the land area between 30°N and 30°S forming two vast belts. In between tropical rain forests and adjacent savannas were reduced to a narrow corridor, in places only a few degrees of latitude wide. (M. Sarnthein, 'Sand deserts during glacial maximum and climatic optimum', *Nature*, 1978, 272, 43–6)

Another potent indicator of the degree of climatic change is provided by the changing history of lake levels in arid areas. In many areas there is considerable evidence that at certain points in the past lakes have been much more extensive than now. Classic examples include the currently highly saline

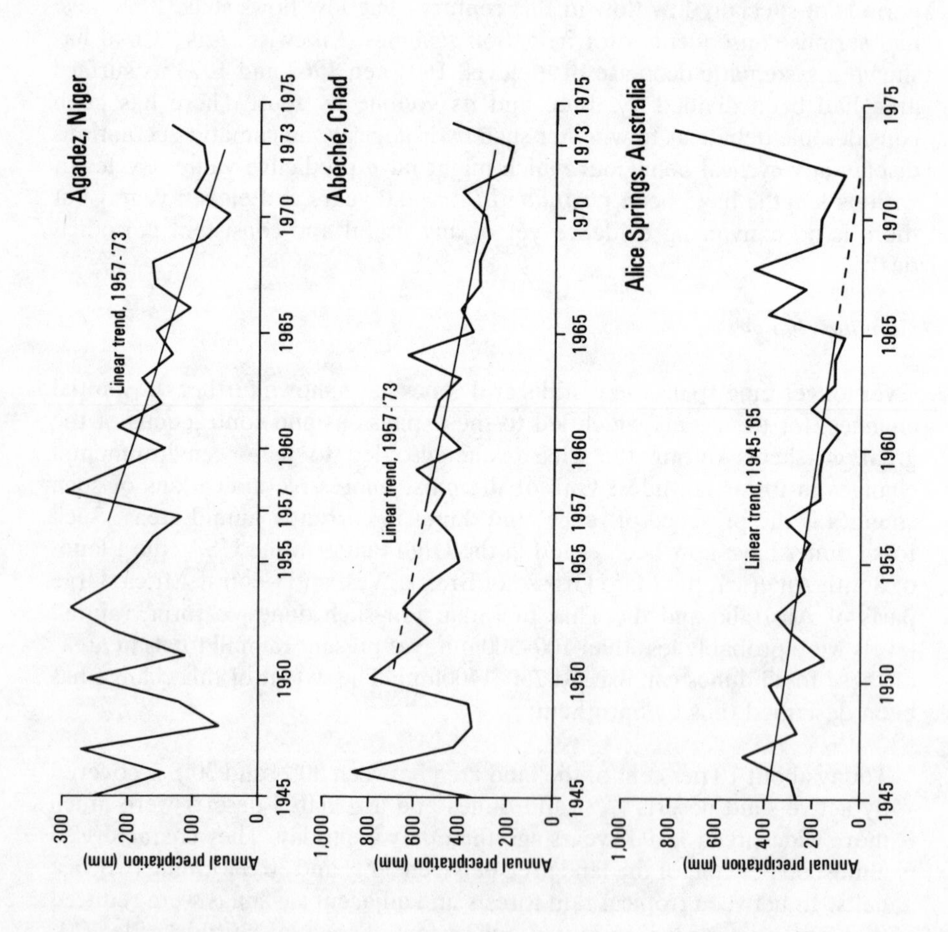
Agadez, Niger
Linear trend, 1957-'73
Annual precipitation (mm)
300
200
100
0
1945
1950
1955
1957
1960
1965
1970
1973
1975
Abéché, Chad
Linear trend, 1957-'73
Annual precipitation (mm)
1,000
800
600
400
200
0
1945
1950
1955
1957
1960
1965
1970
1973
1975
Alice Springs, Australia
Linear trend, 1945-'65
Annual precipitation (mm)
1,000
800
600
400
200
0
1945
1950
1955
1960
1965
1970
1975

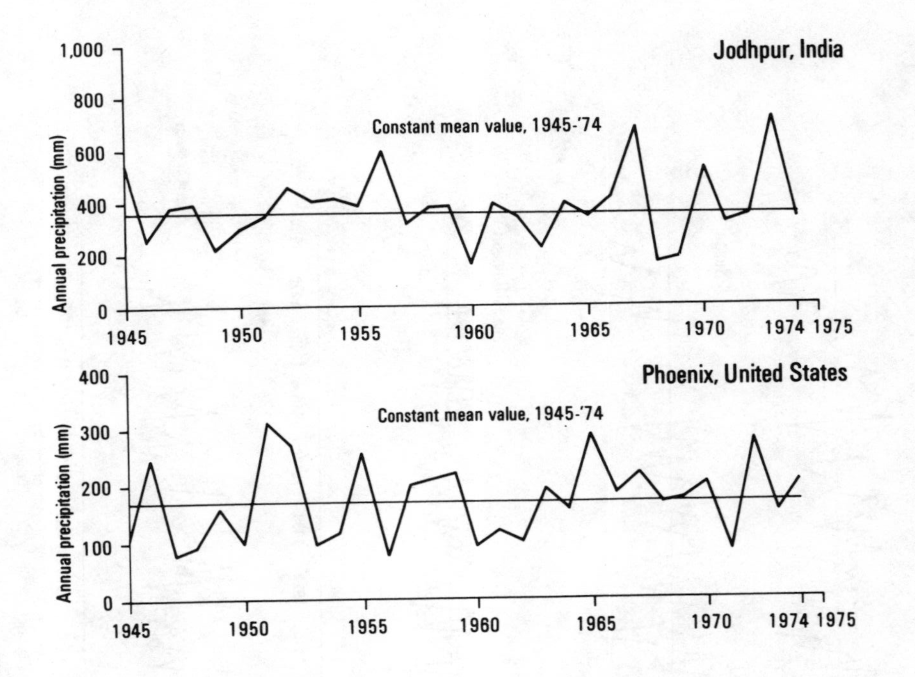

FIGURE 3.4.1 *Rainfall variations at selected arid zone stations since 1945 (From Anon, 1984)*

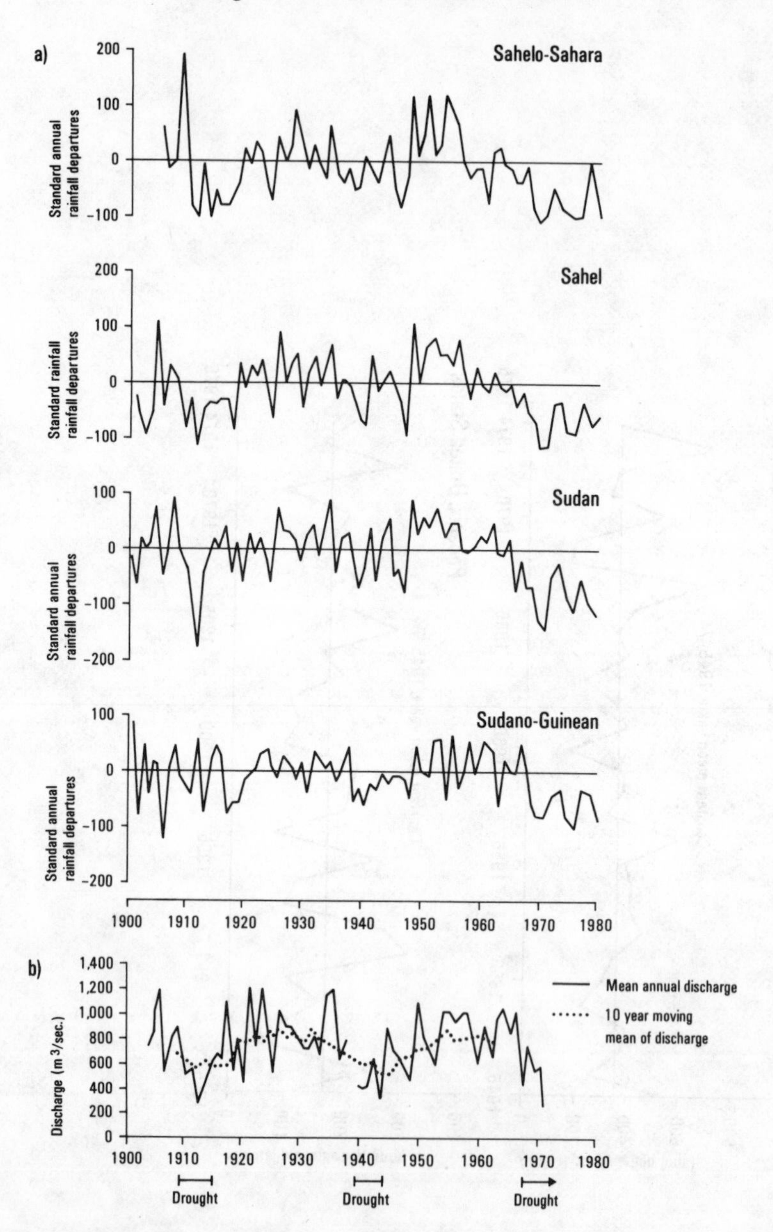

FIGURE 3.4.2 *Rainfall and river discharge fluctuations in West Africa in the twentieth century*

(a) Standard annual rainfall departures for four sub-Saharan zones (after S. Nicholson in Anon, 1984).

(b) The discharge of the Senegal river at Bakel showing the record of mean annual discharge, and the ten-year moving mean of discharge (after Goudie, 1983, fig. 5.11).

basins of the Basin and Range Province in the USA, Lake Chad on the margins of the Sahara, and the Aral–Caspian system in the USSR. The dates of such expanded lake basins vary from area to area (Fig. 3.4.3) and patterns are apparent. In North America lakes appear to have been high during the period of the maximum extent of the last glaciation (i.e. glacial was in a very general sense synchronous with pluvial), whereas in Africa and Australia lake levels were particularly low at the time of maximum glaciation and during the subsequent deglaciation. They reached very high levels in the early Holocene (*c.*9000 years ago).

Within the Holocene itself desert areas have seen certain important changes, and some of these have had social implications. Large areas of the High Plains of the USA were much more arid than today in the Mid-Holocene 'altithermal' and archaeological evidence indicates that human population levels declined over wide areas. Similarly the decline of the Great Indus civilisation around 4000 years ago and the migrations of peoples from Central Asia may have been stimulated by pulsations of aridity.

CHANGE AND HUMAN IMPACT

Desertification

Desertification, which is sometimes used interchangeably with the terms 'desertisation' and 'desert encroachment' is a serious and long-continued problem on desert margins, and has been defined by Dregne (1977) as:

> The impoverishment of arid, semi-arid and subhumid ecosystems by the continued impact of man's activities and drought. It is the process of change in these ecosystems that can be measured by reduced productivity of desirable plants, alteration in the biomass and the diversity of the micro and macro fauna and flora, accelerated soil deterioration, and increased hazards for human occupancy.

Present interest in this theme has been magnified by the persistent Sahelian drought, but in reality the degradation of arid zones has been occurring for five milliennia over a vast territory from the Indus to the Atlantic, and from Scandinavia to the Sahel. Considerable controversy surrounds the causes of desert encroachment. The question has been asked whether the process is caused by temporary drought periods of high magnitude, whether it is due to long-term climatic change towards aridity (either alleged post-glacial progressive desiccation or shorter-term cycles), whether it is caused by man-induced climatic change, or whether it is the result of human activities.

The general consensus is that most desertification is caused by human activities (e.g. overgrazing, overburning, overcultivation and wood cutting) operating on naturally fragile environments which are put under severe stress

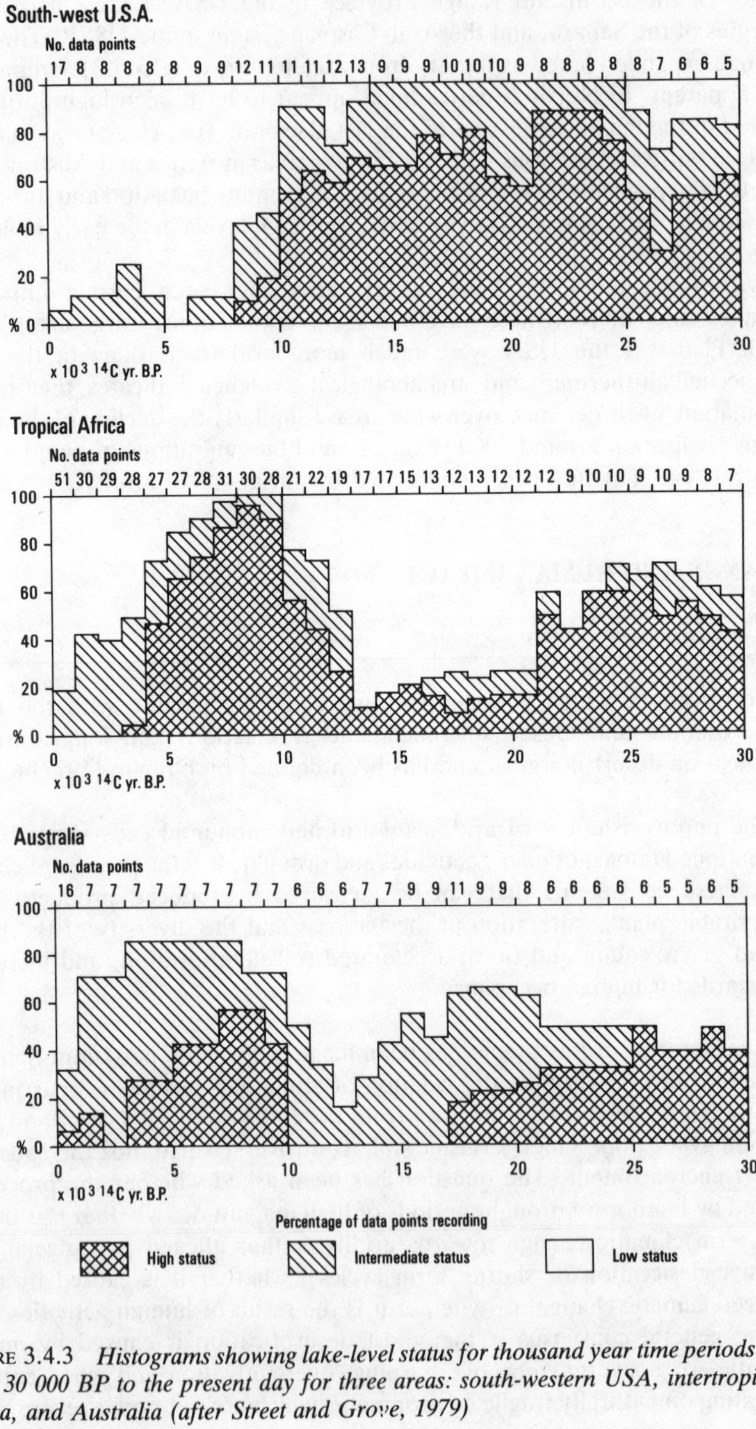

South-west U.S.A.

No. data points

17 8 8 8 8 8 8 9 9 12 11 11 11 12 13 9 9 9 10 10 10 9 8 8 8 8 7 6 6 5

100 80 60 40 20 % 0

0 5 10 15 20 25 30

x 10^3 ^{14}C yr. B.P.

Tropical Africa

No. data points

51 30 29 28 27 27 28 31 30 28 21 22 19 17 17 15 13 12 13 12 12 12 9 10 10 9 10 9 8 7

100 80 60 40 20 % 0

0 5 10 15 20 25 30

x 10^3 ^{14}C yr. B.P.

Australia

No. data points

16 7 7 7 7 7 7 7 7 7 6 6 6 7 7 9 9 11 9 9 8 6 6 6 6 6 5 5 6 5

100 80 60 40 20 % 0

0 5 10 15 20 25 30

x 10^3 ^{14}C yr. B.P.

Percentage of data points recording

High status   Intermediate status   Low status

FIGURE 3.4.3 *Histograms showing lake-level status for thousand year time periods from 30 000 BP to the present day for three areas: south-western USA, intertropical Africa, and Australia (after Street and Grove, 1979)*

by periods of prolonged drought. Certainly, this combination of human and climatic factors can be seen in the case both of the 'Dust Bowl' of the USA in the 1930s, and the Sahel drought of the present time. The former was in part caused by a series of hot, dry years which depleted the vegetation cover and made the soils dry enough to be susceptible to wind erosion. However, the situation was exacerbated by the rapid expansion of wheat cultivation in the Great Plains following the development of the internal combustion engine (tractors, combines and trucks). The Dust Bowl indicates that desertification is not restricted to the less developed countries. Indeed, even today, the USA has one of the largest areas of severe desertification, and much of it is caused by the exploitation of susceptible areas by means of technologically advanced farming methods, such as centre-pivot irrigation.

However, in recent years attempts have been made to delve deeper into the social, economic and political causes of desertification. As Spooner (1982) put it:

> In terms of the conventional wisdom there has been a tendency always to assume implicitly that the immediate human cause of any particular symptom of desertification, such as over-grazing, was the significant cause, to look no further to pursue the reconstruction of a chain of causation, but to prescribe simple remedies in the form of management regimes and expect them to be easily and efficiently implemented.

This traditional view can be equated with the 'dominant paradigm' of natural hazards research which has recently been questioned by Hewitt. Thus the desertification of the Sahel may indeed result from unwise human activities creating stress in drought years, but it can also be seen in terms of a variety of more general issues such as exploitive western capitalism, the decline of traditional social structures and nomadism, core–periphery tensions, the incompetence of certain national and local administrations, the need to grow cash crops to pay for imported manufactured goods and defence equipment, inadequate land-tenure systems, political directives to cultivate new parts of the national territory, restrictions on where people can live and so forth.

Given issues such as these, ultimate solutions to desertification must be long delayed, and there has been a striking lack of progress since the Nairobi Desertification Conference of 1977. There is also some debate as to whether the environmental degradation associated with desertification is itself irreversible. Some experiments and field observations have shown that in certain specific circumstances, recovery is so slow and so limited it may be appropriate to talk of 'irreversible desertification'. In southern Tunisia, for example, the damage caused by military vehicles from the battles of the Second World War is still evident in unregenerated vegetation. Elsewhere, where ecological conditions are favourable because of the existence of such factors as deep sandy soils, or favourable hydrological circumstances, vegetation is able to recover once the excess pressures on it are eliminated. Most desert fauna and flora evolved in an environment where the normal pattern is one of more or less random alternations of short favourable periods and long stress periods.

In developed countries with a desertification problem, solutions may be relatively simple, for population densities are much less, and capital and technical expertise are available, but in the less developed nations the situation is more intractable. Nonetheless, a wide range of strategies has been developed to assist in ecological rehabilitation, including climatic modification; biological recovery through fencing and stock control, semi-natural approaches such as contour terracing or fertilisation, or artificial techniques such as planting and dune stabilisation programmes; and hydrological management.

Salinisation

One major cause of the decline of agricultural productivity in arid lands is the spread of saline conditions. Most arid zones are naturally salty, for they are areas where there is substantial water deficit and thus insufficient water to percolate through the soil and leach soluble materials into the rivers and hence into the oceans. There are also numerous sources of salt. The presence of salt is important for a variety of reasons. First, high salt concentrations render water non-potable for humans and stock. Secondly, plants have only a limited tolerance to salt, through its effects on osmotic pressures of soil moisture and through direct toxicity. Thirdly, soil structure may deteriorate, for as the salinity of soil water becomes more concentrated, calcium and magnesium tend to precipitate as carbonates, leaving sodium ions dominant in the soil solution. The sodium ions tend to be absorbed by colloidal clay particles, deflocculating them and leaving the resultant structureless soil almost impermeable to water and unfavourable to root development.

Human activities are increasing the problems of salinity in a variety of ways. The most important of these is the extension of irrigation, which brings the water table close enough to the ground surface for capillary rise and subsequent evaporative concentration to take place. Concentration also occurs because much water will evaporate directly from the soil surface, or as a result of evaporation from reservoirs. Furthermore, notably in areas of high soil permeability, water seeps laterally and downwards from irrigation canals so that further evaporation takes place.

A second cause of accelerated salinisation occurs in coastal areas as a result of sea water incursion brought about by overpumping. Fresh water is less dense than sea water, such that a column of sea water can support a column of fresh water 2.5 per cent higher than itself. Thus where a body of fresh water has accumulated in an aquifer which is also open to penetration from the sea, it forms a lens on top of the salt water. The corollary of this is that if the hydrostatic pressure of the fresh water falls as a result of overpumping in a well, then the underlying salt water will rise. Given the low rates of recharge in arid areas, the fact that many groundwater reserves may be relicts of pluvial phases, and the very substantial requirements for water, this is a very serious phenomenon in areas like the coasts of Israel, Bahrain, California and the United Arab Emirates.

A third cause of accelerated salinisation is vegetation clearance. In some parts of Australia the removal of the natural forest cover has led to a reduction in rainfall interception and evapotranspirational losses. This causes groundwater levels to rise, creating seepage (and subsequent evaporation) of sometimes saline water in low-lying topographic situations.

The spread of salinisation through these mechanisms is not a new phenomenon and was creating problems for agriculturalists in Mesopotamia around 4000 years ago. Nonetheless, with the implementation of modern irrigation schemes the problem has become increasingly acute. It has been estimated that the percentage of salt-affected and water-logged soils amounts to 50 per cent of the irrigated areas in Iraq and the Euphrates valley of Syria, 30 per cent in Egypt, 23 per cent in Pakistan and over 15 per cent in Iran. Reversal of this trend requires more careful control of water and the installation of improved drainage to reduce groundwater levels. Tubewells are used extensively for this purpose in Punjab and Sind, but the energy requirements and capital costs are high.

Water management problems

A third illustration of the susceptibility of desert areas to human interference is provided by the various ramifications of dam construction. The first dam was probably constructed in Egypt some 5000 years ago, and since that time the number and size of dams has grown exponentially. Dams and their associated reservoirs and canals have many important virtues, including drought and flood prevention, and power generation. However, ill-conceived implementation of major dams can have a whole series of environmental consequences including ground subsidence and increased seismic activity, slope instability, reduction of river sediment downstream, change in water quality, flooding of habitats, alteration of microclimates and ecological disruption.

Other major water management schemes can also have adverse effects. For example, the 300km Jonglei Canal on the Nile in Sudan will, through bypassing the Sudd Swamps in which evaporative losses are high, increase the flow down the river very substantially. At the same time it will remove some major wetland habitats and change the whole ecological and economic basis of the seasonal floodplains. Equally, large-scale groundwater abstraction is causing the water-table level to fall in areas like the High Plains of Texas, and one consequence of this is that ground subsidence may occur. Seven to eight metres of subsidence has taken place in the Central Valley of California and in the basin in which Mexico City lies.

Geomorphological instability and hazards

Our discussion of climatic variability and of landscape changes wrought by man has already indicated that many desert areas are both unstable and hazardous to humans. As the economic development of arid lands takes place, it becomes more apparent that there must first be a careful assessment of potential environmental problems. The list of arid zone geomorphological hazards is long; covering, for example, wind and water erosion, expansive soils and hydrocompaction, dune encroachment and crust formation. It is only possible here to discuss three of these in detail: salt weathering, dust storms and gully incision.

Salt weathering Recent laboratory experiments and field observations have demonstrated that rocks can break down very rapidly in the presence of salt due to two main mechanisms: salt crystallisation and salt hydration. Salt crystal growth may take place in a rock as a result of the evaporation or cooling of a salt solution, and sets up pressures which exceed the tensile strength of most rocks so that disintegration occurs. Salt hydration occurs as a result of the volume expansion that takes place when a crystal changes phase by taking up moisture into its structure.

The power of salt to cause rock disintegration has implications for buildings made of concrete, stone or brick. The Sphinx in Egypt and the ancient city of Mohenjo Daro in Pakistan are examples of archaeological sites that are suffering from the ravages of the sulphates of magnesium and sodium respectively. Of rather greater significance, however, is the rapidity with which modern buildings may deteriorate, especially if their foundations penetrate the zone of groundwater and act like a wick. The problem may, however, be alleviated, by planning restrictions in areas where groundwater is saline and close to the ground surface, by making sure that the concrete is made with materials that do not contain large quantities of deleterious salts, and by selecting aggregate that is resistant to salt attack. In the case of Mohenjo Daro attempts are being made to reduce groundwater levels beneath the site by the installation of an expensive ring of tube wells.

Dust storms The Dust Bowl in the USA in the 1930s is possibly the most famous example of the importance of dust storms in the arid environment. Deflation of susceptible surfaces, many of which had been rendered unstable by ploughing, caused dust storms to occur with great regularity during the 1930s. Such dust storms have many environmental consequences which are disadvantageous to man. As more meteorological data become available, and as more dust storms can be recognised and tracked by means of meteorological satellites, it is becoming possible to identify those areas which are especially prone to this process. It is also becoming possible, using long-term meteorological data, to see whether human influences are causing trends in dust storm occurrence (Figure 3.4.4). As Table 3.4.4 shows, there are many

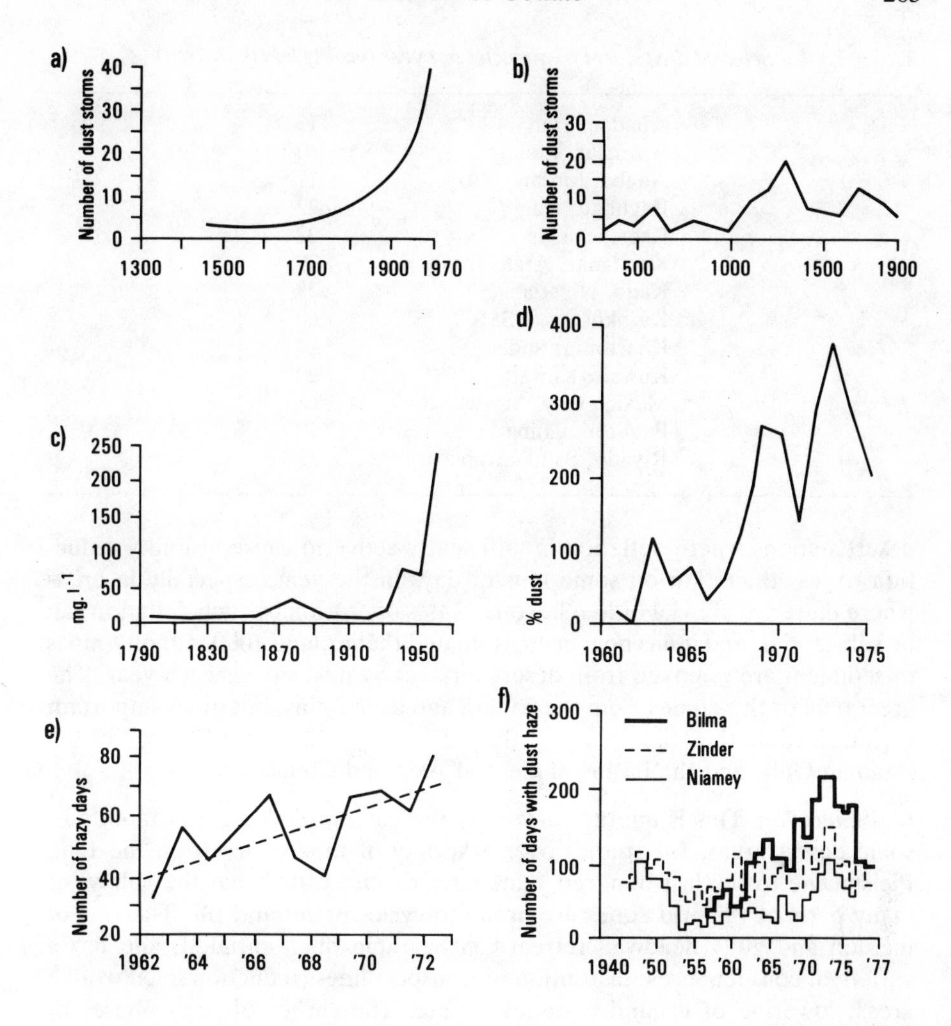

FIGURE 3.4.4 *The changing frequency of dust storm activity through time (from various sources in Goudie, 1983, fig. 13)*

(a) based on Russian data since the late fifteenth century.
(b) China since AD 300.
(c) Dust in glacier ice in Kazbek, USSR.
(d) Number of dust storms at El Fasher, Sudan since 1960.
(e) Number of hazy days at Samaru, Nigeria from 1962–1973.
(f) Number of hazy days at miscellaneous West African stations since the 1940s.

TABLE 3.4.4 *Selected dust storm frequencies per year (visibly less than 1km)*

Abadan, Iran	13
Amritsar, India	10
Aqaba, Jordan	11
Baghdad, Iraq	22
Beersheva, Israel	27
Kandahar, Afghanistan	79
Kano, Nigeria	23
Kazakhstan, USSR	60
Khartoum, Sudan	24
Kuwait, Kuwait	27
Mexico City, Mexico	68
Paoutou, China	19
Riyadh, Saudi Arabia	13

desert stations where deflation is sufficiently active to cause visibility reduction to less than 1km on some tens of days in the year, especially in areas where there are alluvial or loessic soils. Rates of erosion by such deflation can be substantial, and it has now been estimated that around 1000 million tonnes of sediment are removed from desert surfaces by dust storms each year. The great bulk of this comes from the Sahara and its margins, but other important regions include the High Plains of the USA, the lowlands at the head of the Arabian Gulf, and the basins of Central Asia and China.

Gully incision This is another aspect of the geomorphological instability of some desert areas, for studies over a variety of time scales from the Late Pleistocene to the last hundred years have demonstrated that the valleys of many parts of the arid zones are prone to cycles of cut and fill. The rate of incision and gully headwall retreat can be rapid once initiated, and has a variety of consequences: disruption of transport lines, reduction in cultivable areas, lowering of groundwater levels, etc. The causes of such phases of incision have been the subject of considerable controversy, and a large number of possible mechanisms have been postulated (Fig. 3.4.5). Particular interest has surrounded the question whether the incision results from human activities, from changes in climate, or from some geomorphological threshold being crossed. Recognition of valley bottoms that may be approaching some threshold (see Fig. 3.4.6) may have some predictive value in terms of trying to identify those areas upon which soil conservation measures should be concentrated. However, the whole problem of gully or arroyo initiation is a neat illustration of the principle of equifinality, demonstrating that the plausible explanations may differ from area to area and from gully to gully. For example, it has been demonstrated that in the 1880s there was a widespread phase of gully development in the American West, and that the cause in southern Arizona was a change in rainfall type while the cause in coastal California was land-use change.

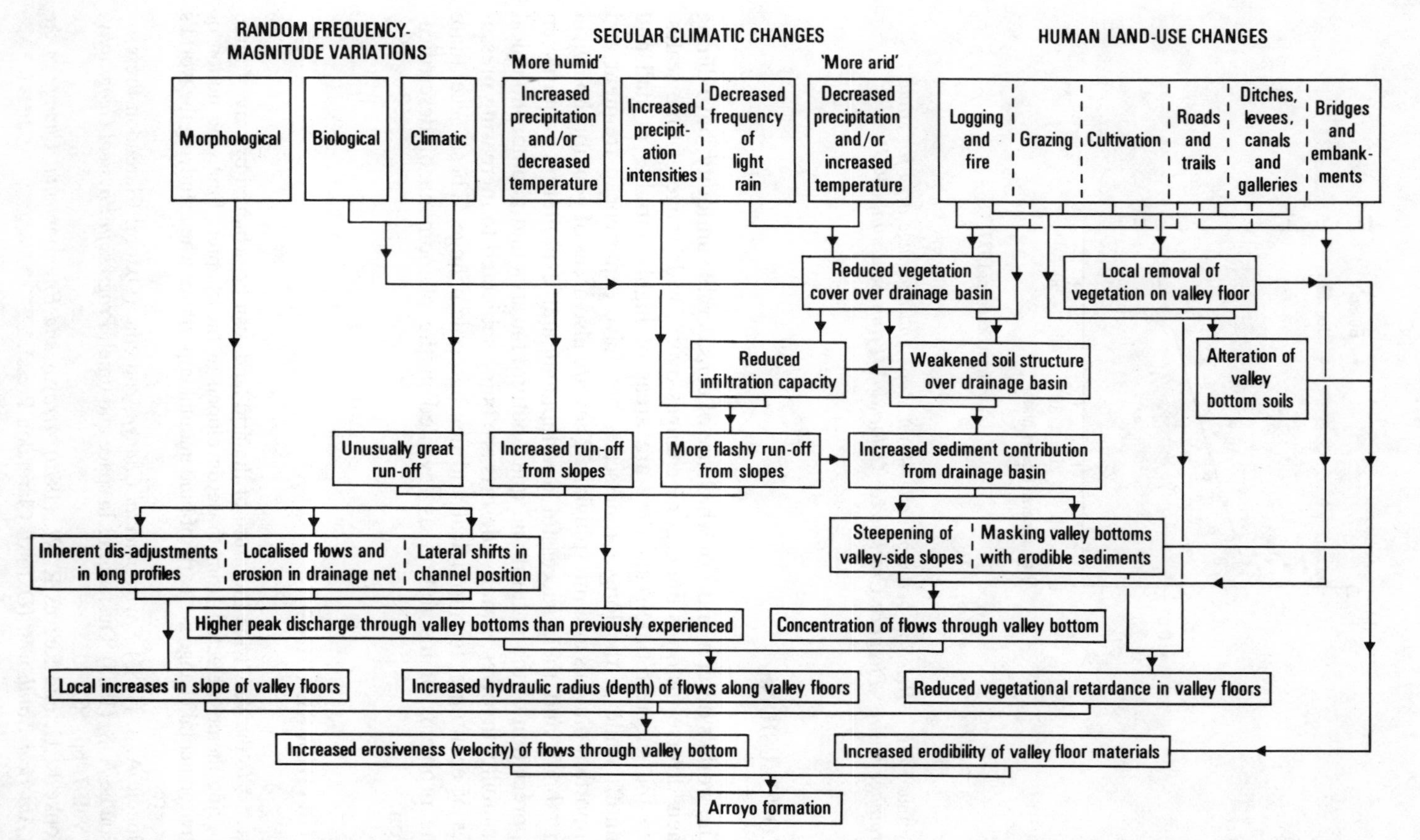

FIGURE 3.4.5 *The complexity of factors that can cause gully (arroyo) incision (from Cooke and Reeves, 1976)*

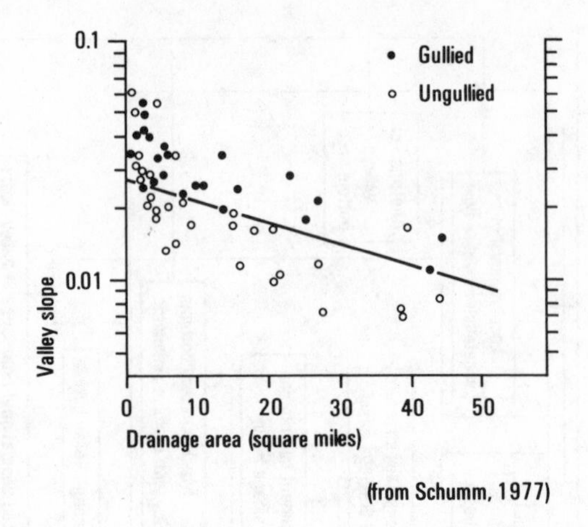

FIGURE 3.4.6 *Relation between valley-floor slope and drainage area for small drainage basins, Piceance Creek area, Colorado. (From Schumm, 1977)*

CONCLUSIONS

The world's deserts, most of which are of considerable antiquity, have during their diverse histories undergone changes over a wide range of time scales. On the shortest time-scale they are areas of highly variable rainfall and runoff, while at the longer time-scale they have experienced frequent and important pluvials and interpluvials. They are also areas of geomorphological instability and their successful development and exploitation requires an appreciation of the natural and quasi-natural hazards that geomorphological instability involves. Finally, deserts are being subjected to increasing pressures of economic, technological and demographic change. These accentuate the problems of instability, as revealed in the phenomena of desertification.

FURTHER READING

Access to the very wide literature of the arid lands can be achieved by way of three specific themes selected from those developed in the chapter. First, the notion of environmental change is clearly of fundamental importance when studying the world's deserts:

Goudie A. S. (1983) *Environmental Change*, 2nd edn. (Oxford: Clarendon Press).

Goudie A. S. (1983) 'Dust storms in space and time', *Progress in Physical Geography*, vol. 7, pp. 502–30.

Cooke R. U. and Reeves R. W. (1976) *Arroyos and Environmental Change in the American South-west* (Oxford: Clarendon Press).

Graf W. L. (1983) 'The Arroyo problem – palaeohydrology and palaeohydraulics in the short term', in Gregory K. J. (ed.) *Background to Palaeohydrology* (Chichester: John Wiley) pp. 279–302.

Secondly, change can be related very specifically to the complex but fundamental concept of desertification:

Uncod (1977) *Desertification: Its Causes and Consequences* (Oxford: Pergamon Press).

Worster D. (1979) *Dust Bowl* (New York: Oxford University Press).

Spooner B. and Mann H. S. (eds) (1982) *Desertification and Development: Dryland Ecology in Social Perspective* (London: Academic Press).

Dregne H. (1977) 'Desertification of arid lands', *Economic Geography*, vol. 53, pp. 322–31.

Thirdly, the arid lands today present many challenges to the geographer, both pure and applied. From many possibilities we might select:

Heathcote R. L. (1983) *The Arid Lands: Their Use and Abuse* (London: Longman).

Cooke R. U., Brunsden D., Doornkamp J. C. and Jones D. K. C. (1983) *Urban Geomorphology in Drylands* (Oxford: Oxford University Press).

Worthington E. B. (ed.) (1977) *Arid Land Irrigation in Developing Countries: Environmental Problems and Effects* (Oxford: Pergamon Press).

Butzer K. W. (1974) 'Accelerated soil erosion: a problem of man-land relationships', in Manners I. R. and Mikesell M. W. (eds) *Perspectives on Environments* (Washington DC: Association of American Geographers).

Gray W. (S.) (1983) '[illegible] problems—palaeohydrology and palaeohydraulics in [illegible]', in Gregory K. J. (ed.) Background to Palaeohydrology (Chichester: John Wiley), pp. 279–302.

Secondly, change can be studied very specifically in the complex but fundamental concept of desertification:

[illegible] (1977) Desertification: Its Causes and Consequences (Oxford: Pergamon Press).
[illegible] (1978) [illegible] (Oxford University Press).
Spooner B. and Mann H. S. (eds) (1982) Desertification and Development: Dryland Ecology in Social Perspective (London: Academic Press).
Dregne H. E. (1977) 'Desertification of arid lands', Economic Geography, Vol. 53, pp. [illegible].

Thirdly, the arid lands today present many challenges to the geographer, both pure and applied. From natural hazards to [illegible]:

Heathcote R. L. (1983) [illegible] (London: Longman).
Cooke R. U., Brunsden D., Doornkamp J. C. and Jones D. K. C. (1982) Urban Geomorphology in Drylands (Oxford: Oxford University Press).
[illegible] (1979) [illegible] (Oxford: Pergamon Press).
Butzer K. W. (1974) 'Accelerated soil erosion: a problem of man-land relationships', in Manners I. R. and Mikesell M. W. (eds) Perspectives on Environment (Washington DC: Association of American Geographers).

Section 4
Management of the Physical Environment

4.1
Geomorphology and Environmental Management

Ronald U. Cooke
University College London

The previous chapters have revealed both the changing nature of physical environments and the changing roles of physical geographers in studying them. In approaching the links between such studies and environmental management, two perspectives are fundamental: first, the perspective of the specialist physical geographer who seeks to serve the needs of a range of management agencies; second, the perspective of management agencies on what the physical geographer can provide. Both perspectives are commonly partisan, but they can be explored in a relatively detached way through an appraisal of recent geographical contributions. In this chapter the perspective of the specialist physical geographer is explored through the manifold links between geomorphology and the hierarchy of management agencies in which geomorphological work can and often does play a role. Geomorphology is not alone amongst the systematic branches of physical geography in seeking to develop some of its research in the context of management, but it has made useful, perhaps exemplary progress in recent years.

VIEWPOINTS

Figure 4.1.1 shows a relatively unpretentious desert plain in southern Peru – a fairly flat stony surface underlain by a primitive soil and a thick accumulation of calcium carbonate, fringing escarpments, and numerous small isolated dunes. This view, like so many, can inspire in the geomorphologist a plethora of research questions. Take the small barchan dunes, for example: how are they distributed, what are they composed of, and what are their geometrical properties? Are they moving, and if so, in what directions, when and how quickly? Why does their colour differ from that of the prevailing surface material? And how do all the dune characteristics relate to wind patterns and sand sources at present and in the past? All such questions

FIGURE 4.1.1 *A scene in southern Peru*

are scientifically respectable and they can be answered by well-established methods of geomorphological analysis; the answers can illuminate an understanding of this piece of landscape and desert landscapes in general, and they may contribute to the growing body of geomorphological theory.

But this same view can be seen in a different light. To a European consultancy, part of this plain appeared to be suitable for the development of irrigation agriculture. As a result of the consultants' report, open canals were built to supply the area with water from the Andes, and the encumbrance of a stony surface was removed to expose a cultivable soil. A good crop of vegetables was grown in the first year of production, but problems soon emerged. The sand dunes happen to be moving quickly across the stony surface – when they encounter a canal, it traps them and the supply of water is seriously disrupted. Removal of the stony surface meant that the underlying silt was prey to quick and easy erosion by the wind and, in places, the calcium carbonate in the soil profile (wrongly identified by the consultants as a marine deposit) became exposed. Productivity and enthusiasm for the development rapidly declined. The developers began to ask questions, such as those posed by the geomorphologist, but they asked them too late.

On the other side of the earth, London was rapidly running short of aggregates for the building industry during the great phase of development and reconstruction after the Second World War. As the prime aggregate resources of the River Thames terraces were depleted, or held hostage by construction on top of them, where were new supplies to come from? Geomorphologists had shown that in the Vale of St Albans there was a major gravel train associated with a little known, much older course of the River Thames east of Rickmansworth. This research, carried out largely because the history of the River Thames and the evolution of the landscape of south-east England is one of great intrinsic complexity and interest, provided the key to future sources of aggregates around north London. The same gravel train is still being worked today, for example in connection with the building of the M25 motorway.

These very different anecdotes provide a single message – geomorphological studies may be intrinsically interesting in their own right; but the same work may also be useful in environmental management.

THE RISE OF GEOMORPHOLOGY IN ENVIRONMENTAL MANAGEMENT

It is as well to remember that some very substantial and theoretically fundamental geomorphological contributions in the past have been stimulated by the demands for solutions to environmental problems. Leonardo da Vinci's observations on fluvial processes were certainly provoked, in part, by his appreciation of the ravages of floods and alluviation. G. K. Gilbert's seminal studies of debris transportation in running water were undertaken in the context of his efforts to understand the effects of hydraulic mining on the Sacramento River. More recently, the fundamental work of Robert E. Horton on the drainage basin and soil erosion by water, and that of W. S. Chepil and his collaborators on soil erosion by wind were certainly sustained by the perceived seriousness of the soil erosion problem in North America from the 1930s onwards.

During the philosophical turmoil within geography in the early 1960s, applied work in general did not figure prominently. Although some of the inspiration for the changes at that time undoubtedly came from the work of Robert Horton, Luna Leopold, Walter Langbein and others, all of whom had conspicuously succeeded in applying their geomorphological research to environmental problems, of the widely used texts available only Thornbury's fading volume on the *Principles of Geomorphology* included an 'applied geomorphology' chapter. Since that time, both geomorphology and its applications to environmental management have begun to flourish, and for several reasons.

First, the change of emphasis within geomorphology in recent decades towards the study of contemporary processes and changes at the earth's surface has placed part of the research frontier firmly in those areas of the subject of greatest potential value to resource and hazard appraisal, planning and civil engineering. In the United Kingdom, this change was provoked, in part, by the 1960s' wish to be more scientific and quantitative. This promoted the interest in dynamics, systems and methods of evaluating complex spatial and temporal interrelationships, and encouraged the objectives of defining laws and gaining respectability in other cognate disciplines. It was a change forced mainly by academic dissatisfaction with the results of earlier geomorphological work, in which long-term landform evolution was the focus of attention, rather than by any stimulus external to the academic world.

Secondly, this change within geomorphology was accompanied by the development of more precise techniques for mapping the landscape, monitoring change and rigorous analysis, so that geomorphological advice could be offered in a form that was both intelligible and acceptable to those outside the subject (especially planners and engineers) who sought it.

Thirdly, the changes coincided with a resurgence of international interest in environmental problems and terrain resource evaluation. As the colonial era closed, newly independent countries sought details of their natural environment as a basis for planning and development, in much the same way as countries such as Australia and the USSR had attempted in previous decades. Land was a major focus of interest in these appraisals, and geomorphology was perceived to have an important contribution to make. Growing international concern with the environment was also generated by pressure groups, perhaps by a growing political awareness of the finite limits to some resources, and especially by a growing number of crises and catastrophes. In the desert realm, for example, desertification became an issue of alarming proportions and of vital concern to many poor countries; and it is one that has generated discussions and research in which geomorphologists have contributed substantially, as is exemplified in Chapter 3.4. Elsewhere, specific disasters enhanced public awareness of geomorphological problems: the Aberfan disaster in 1966, when the failure of a coal waste tip killed 115 people; the permafrost problems associated with the trans-Alaska oil pipeline in the late 1960s; the Yungay debris avalanche in Peru that killed 25 000 in 1970.

The response to these concerns was profound, and it was often accompanied by legislation. For instance, after the passing of the National Environmental Policy Act by the US Congress in 1969, the need for 'environmental impact assessments' prior to development became widely accepted in the United States and many other countries, and such assessments commonly required a geomorphological contribution. In the UK, the newly established Department of the Environment did not propose similar legislation, but it does now frequently seek geomorphological advice, for example in the context of highway planning, aggregate appraisal and slope stability problems.

Finally, and perhaps most significantly, not only has the number of geomorphologists grown in the 1960s and 1970s (associated especially with expansion of geography in British universities and with the development of engineering geology in the United States), but an increasing number began more deliberately to try to serve the needs of environmental managers. Their efforts enhanced the contacts between geomorphologists and environmental managers, and with scientists in cognate fields in the context of common problem solving, and thus provided a very valuable stimulus to, and gave a higher public profile to, geomorphological research. As a result, geomorphologists have persuaded with increasing success a wide range of planners and engineers – most of whom are hard task masters – that they can contribute effectively to the solutions of their problems, and often cheaply and at relatively short notice. Although applied geomorphological work has become technically much more sophisticated and more expensive in recent years, it is still often relatively cheap by comparison with, for example, some types of geotechnical surveys; also, low-technology research based on low-cost field mapping and the analysis of available data is seen as being efficient, quick and adequate in many developing countries with limited funds for environmental appraisals.

Not all evidence of recent success is obvious. Certainly there are now several books that draw heavily on the experience, one company in the UK now provides a professional service, and a small but rising number of geomorphologists is finding its way into engineering and planning companies, consultancies and public agencies. But the main, yet largely invisible success is in the number of organisations concerned with environmental management that have actually sought and used geomorphological advice in recent years. Of these groups, those concerned with planning, regional development and engineering are the most important. Many institutions in the UK currently employ geomorphologists as do government departments, international agencies and foreign governments.

THE WORK OF APPLIED GEOMORPHOLOGISTS

The needs of clients

Most geomorphological work for environmental management serves the requirements of clients who are almost certainly not geomorphologists. The research problems are posed by the clients, and the answers must be provided in a form that the clients can understand. As a result, most such applied geomorphological research – while it may be innovative in terms of methods and may ultimately feed back into geomorphological theory – must be cast in a form that is imposed from outside. Thus, while the work is undertaken using geomorphological methods, it normally has to be translated for clients and critically tested by them. This process of translation and testing is fundamental, and one of the great attractions of applied work.

The range of clients is now extensive. In the context of planning (Fig. 4.1.2), there may be a need for geomorphological advice at different scales before, during and after planning procedures. Each scale has its own range of management agencies, and each poses distinctive questions arising from its particular perspectives and responsibilities. At international and national scales, governments and supranational bodies commonly seek exploratory or reconnaissance surveys, resource inventories and systematic mapping at a small scale. FAO's recent use of Landsat imagery to assess the global pattern of soil degradation, a study that involved geomorphologists, is a good example. At a regional scale, data are often required to help select sites for particular purposes within a range of possibilities, and to evaluate relative terrain qualities. The appraisal of route corridors prior to road construction, provides a good illustration where geomorphologists have worked successfully in recent years.

Locally, the appraisal of hazards and resources is normally required in the context of major developments of towns, and industrial and agricultural projects. For example, as part of a survey of the environs of a town in Saudi Arabia, a geomorphological map of the area was produced and then translated in terms of the potential of different types of desert surface to supply sand and dust to the urban area. The resulting relative hazard maps – qualitative, approximate but readily understood – served as one part of the assessment of the aeolian hazards threatening the expanding settlement. Finally, site planning requires evaluation of such specific problems as slope failure and controls on land engineering. The work of numerous engineering geomorphologists on landslides affecting property in Los Angeles provides one example, and the geomorphological appraisal of reservoir sites in the context of sediment yield and reservoir longevity provides another. In such studies, both site conditions *and* regional context of sites tend to be

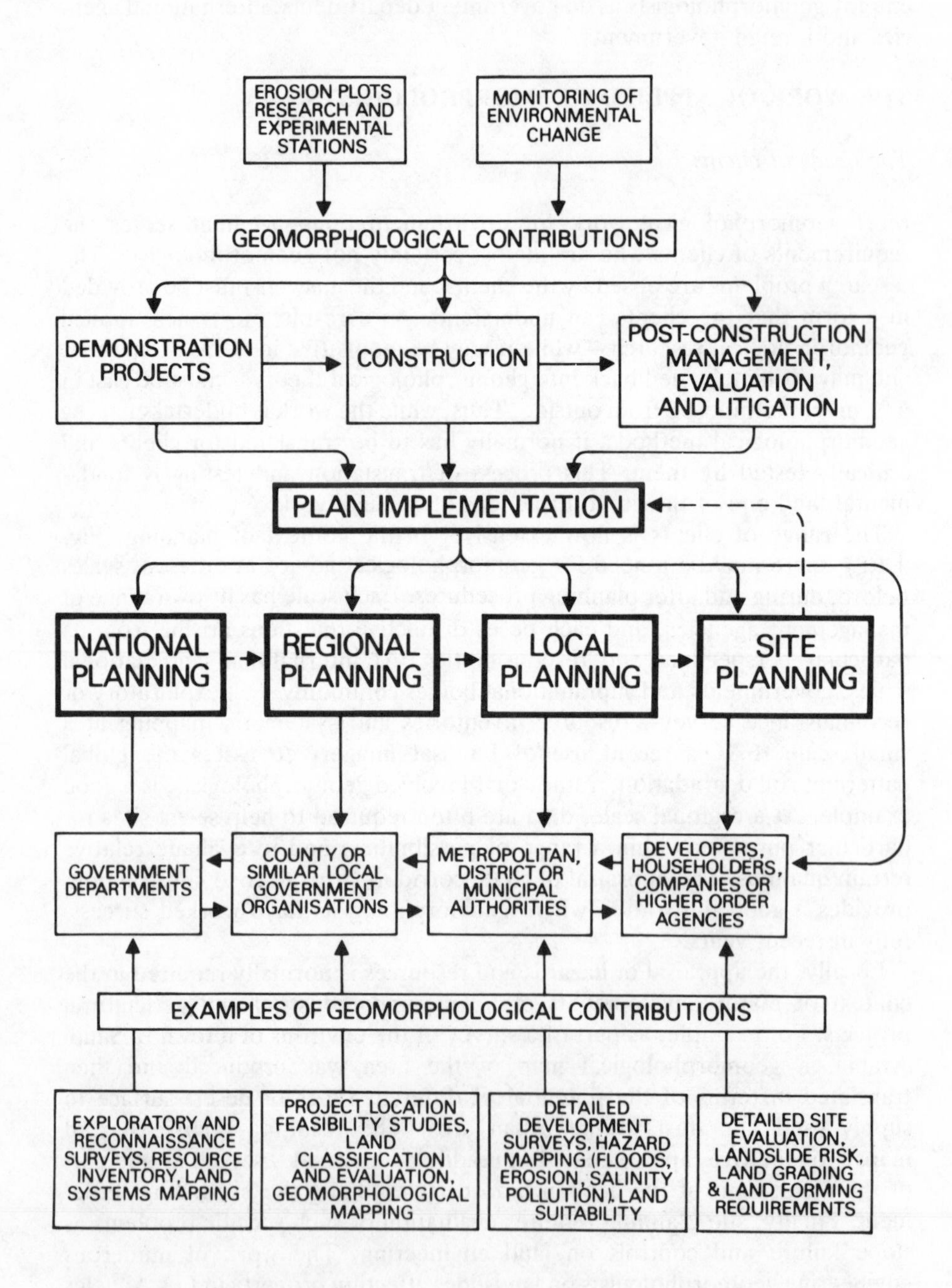

FIGURE 4.1.2 *Relations between geomorphology, the hierarchy of management agencies and scales of planning*

important, and it is the latter in which geomorphological assistance is often most effective.

During the phase of development, most geomorphological queries are raised by consultant engineers or site contractors, usually when unexpected problems are encountered. Of numerous examples, the infamous case of the Sevenoaks (Kent) bypass – in which excavation for the road led to the reactivation of a 'fossil' solifluction lobe – is one of the best known. After construction is complete, the evaluation of policies and planning decisions may be required, and contributions to litigation may be sought, especially following damage caused by floods, storms and other hazardous events; again, for example, geomorphological expertise has been widely used for these purposes in Los Angeles.

The relations shown in Figure 4.1.2 are highly generalised. A specific example (Fig. 4.1.3) reveals how problems, and responsibilities for them, can become disseminated through a hierarchy of management agencies, often in a most complex way. In the case of slope failure, flooding and sediment hazards in Los Angeles, the range of agencies is very wide and the number of agencies is extremely large: some have an interest in survey, planning, assessment and insurance. Most of these agencies have a need for geomorphological advice: some of them employ engineering geologists; others obtain their requirements through contracted research.

Within Figure 4.1.3, each relationship described by a blob-in-a-box provides opportunities for research on the relationship itself and for a geomorphological contribution to it. For example, Figure 4.1.4 develops this relationship with reference to geomorphological objectives and responses in Los Angeles County.

The geomorphologist's skills

Requests for geomorphological assistance in environmental management usually arise from a need to appraise the nature – and especially the distribution and changes over time – of hazards and resources. The most common requirements are to map a landscape, or selected attributes of it, and/or to monitor the nature and causes of change. It is valuable to emphasise some of the analytical approaches the geomorphologist brings to these critical tasks of mapping and monitoring, because they arise directly from geomorphological training, and they involve skills that remain a part of most geography curricula.

Most importantly, the skill of landscape interpretation in the field undoubtedly underlies the success of much recent geomorphological work in environmental management. It is not a new skill, and it is not one that the philosophical changes in geography during the 1960s particularly encouraged; indeed it is a traditional skill that may have been weakened within the subject by the fact that so many geomorphologists became preoccupied with the minutiae of contemporary processes to the exclusion of a broader view of

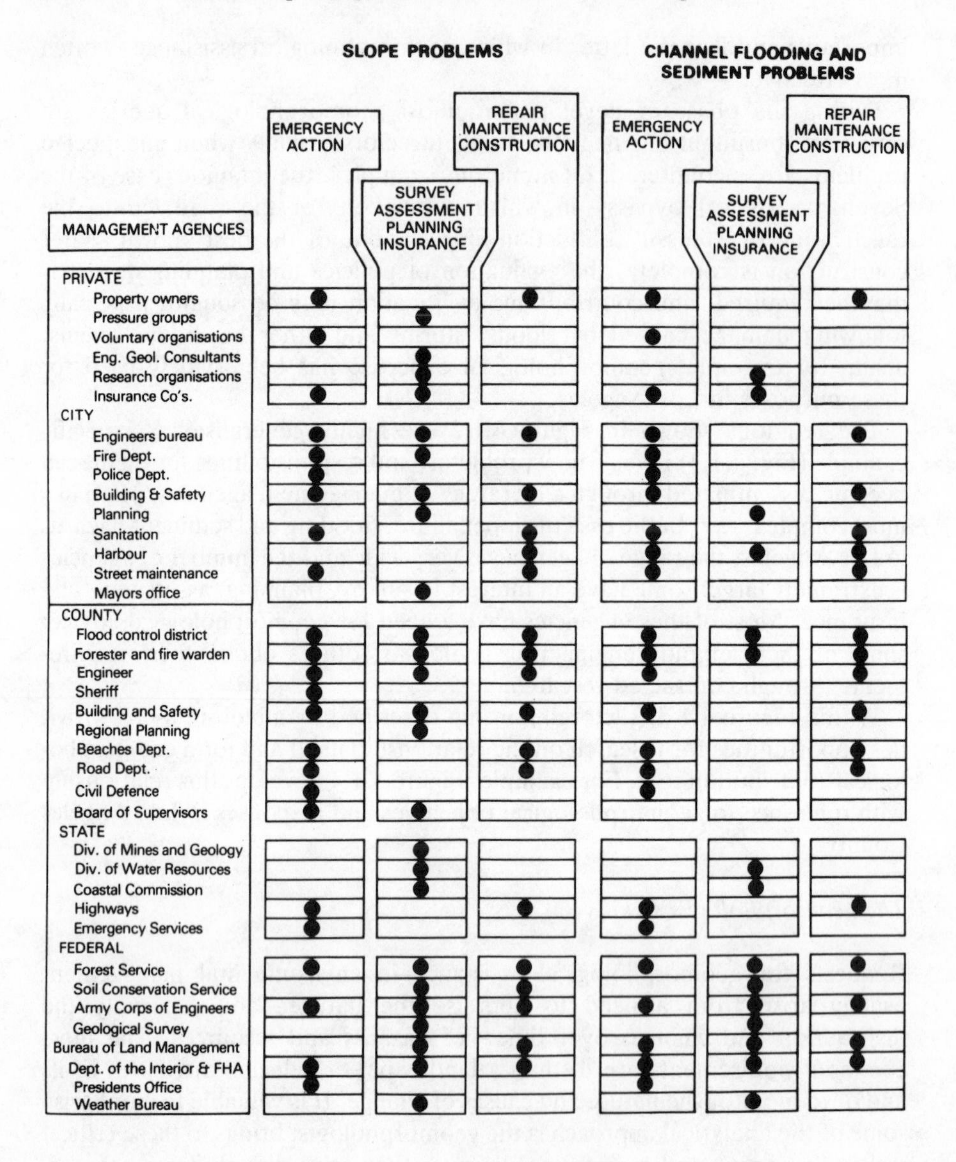

FIGURE 4.1.3 *Some relations between the hierarchy of management agencies and slope, channel-flooding and sediment problems in Los Angeles County*

MAJOR OBJECTIVES	METHOD	PRINCIPAL ADOPTING AGENCIES
AVOIDANCE	I. Control of location, timing and supervision of development	
	A. Land use restrictions	CLABC CBL UBC
	B. Geology / soil surveys	CLABC CBL UBC
	C. Seasonal limitations to slope development	CLABC CBL
	D. Relocation or avoidance of failed sites	CLABC CBL *
REDUCE SHEAR STRESS	II. Control of cut and fill	
	A. Limit / reduce slope inclination	CLABC CBL UBC
	B. Limit / reduce slope length	CLABC CBL UBC
	C. Remove unstable material	*
	D. Expansive soil provisions	CBL
REDUCE SHEAR STRESS AND INCREASE SHEAR RESISTANCE	III. Improve drainage	
	A. Surface	
	1. terrace drains	CLABC CBL UBC
	2. drains	CLABC CBL UBC
	3. others	UBC
	B. Sub-surface	
	1. drains	UBC
	2. drain wells	
	C. Control irrigation	CLABC
INCREASE SHEAR RESISTANCE	IV. Retaining structures	
	A. Buttress at slope foot	CLABC *
	B. Cribs or retaining walls	UBC *
	C. Piling, tie-rodding or other foundation engineering controls	CLABC *
CHIEFLY INCREASE SHEAR RESISTANCE	V. Protect surface	
	A. Control vegetation cover	CLABC CBL UBC
	B. Harden surface (e.g. concrete cover)	CLABC *
	VI. Compaction	
	A. Control fill compaction	CLABC CBL UBC

CLABC City of Los Angeles Building Code
CBL Los Angeles County Building Laws
UBC Uniform Building Code
* methods not often incorporated in codes

FIGURE 4.1.4 *The major geomorphological objectives of building code legislation, methods adopted, and the principal building codes adopting the methods in Los Angeles County*

specific landscapes and their unique historical antecedents. It is a highly prized skill in environmental management, with its requirements of four-dimensional thinking at different scales and a great deal of often strenuous field tramping, especially because it is practised by so few. At one level it allows the accurate recording of geomorphological data in the form of maps that can often serve as the basis for a whole range of environmental surveys. Geomorphological mapping can provide a good basis for such surveys because landforms comprehensively cover the landscape (unlike, say, vegetation). Landforms are also capable of hierarchical classification (e.g. elements, facets, land systems, etc.) so that data can be recorded at the scale appropriate for the level of the management hierarchy to which it is directed. In addition, geomorphological patterns are often related in a predictable way to other environmental variables, such as soils. Table 4.1.1, which is based on a recent survey by geomorphologists for the Natural Environment Research Council, lists some recent examples of applications by British geomorphologists of geomorphological mapping in hazard assessment and resource evaluation.

TABLE 4.1.1 *Selected examples of geomorphological mapping applied to hazard assessment and resource evaluation*

I. **HAZARD ASSESSMENT**

Country	*Project*	*Hazard*	*Client*
UK	Taf Vale trunk road	Landsliding	Welsh Office through Rendel, Palmer & Tritton
	Thame–Stevenage bypass	Landsliding Flooding	Howard Humphreys
	S. Wales gas pipeline	Landsliding	Howard Humphreys
Cyprus	Evreton Dam	Siltation	Sir William Halcrow & Partners
Egypt	Suez urban development	Flooding Aggressive saline soils Aggressive groundwater	Sir William Halcrow & Partners
Saudi Arabia	Urban development	Blown sand & dust	Sir William Halcrow & Partners
	Steelworks site	Flooding & bank erosion	British Steel Corporation
	Irrigation scheme		Sir William Halcrow & Partners
Bahrain	Regional development	Aggressive saline soils	Ministry of Works

TABLE 4.1.1 *contd.*

Country	*Project*	*Hazard*	*Client*
Dubai	Airport sites	Sand dune migration Aggressive saline soils	Halcrow Middle East
Ra's Al Khaimah	Regional development	Flooding Aggressive saline soils	Halcrow Middle East
Nigeria	Road development	Flooding	Engineering Geology Ltd
Nepal	Route assessment Road design Irrigation scheme	Landslides Flooding Landslides Sediment transport	Rendel, Palmer & Tritton Goode & Partners
Sri Lanka	Kotmale Reservoir	Landslides Siltation Soil erosion	Central Engineering Consultancy Bureau
Papua	Rouna 4 HEP	Landslides	Sir William Halcrow & Partners

II. **RESOURCE EVALUATION**

Country	*Project*	*Hazard*	*Client*
Libya	Harbour construction	Rock	Hyundai (S. Korea)
Bahrain	Regional development	Construction materials Soils for agriculture	Ministry of Works
Dubai	Coastal zone development	Construction materials	Wimpey
Ra's Al Khaimah	Regional development	Aggregate sources	Halcrow Middle East
Nigeria	Road construction	Borrow material	Engineering Geology Ltd
Brazil	Diamond mining	Alluvial diamonds	Rio Tinto-Zinc

The skill of field interpretation is also fundamental to the analysis, monitoring and prediction of environmental change. Because environmental managers often require knowledge of past and present changes and prediction of future trends, and need to know quickly without giving any opportunity for long-term monitoring, it is frequently necessary to deduce change and process from the field interpretation of the surrogate evidence of landform and surface materials. Thus, a field survey of the geomorphology of the area around the city of Suez in Egypt provided the data necessary for the preparation of a relative flood hazard map – despite the fact that no runoff data were available (Fig. 4.1.5). The estimated potential flood hazard was based on an assessment of the alluvial fans, their degree of entrenchment, the relative age and stability of the stone pavement surfaces on them, the size of catchment areas, and evidence of flow types and competence revealed by sediments from recent flow events.

The field skill and experience is also relevant to the complementary skill of indirect monitoring of change and the mapping of landscape patterns from aerial photographs and satellite photography. Here, geomorphologists have been in the forefront of those seeking to exploit the potential of remote sensing technology for environmental and other applied purposes because landforms are often clearly displayed on remote sensing imagery, and successive images of the same locations can reveal changes not easily monitored on the ground.

The applied geomorphologist also often calls upon other skills that now figure in the training of geographers. Direct measurement and monitoring of processes in the field have been developed substantially in recent decades, especially in drainage basins, along coasts and in aeolian environments. Closely allied to these skills are those related to field measurements of variables of predictive value in studying process (such as salinity in the context of salt weathering or soil properties in the context of soil erosion). Indirect monitoring of change by the analysis of successive remote sensing images finds an important counterpart in the growing efforts to monitor and explain environmental change through the use of historical archives – efforts that require skills closely allied to those of the historical geographer.

The ability to use hardware models in the analysis of processes and complex environmental situations under controlled laboratory conditions requires another set of skills of great applicability, as shown by recent studies in flumes, wind tunnels and climatic cabinets. Finally, the use of computers is clearly fundamental, for instance in creating digital simulation models, in analysing digital tapes of satellite data, and in preparing maps and other derivatives from field information.

Beyond contractual contributions

Most geomorphological work in environmental management is undoubtedly sponsored by clients with specific objectives and most of it uses the techniques

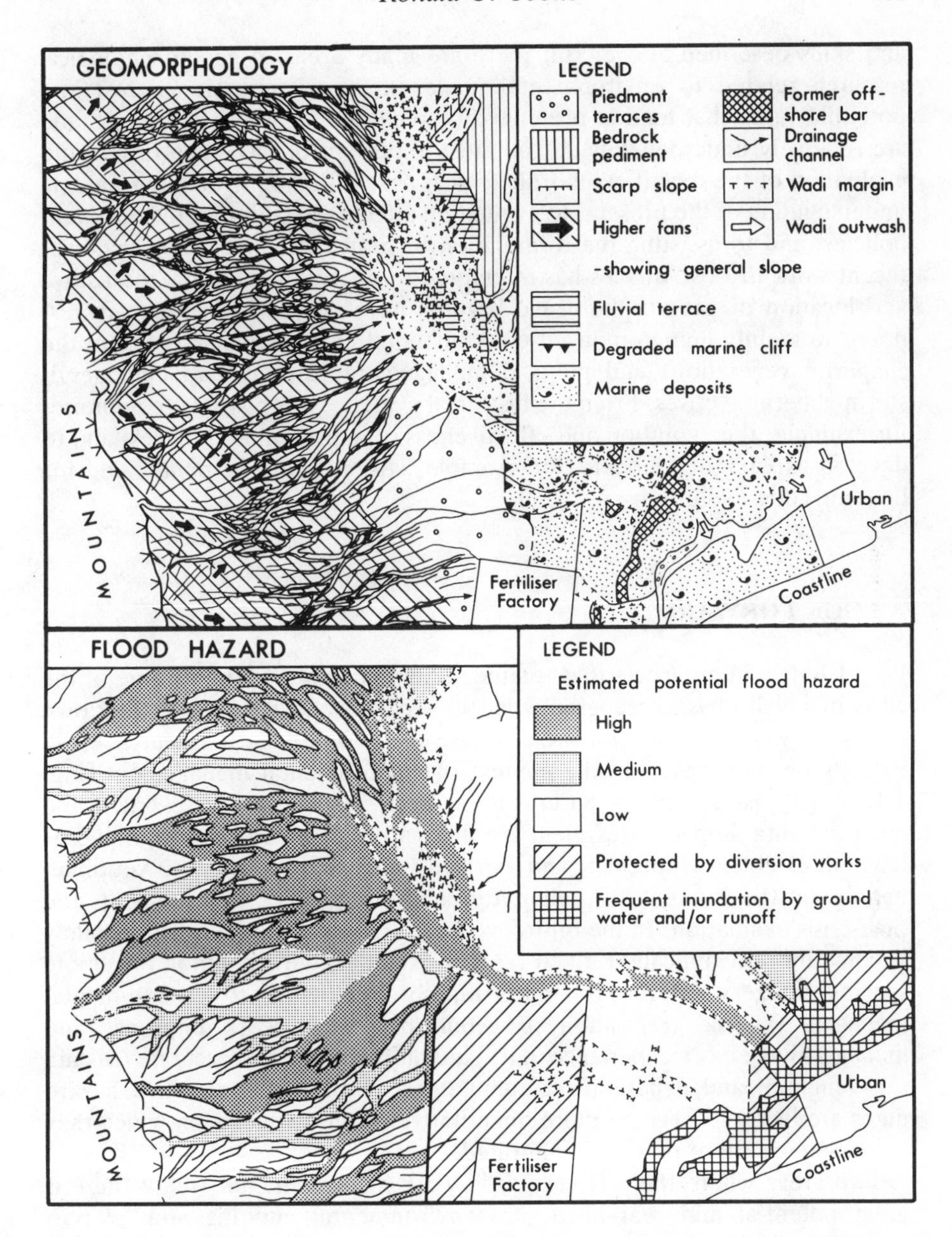

FIGURE 4.1.5 *An example of a relative hazard map (lower diagram) based on the qualitative translation of data derived from field mapping of sediments and landforms (upper diagram)*
These maps formed part of flood-hazard study of the Suez City region undertaken by the author, Denys Brunsden, Peter Bush, John Doornkamp and David Jones for Sir William Halcrow & Partners.

and skills described above. But there are many areas of geomorphological research related to environmental management of intrinsic interest and potential value that tend at present to be regarded as purely 'academic' and are relatively underdeveloped. The most important of these concerns is the evaluation of the recent geomorphological past with a view mainly to aiding understanding of the present, to evaluating effectiveness of past management policies, and to assisting the formulation of future strategies. For example, recent work in Los Angeles has attempted to show in detail how the nature and location of geomorphological hazards have changed over time in response to mainly human-induced environmental changes (such as fire in the chaparral vegetation) and political management responses to ephemeral, storm-generated crises. From the historical evidence it is possible to appraise, for example, the evolution and effectiveness of management policies; and to develop explanatory models, for example, storm-hazard response, and for building code evolution.

A LOOK FORWARD

David Jones, in an article that reminded human geographers of some of the ways in which physical geography is still very relevant to some of their own concerns, identified several areas where geomorphological research could usefully be undertaken in the context of environmental management (Fig. 4.1.6). The assessment of natural hazard effects, resource assessment and environmental impact assessment are three areas in which geomorphologists have an established interest, and work in these areas is likely to continue, mainly in the context of policy formulation, project development and post-crisis evaluation. In the future, geomorphological methods will certainly become more technically refined (as, for example, the resolution of remote sensing imagery improves), more rapidly intelligible to environmental managers (as the translation of geomorphological work improves and managers have more experience of it), and more precise and accurate (as data sets lengthen and experience accumulates). In particular, relative hazard maps are likely to become more quantitative as the variables on which they are based are more precisely calibrated.

Two areas where there is very little work at present, but where there is great potential and real need, are environmental auditing and 'ex-post assessments'. The former involve the monitoring of environmental change. Geomorphological monitoring exercises contribute to this work. Indeed, many such exercises have now been carried out for more than two decades and are beginning to yield data sets of real predictive value. Unfortunately, the exercises tend to be both expensive and time-consuming, and for them to be successful often requires a very high degree of independent, sustained motivation, or rather rarely forthcoming sponsorship from an interested environmental agency. Also, the monitoring usually needs to involve more

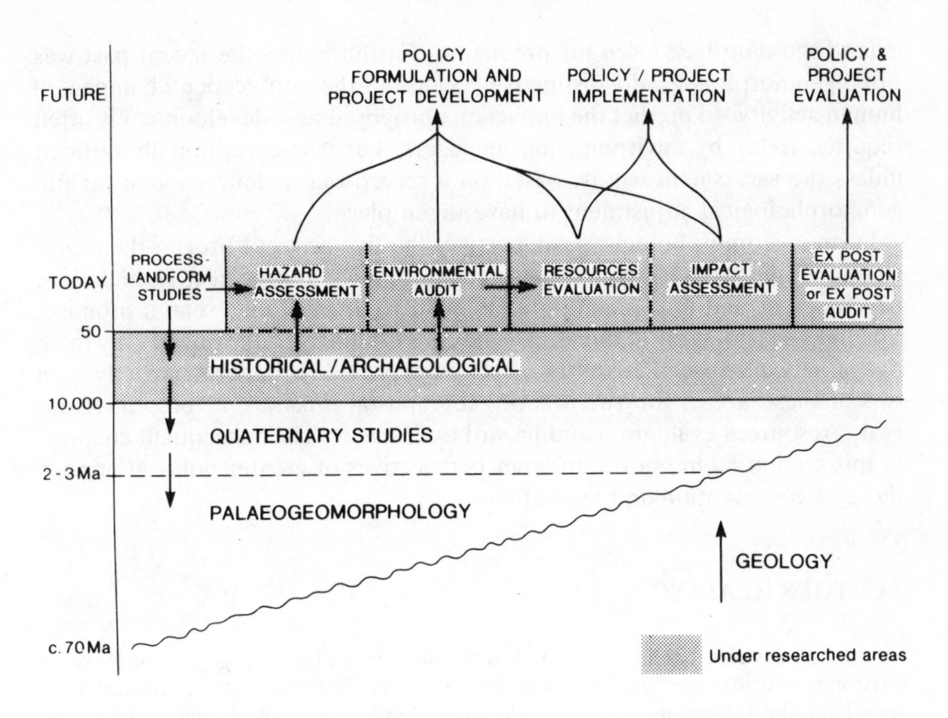

FIGURE 4.1.6 *Under-researched areas in geomorphology with potential application to environmental management (after D. K. C. Jones, 1983)*

than measurements of the landforms, materials and processes – it often requires cognate data on, for example, land use and land use practices, public policy and land ownership, which is often not easily acquired.

Ex-post assessments are designed to evaluate the success of earlier predictions undertaken before developments were initiated, and the effectiveness of management policies. Here the research questions may pose a political or even an academic threat, as with such questions as 'How accurate were Professor *X*'s predictions of channel changes following the construction of the reservoir *Y*?', or 'How successful has the US policy for reducing soil erosion on agricultural land been?'. Perhaps this explains why funds for the assessments are scarce. Nevertheless, such studies are essential as a means of improving management practices. Independent research scholars in universities may well see this as a very significant area of research in the future.

There is also an important future for research into the recent geomorphological past. The recent past provides the stage upon which present day processes act: much contemporary monitoring is of landforms that were initiated or profoundly modified by unique environmental changes over the past 150 or so years. Many monitoring studies are undertaken over too short periods to allow sensible extrapolation of process rates and to permit realistic prediction for planning purposes; an analysis of historical records may help to extend the data base used for prediction. Furthermore, the recent past was often geomorphologically distinctive because of the unprecedented impact of human activity: to predict the impact of a proposed new development is often required today by environmental managers, but it is exceptionally difficult unless the assessment can be based on a record that is long enough for full geomorphological adjustment to have taken place.

From a geomorphologist's perspective, the five areas of process–landform studies of most relevance to environmental policy and project formulation, development and implementation (Fig. 4.1.6) are full of research promise. The links are not wishful thinking – they are established; they need only to be strengthened by use. The following chapters provide broader perspectives on two of these areas, the two where geographical links are at present strongest – resources evaluation and hazard assessment. The subsequent chapters in this section address even broader perspectives of greater political import, those of conservation and legislation.

FURTHER READING

There are several books concerned with the application of geomorphology in environmental management. The first represents the first effort at a systematic view seen from the standpoint of geomorphologists (and is available in paperback). The second complements this with a series of extended essays on selected themes, whilst the third provides a detailed examination of the application of geomorphological surveys for environmental development. Finally, a series of case studies derived from a symposium yield the fourth book:

Cooke R. U. and Doornkamp J. C. (1974) *Geomorphology in Environmental Management* (Oxford: Oxford University Press).
Hails J. R. (ed.) (1977) *Applied Geomorphology* (Amsterdam: Elsevier).
Verstappen H. Th. (1983) *Applied Geomorphology* (Amsterdam: Elsevier).
Craig R. G. and Craft J. L. (1982) *Applied Geomorphology* (London: Allen & Unwin).

A number of volumes develop particular aspects of the relations between geomorphology and environmental management. The first two concentrate on geomorphological contributions to recent urban growth in deserts, and to the detailed management of hazards in a rapidly urbanising metropolitan county. The last provides a rural counterpart in a collection of papers on soil erosion by wind and water:

Cooke R. U. *et al.* (1982) *Urban Geomorphology in Drylands* (Oxford: Oxford University Press).
Cooke R. U. (1984) *Geomorphological Hazards in Los Angeles* (London: Allen & Unwin).
Kirkby M. J. and Morgan R. P. C. (eds) (1980) *Soil Erosion* (Chichester: John Wiley).

Four recent systematic texts of relevance to this chapter follow. Their titles are self-explanatory:

Gardiner V. and Dackombe R. (1983) *Geomorphological Field Manual* (London: Allen & Unwin).

Goudie A. (1981) *Geomorphological Techniques* (London: Allen & Unwin).
Hooke J. M. and Kain R. J. E. (1982) *Historical Change in the Physical Environment* (London: Butterworths).
Verstappen H. Th. (1977) *Remote Sensing in Geomorphology* (Amsterdam: Elsevier).

Focuses for Environmental Management

4.2
Evaluating Environmental Potential

Vincent Gardiner
University of Leicester

Environmental potential has helped to shape the course of history. In Britain, geology and climate provide abundant coal, iron and water which helped stimulate the Industrial Revolution. The demise of Hitler's Reich was accelerated by the lack of key resources within its boundaries, and contemporary geopolitics are greatly influenced by the distribution of oil reserves. The physical environment is a resource which can be used for many purposes; their reconciliation within a spatial framework, as conditioned by social, economic, political, moral and aesthetic constraints, is the challenge faced by environmental management. This necessarily requires evaluations of environmental potential for particular purposes, and their synthesis into an overall management framework.

Society's earliest concerns for environment are evidenced by man's deification of Earth, Air, Fire and Water. In the eighth century AD land classification was carried out in Japan, by Imperial edict. A description of the village of Harima was accompanied by an account of the 'soil', this being classified into nine grades (Fig. 4.2.1). 'Soil' meant land productivity, and its assessment therefore embraced the entire physical environment.

As technology has enabled greater potential for environmental impact, concern for environmental issues has increased with, for example, Environmental Impact Statements now forming part of planning procedures in the United States and elsewhere. Techniques for environmental evaluation have evolved in many disciplines, including geography, environmental sciences, ecology, landscape architecture and planning. In addition, individual aspects of the environment have been evaluated by geologists, climatologists and agronomists, amongst others. Although the spatial viewpoint is implicit throughout, only in geography is it explicit. Evaluation of environmental potential within a spatial framework demands an appreciation of virtally all aspects of geography, as well as cartography, and therefore provides an important integrating element within the discipline, as well as an important application of it.

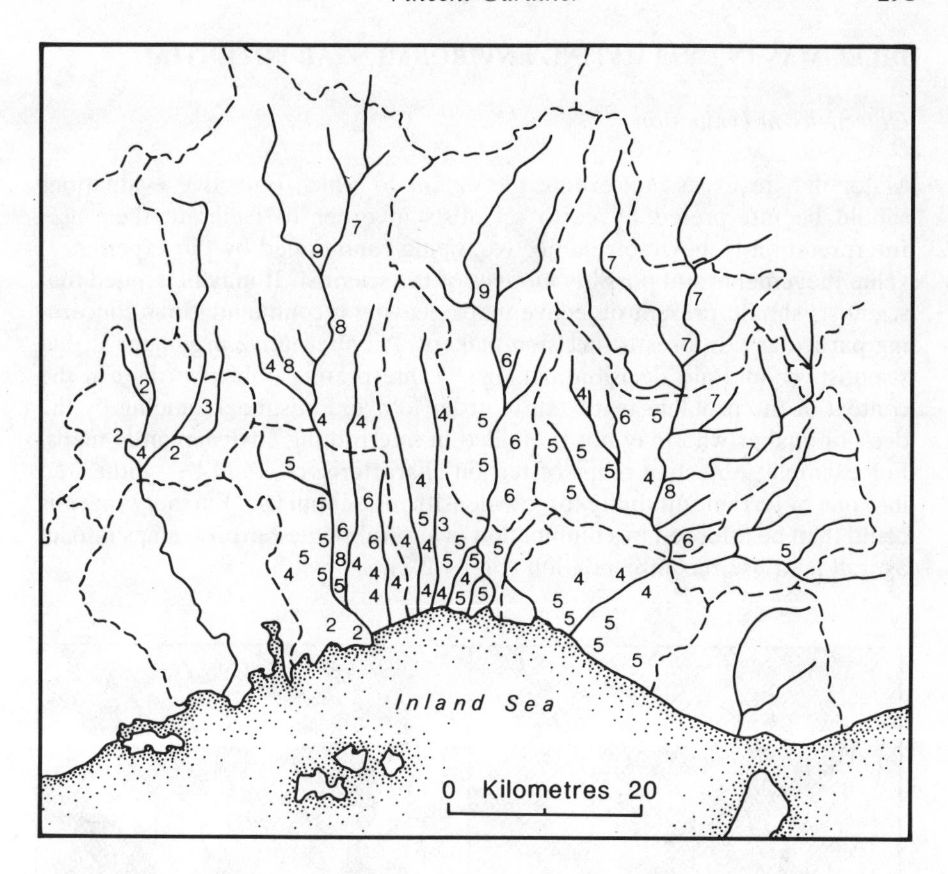

FIGURE 4.2.1 *An early evaluation of environmental potential*
The distribution of productivity in Harima, south-west Japan, in the eighth century. Observations are numbered from 1: 'best of best' through 4: 'best of medium' to 9: 'worst of worst' (from M. M. Yoshino (1975) *Climate in a small area* (Tokyo)).

Maps are the usual means by which evaluations of environmental potential are presented. They are an extraordinarily efficient means of data storage, and also facilitate rapid and effective communication. By comparison of several maps, relationships can be readily identified between aspects of the physical environment or patterns of human activity, and maps showing these can be synthesised. Finally, it is possible to envisage a range of maps becoming progressively more evaluative and less descriptive, thereby allowing particular kinds of maps to be produced for particular purposes and audiences. Effective environmental evaluation therefore demands some acquaintance with cartographic concepts and techniques; it is, however, an essentially geographical task rather than one of environmental survey.

DILEMMAS IN EVALUATING ENVIRONMENTAL POTENTIAL

Objectivity in evaluation

Major debate exists concerning the extent to which objective evaluations should be interpreted by earth scientists in order to facilitate their use. Interpretation is inevitably subjective, being conditioned by the experience, value judgements and possibly motives of the scientist. It may be argued that scientists should present objective maps, leaving recommendations concerning particular purposes to decision-makers. An alternative argument is that scientists should aid decision-making, by interpreting objective data in the context of the problem involved, in order to avoid misunderstanding by the decision-maker who may not be skilled in interpreting environmental maps. For example, objective maps of terrain characteristics could be synthesised into one of terrain suitability for low-density residential use. Further synthesis could then be afforded by combination of several single-purpose maps into an overall land-use recommendation (Fig. 4.2.2).

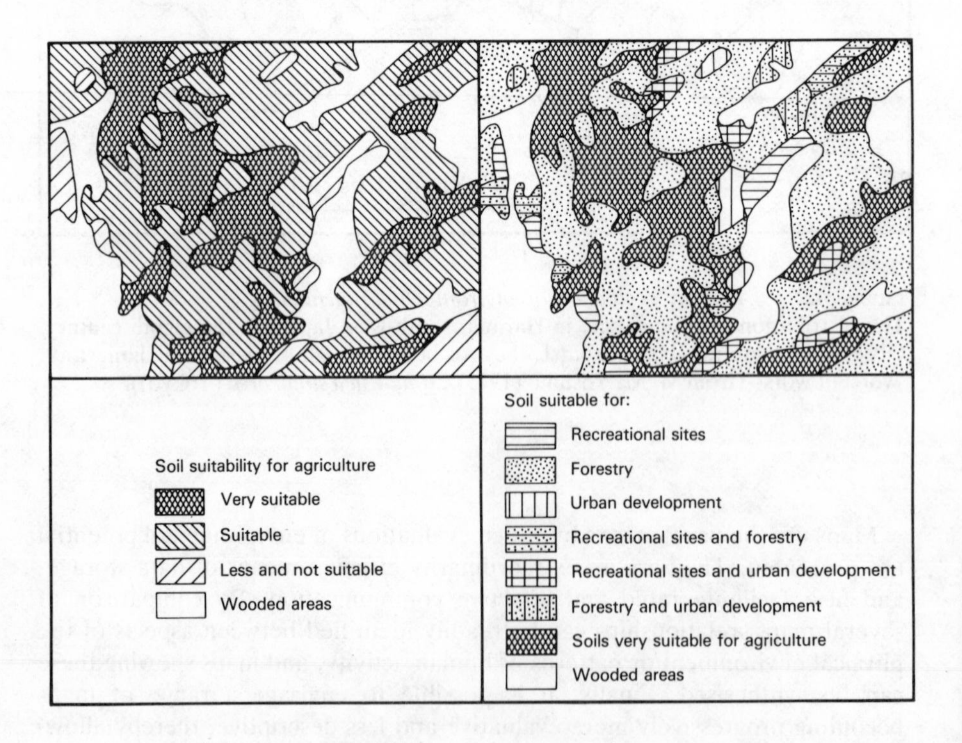

FIGURE 4.2.2 *A single-purpose evaluation and a multipurpose evaluation formed by synthesis (from J. C. F. M. Haans and G. F. W. Westerveld (1970)* Geoderma 4, *279–309)*

Bases for evaluation

The number and kind of attributes embodied in land evaluations are also debated. A single terrain characteristic might be all important. For example, building foundation conditions are dependent upon geological characteristics of the terrain. However, environmental potential is commonly dependent upon many factors; the logical extension of this is that evaluation is best achieved by considering the fullest range of environmental attributes, preferably by an integrative method.

A further dilemma concerns criteria by which evaluations might be judged. Economic values are the most obvious, but it is not always possible to cost social and aesthetic issues. For instance, when evaluating proposed reservoir sites it is impossible to cost in monetary terms the landscape improvement (or despoilation) caused.

Reconciliation of information

Finally, conceptual and technical difficulties occur because of the necessity to reconcile information on various aspects of the environment, this often having different temporal and spatial scales and resolutions. Thus slope angle data may be ordinal scale point measurements, soil type a nominal scale areal classification. Static characteristics such as these may have to be combined, and then reconciled with dynamic quantities such as precipitation and slope erosivity in evaluating agricultural potential.

METHODS FOR EVALUATING ENVIRONMENTAL COMPONENTS

Methods for evaluating environmental potential are briefly examined below according to the major terrain characteristics embraced by particular methods. A progression is followed from characteristics for the evaluation of which many techniques have been developed, such as geology and relief, to those which in relative terms have received much less attention, such as processes.

Geology

Knowledge of the spatial distribution of geological phenomena may be essential in choosing between alternative routes for roads or sites for buildings and dams, or in exploiting mineral deposits. However, earth scientists are not always included in bodies responsible for making decisions, and reference is often made to geological maps without necessarily a full understanding of the bases upon which the map rests, or its accuracy. For example, 'boulder clay' is a genetic rather than lithological term; most so-called boulder clays contain few boulders, and many are not clays!

The apparent complexity of geological maps arises because they show two three-dimensional distributions; the topography, usually by contours, and the geological structure, shown implicitly by the outcrop pattern. Geological maps have severe limitations arising from field survey. At the field mapping scale of perhaps 1:10 000 it is impossible to show without exaggeration very small features. However, valuable mineral veins may be only a few metres across. Linear features can be exaggerated in thickness but small areas cannot be shown without distorting the map. Geological field survey is also a partially subjective procedure, being limited by available exposure and the extent to which secondary evidence of topography, vegetation and soil type may be used.

Geological maps are based upon stratigraphy; their purpose is to depict the relative ages of rocks. Although litho-stratigraphic units may be used by the field surveyor, lithology is a secondary consideration on the map. This renders geological maps difficult to use in evaluating environmental potential, where lithology irrespective of age is the main property of interest. Geological maps may be recast into lithological maps or indeed evaluative derived maps in terms of any criterion, such as rock strength, thickness of a particular formation or bed, or particular aspects of geology such as structural or geochemical characteristics.

This process of abstraction may be extended to evaluation of terrain in terms of geological suitability for a particular purpose (Fig. 4.2.3). For example, mineral assessment maps evaluate volumes of economically valuable rocks, as a basis for development of mineral extraction, and engineering geological maps abstract information on the physical properties of rocks, rather than their relative ages.

Application of geological evaluation in Grant County, Indiana, USA, is illustrated by Figure 4.2.3. From the basic inventory maps of solid and drift geology, maps of potential mineral resources and areas with impermeable surface materials or other drainage problems were extracted. Evaluative maps were then produced, assessing the terrain in terms of suitability for aspects of urban development, such as septic systems, sanitary landfills and storage lagoons.

Soil

It is essential to assess the agricultural potential of soils; in addition soil properties can be of great significance in many non-agricultural applications. Like geological maps, soil maps are prepared by detailed field survey followed by compilation and generalisation to a reduced scale. The soil surveyor, too, has to rely on indirect information such as vegetation type in completing the map. The most severe difficulty surrounding preparation and use of soil maps is that concerning the nature of soil. Soils do not occur as discrete parcels which can be confidently identified and delimited. Soil

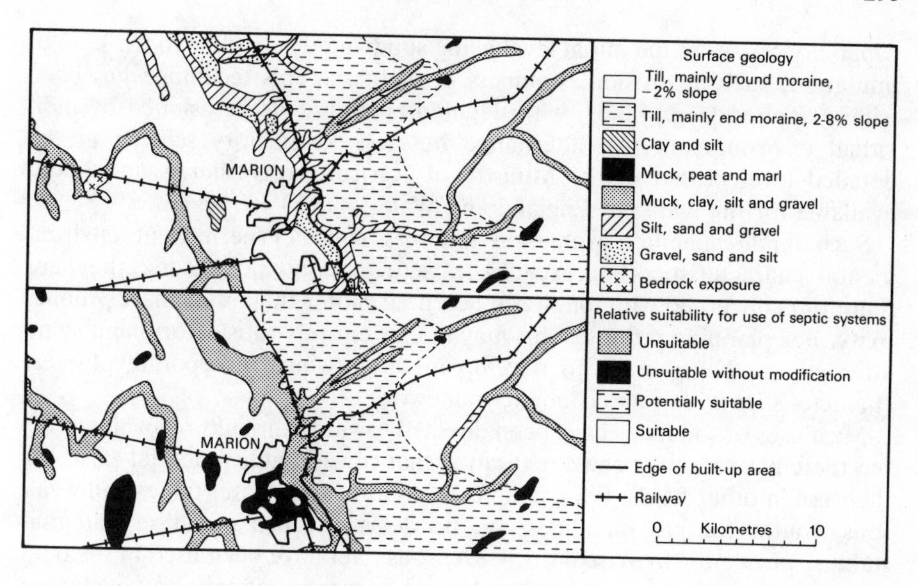

FIGURE 4.2.3 *Land evaluation on the basis of geology, for residential development* (From E. J. Hartke (1982) Environmental Geology of Grant County Indiana, Special Report 23. Indiana Department of Natural Resources, Geological Survey.)

properties change gradually horizontally and vertically, some change through time, and change in one property is not necessarily accompanied by changes in others. Soil maps are therefore based upon a conceptual model of how soils occur in the landscape, with real soil areas defined by mapping units such as soil series; these may contain considerable areas of 'impurities' in relation to the conceptual model upon which the soil mapping is based.

Soil surveys are expensive, and even for developed countries only a limited coverage is available at other than a reconnaissance level. A rather restricting drawback of soil maps is that general purpose maps have very limited capabilities for prediction of certain soil properties. Derivative and evaluative soil maps are thus of great value in particular applications. Again, a sequence of evaluations exists, from basic soil maps through maps extracting single aspects of soil to maps attempting soil-based land evaluation for various purposes. Derivative maps of single elements, such as soil thickness or calcium carbonate content, may be prepared by using data from the memoirs accompanying general purpose maps. These may be further developed to produce evaluative maps by considering the physical properties of soils in relation to requirements for particular purposes.

Evaluation of agricultural land potential is a basic tool. The erosional problems evident in the USA during the 1930s stimulated development of the US Department of Agriculture Land Capability Classification Scheme. This assessed limitations to land use imposed by permanent terrain characteristics, capability classes at the initial level being subdivided according to the kind of limitation, such as erosion, stoniness or wetness. This technique has been widely applied (Fig. 4.2.4), with detailed amendments occasioned by individual environments. An alternative but less satisfactory scheme at the detailed level is that of the Ministry of Agriculture, Fisheries and Food, available for the whole of England and Wales at 1:63 360 scale.

Such land capability evaluation schemes consider permanent environmental characteristics but exclude transitory economic factors; they are therefore concerned with long-term potential rather than immediate productivity. For planning purposes this may not be entirely satisfactory, and some attempts are being made to develop methods which incorporate physical characteristics within an economic framework.

Most uses of soil maps have been directed towards agricultural evaluations, but there has recently been a realisation that considerable potential exists for their use in other fields. For example, in areas of high water tables and weak soils, routes for gas pipe-lines can be devised which minimise potential stability problems. In Wisconsin, USA, soils data have been used in regional planning. A study established the physical properties of soils and evaluated their spatial distributions for various purposes. These interpretations have been incorporated into local zoning ordinances by the use of zoning districts and land use regulations, and by delimiting special hazard zones.

Relief

Much human activity depends upon the shape of the underlying land. Terrain evaluations in terms of relief can again be viewed as a continuum from objective depictions to evaluations for particular purposes. Many methods of relief depiction exist, including spot heights, contours (Fig. 4.2.5A), hachures, and their developments into layer shading, inclined contours, plastic shading and other more sophisticated devices. Experiments into perceptual processes involved in reading relief maps suggest that there is no single ‘best’ method, different kinds of representation being preferred for different kinds of information extraction task.

Morphological maps (Fig. 4.2.5B) show the shape of the landscape. These were developed from geomorphological work and are based upon recognition of breaks and changes of slope, which may be convex or concave in nature. The slope facets identified should theoretically have uniform slope angles but in practice some variation is permitted. Morphological mapping is an excellent general method for portraying relief, and affords a good basis for further evaluative developments. However, some reservations have been expressed concerning the apparently high degree of subjectivity inherent in it. Morphological maps may be supplemented by slope gradient information, as a partial progression towards further evaluation. Slope maps are very useful evaluation tools in fields such as land capability assessment, forestry and engineering.

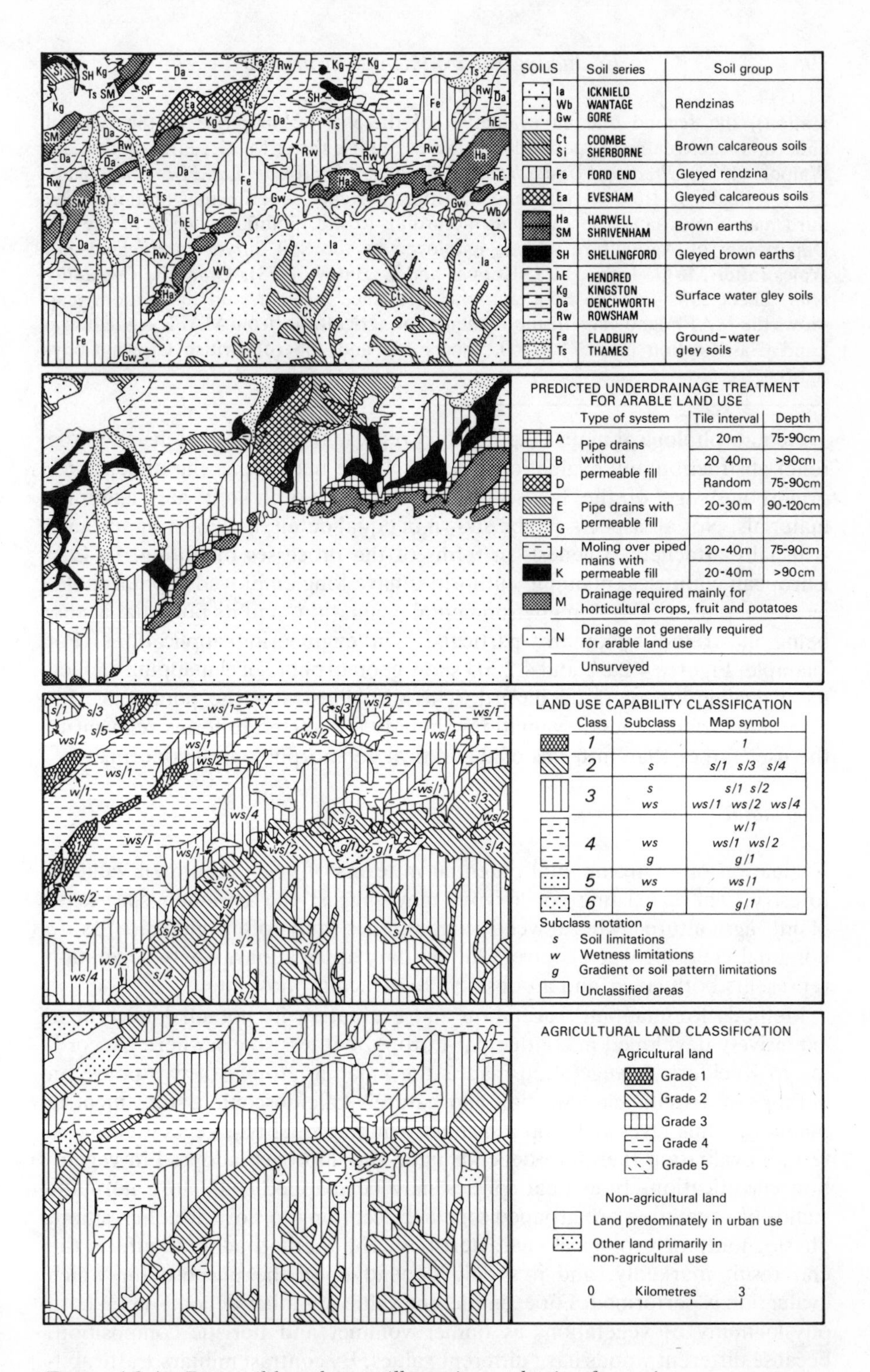

FIGURE 4.2.4 *Four soil-based maps illustrating evaluation for various purposes*
The top map shows the distribution of soil series (after M. G. Jarvis and R. A. Jarvis (1973). *Soil map*. Part of 1:63360 sheet 253. Harpenden: Rothampstead Experimental

Station), the second one an evaluation of these series in terms of the type of underdrainage treatment required for arable land use (after M. G.; Jarivs and R. A. Walpole (1973). *Predicted Underdrainage Treatment for Arable Land Use Map*. Sheet 253 (Abingdon), 1:63360. Harpenden: Rothampstead Experimental Station). The third map shows land use capability according to the modified version of the US Department of Agriculture scheme employed by the Soil Survey of England and Wales (after M. G. Jarivs, L. J. Hooper and T. Batey (1973), *Land Use Capability Map*, Sheet 253. 1:63360. Southampton: Ordnance Survey), and the bottom map shows the MAFF land classification scheme for agriculture (from MAFF Agricultural Land Classification maps, 1:63360, sheets 157 and 158 (c) British Crown Copyright).

Geomorphological maps (Fig. 4.2.5C) represent a further extension in the relief abstraction and evaluation process. These may depict the dimensions, origins and age of the landscape, as well as morphology and landform materials. Not all aspects can be shown in equal detail, and most importance is usually attached to genesis in terms of the processes operating. In East European countries, planning authorities have long employed geomorphological mapping as a tool in environmental evaluation, and its potential is now being illustrated by others, particularly in engineering applications. For example, Figures 4.2.5C and 4.2.5D show geomorphological mapping applied to an evaluation of flood hazard. This also illustrates the point that, for maximum effectiveness, evaluations should be depicted in terms relevant to the user rather than in terms of theoretical geomorphological concepts.

Vegetation

Evaluation of environmental potential in terms of vegetation is of value in forestry, where existing cover-types may suggest the optimum species to plant; agriculture, where weed communities may offer guidance as to potential yields or weed control; and military applications, where vegetation represents both cover and an obstacle to vehicular movement.

Methods for mapping vegetation and evaluating its potential have been extensively developed in continental Europe, but less so elsewhere. According to Kuchler: 'A vegetation map is the meeting ground of two poles: an author's systematic classification and nature's kaleidoscopic arrangement of plants . . . this gives the map a distinctly human quality.' Vegetation-based terrain evaluation therefore depends upon the approach adopted to vegetation classification. Individual species may be mapped, but more normally stands of vegetation are grouped together according to floristic composition, physiognomy or ecological characteristics. The method adopted affects the end result markedly, and it should depend upon the purpose for which evaluation is performed. For example, the value of timber depends upon both physiognomy of vegetation, as timber volume, and floristic composition, because different woods have different values. By contrast military trafficability through woodland is primarily dependent upon vegetation physiognomy, the species of tree being largely irrelevant.

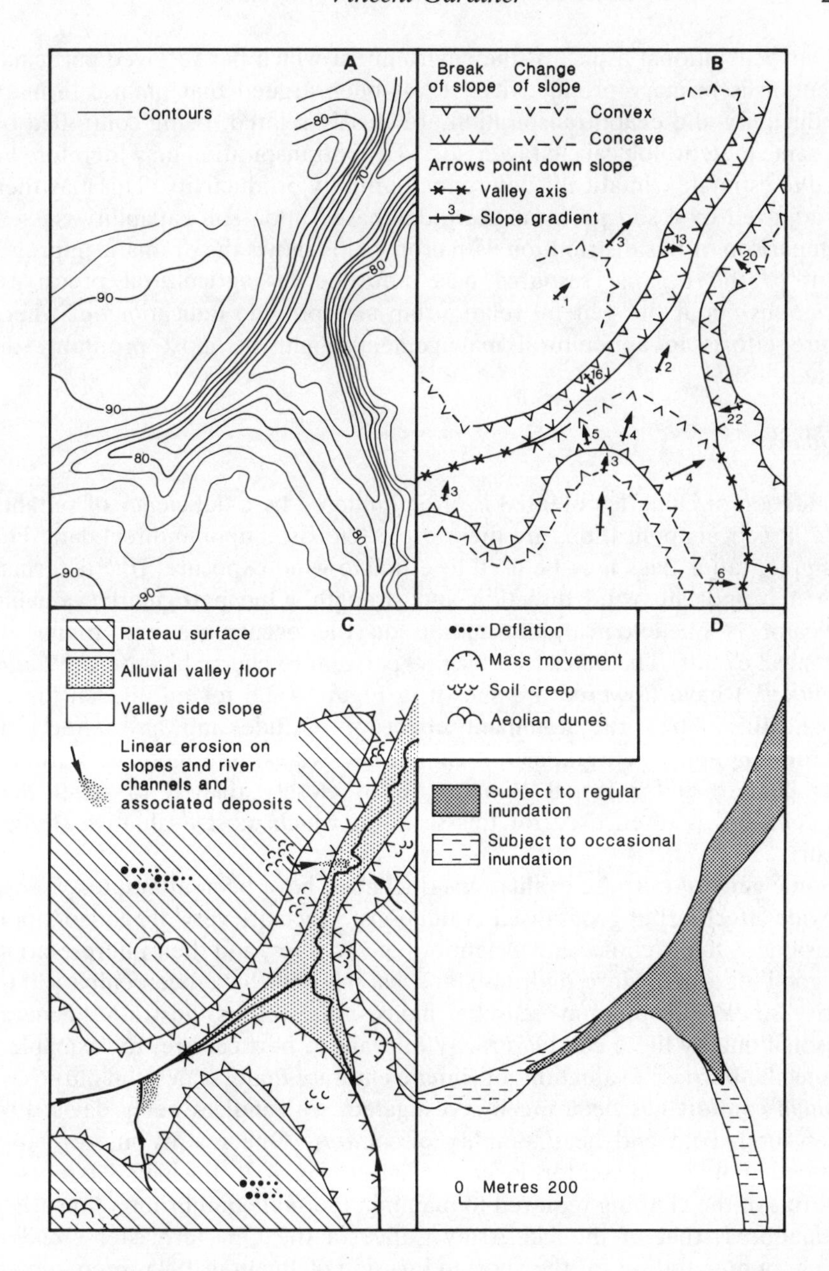

FIGURE 4.2.5 *Morphological and geomorphological representations of an area in the Campine of Belgium*

A Contour map
B Morphological map
C Geomorphological map
D Flood hazard map

One vegetational aspect of the environment which has received particular attention is primary productivity. It has been argued that natural primary productivity and evapotranspiration are closely related, being controlled by the same meteorological characteristics. Evapotranspiration may therefore be used to estimate climatically determined primary productivity. This may then be adjusted by a soil performance index based upon soil capability classes. Comparison of this distribution with crop yields shows the dramatic improvement to the organic resource base afforded by agricultural processes; variations about the general relationship also provide indications of where future efforts in agricultural management might be most profitably directed.

Climate

Evaluation of climatic potential is often hindered by a deficiency of suitable data. Practical applications are therefore often based upon indirect data. For example, tatter flags may be used to estimate wind exposure, tree deformation may indicate wind direction and strength. One particularly valuable indicator is phenological information on the occurrence and timing of botanical events. The dates by which 80 per cent of cherry blossoms (*Prunus subhirtella*) have flowered are shown in Figure 4.2.6 for an alluvial fan in Japan; this shows the dominant effect of altitude and also effects of microtopography. Variations in climate at scales intermediate between microclimate and regional climate are so closely related to relief that 'topoclimate' is often used for this scale, at which most evaluative studies occur.

Some general purpose evaluations of climate have been attempted. These provide effective but generalised evaluations, which are very dependent upon decisions as to the choice and weightings of variables, and their interpretation as 'good' or 'bad'. More rigorous statistical methods have been employed to derive so-called 'optimum' seasonal indices. As well as illustrating general seasonal quality these can be directly relevant to horticulture, for example.

Specific-purpose evaluations of climate embrace many scales and purposes. Human comfort has been much investigated, with indices being devised to assess both cold and heat/humidity discomfort. One readily interpretable expression of human comfort is the 'clo' index (Fig. 4.2.7), which is expressed in terms of the clothing required to maintain thermal equilibrium. An earlier evaluation is that of the US Army Office of the Quartermaster General which, in preparation for the aborted invasion of Japan in 1945, prepared an item-by-item and month-by-month almanac of military clothing requirements.

Agricultural evaluations of climate have also been the subject of much work. For example, the Soil Survey of Scotland has classified climate according to accumulated temperatures in day-degrees in excess of a threshold value, potential water deficit, exposure, and accumulated frost as degree-days below freezing.

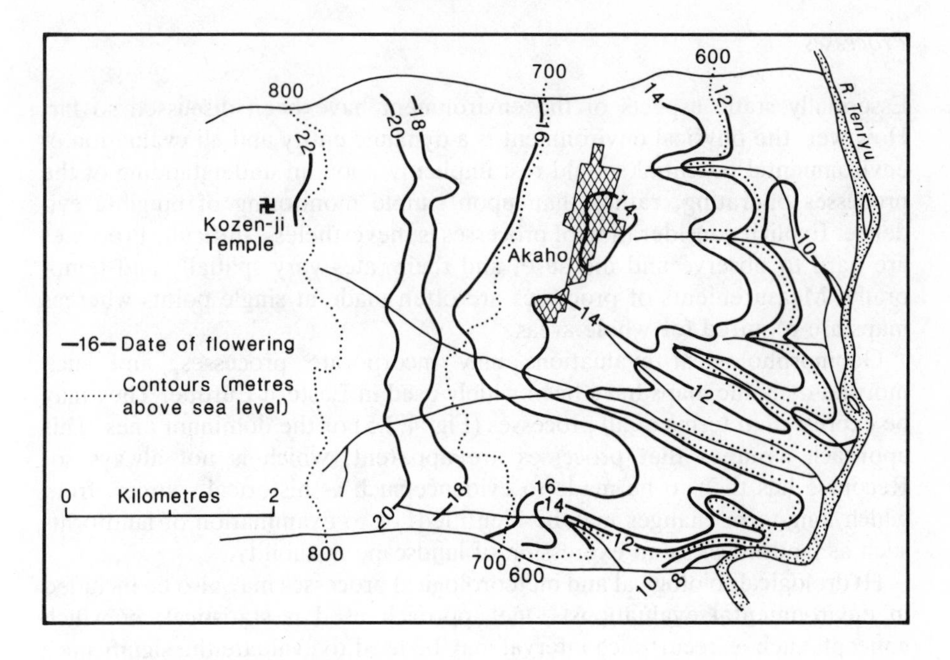

FIGURE 4.2.6 *Phenological data as indirect evidence of topoclimates*
The distribution of flowering dates of cherry blossoms (*Prunus subhirtella*) in Nagano Prefecture in central Japan (after T. Sekiguti (1950) *Geoph. Mag.* (Tokyo) 22,

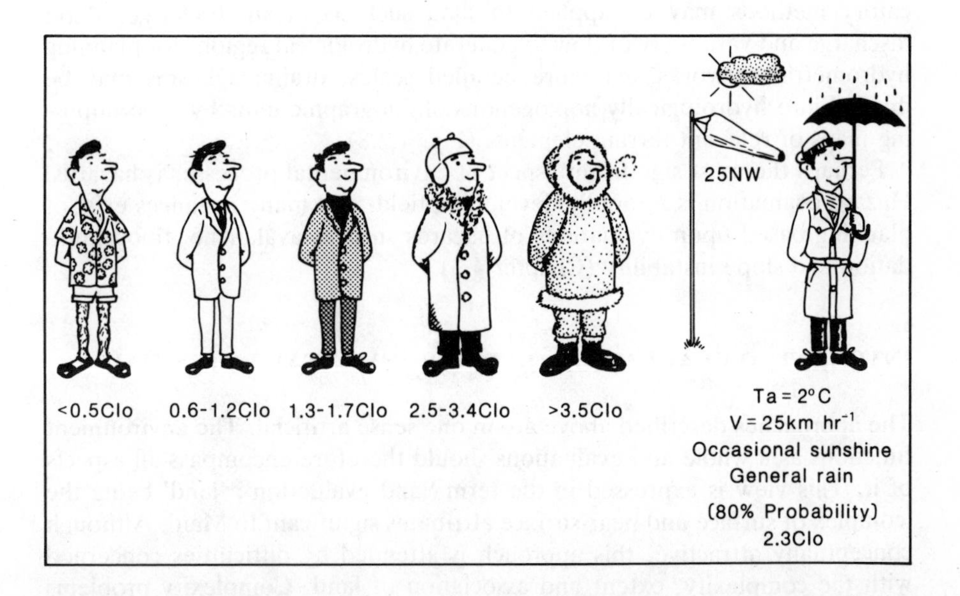

FIGURE 4.2.7 *Expressions of the Clo index and a hypothetical weather forecast*
(From A. Auliciems and F. K. Hare (1973), 'Visual presentation of weather forecasting for personal comfort', *Weather* 28, 478–480.) One Clo is the insulating value of a normal 'business' suit.

Processes

Essentially static aspects of the environment have been discussed so far. However, the physical environment is a dynamic entity and all evaluation of environmental potential should rest implicitly upon an understanding of the processes operating, rather than upon simple monitoring of tangible evidence. Explicit consideration of processes is, nevertheless, difficult. Processes are hard to observe and measure, and their rates vary spatially and temporally. Measurements of processes are often made at single points whereas maps are required for whole areas.

Geomorphological evaluations may incorporate processes, and such morpho-dynamic maps have been widely used in Eastern Europe. They may be expressed in terms of all processes (Fig. 4.2.5) or the dominant ones. This approach assumes that processes are apparent, which is not always so. Recourse has then to be made to evidence such as historical sources, from which long-term changes may be identified, or to examination of landforms such as landslides, as an expression of landscape instability.

Hydrological, biological and meteorological processes may also be included in environmental evaluations. One approach used is statistical, in which concepts such as recurrence interval may be used to evaluate the significance of particular events. One virtue of this is that it allows comparisons between different locations by comparing the magnitudes of, for example, the 'ten year flood'.

Processes can also be regionalised. At regional scales quantitative classificatory methods may be applied to data such as mean discharge, flood discharge and variability of flow to generate hydrological regions for planning hydrometric networks. At more detailed scales, drainage basins may be divided into hydrologically homogenous physiographic units by superimposing maps of relevant terrain elements (Fig. 4.2.8).

Perhaps the most significant aspect of environmental processes is hazards. Hazard evaluation is a rapidly developing field, and many instances exist of planning based upon evaluations of hazards such as avalanche, flood inundation and slope instability (Chapter 4.3).

INTEGRATED EVALUATION OF ENVIRONMENTAL POTENTIAL

The approaches described above are in one sense artificial. The environment functions as a whole and evaluations should therefore encompass all aspects of it. This view is expressed in the term 'land evaluation'; 'land' being the 'complex of surface and near-surface attributes significant to Man'. Although conceptually attractive, this approach is attended by difficulties concerned with the complexity, extent and association of land. Complexity problems arise from the notion of uniqueness, whereby it is argued that each point is essentially unique; delineation of boundaries and classification in regions are therefore artificial. However, uniqueness and absolute similarity are theoretical extremes; for practical purposes there is a fundamental spatial organisation produced by the association between environmental attributes. This renders regionalisation viable in pragmatic terms. Extent problems stem from the differing spatial geometries of various environmental characteristics; for example, soil pH is a point measurement whereas soil type applies to areas. Association problems arise because identification of regions inevitably emphasises boundaries rather than interactions between regions, as transfers of energy and mass.

One way by which at least partially integrated evaluations of land are produced is by the superimposition of single element surveys, as illustrated in Figures 4.2.8 and 4.2.9. This is possible because of the inherent co-variation of environmental characteristics, although judicious cartographic generalisation of outlines is inevitable. This approach is made more feasible by the existence of computer data bases.

A second approach is to carry out related surveys in the first instance. For example, in West Africa pedogenesis is closely related to landform and flood inundation, and teams of geomorphologists and soil scientists have co-operated to assess agricultural potential of floodplain areas. Single integrative indices of land character may also be of value. Soil is produced by interaction amongst all other environmental attributes, and its nature is therefore indicative of them to some extent. Indeed, concepts such as 'soil landscape' have been advanced to express this relationship. It has also been claimed that vegetation and drainage density may be used in this manner.

However, the most popular methods of integrated land evaluation is the 'landscape approach'. This is hierarchical in nature and is best applied to reconnaissance-scale surveys. Small topographic units called facets, which are homogenous in terms of most environmental characteristics, are recognised. Areas with facets in a recurrent pattern are called Land Systems. Surveys are usually based upon field sampling of terrain characteristics along transects, planned from provisional Land Systems identified from aerial photographs. Satellite remote sensing imagery is now often employed, and quantitative methods have been introduced. The method has been criticised as being static in nature, and of limited accuracy. However, it has afforded a rapid and effective means of reconnaissance survey of physical resources in underdeveloped areas, as employed by CSIRO in Australasia and the ODA in Africa.

FUTURE DIRECTIONS AND PROSPECTS

Environmental evaluation is a rapidly developing field. It is clear that much greater use will be made of aerial photographs and satellite remote sensing data, often linked to computerised methods of analysis and presentation. Greater emphasis will be placed upon evaluation of environmental processes

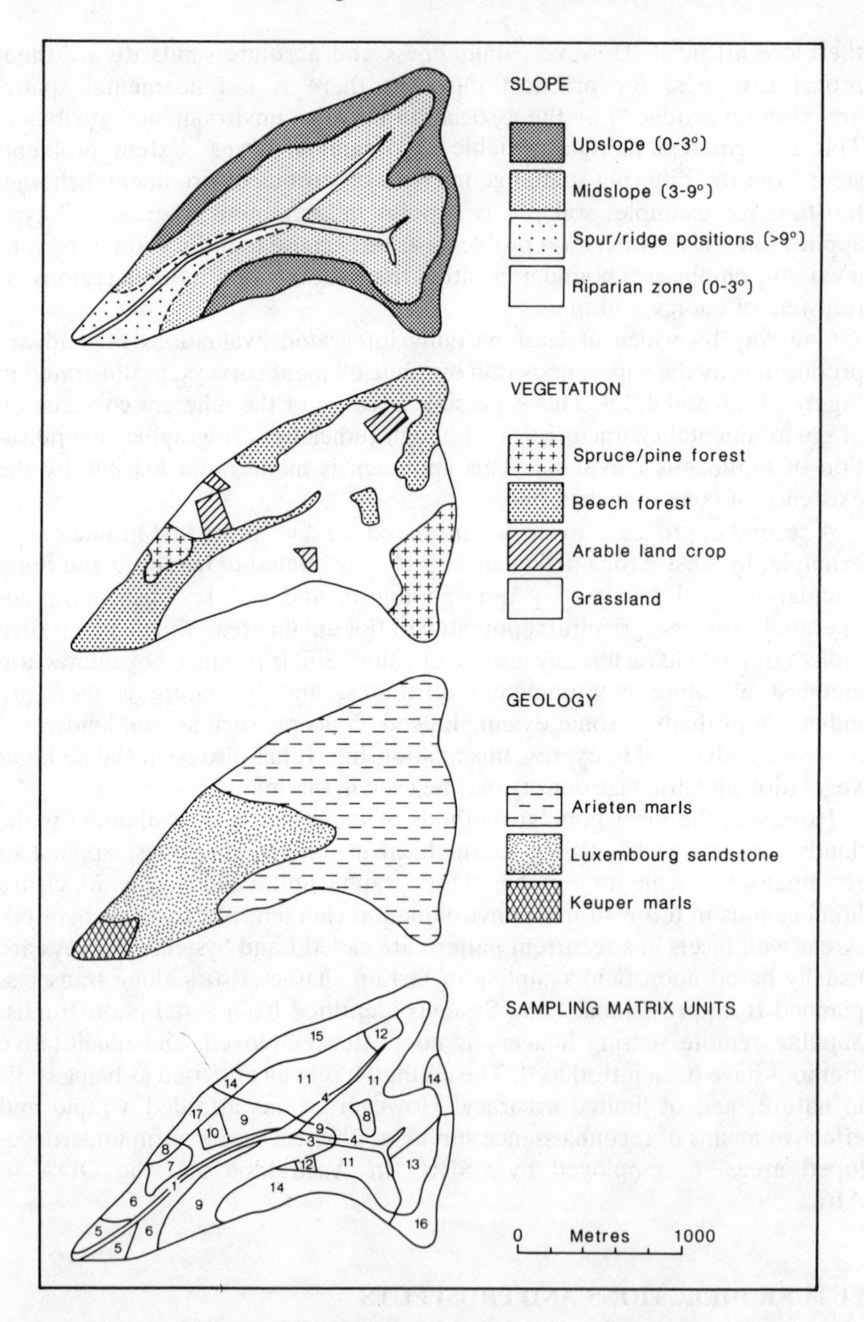

FIGURE 4.2.8 *Derivation of homogeneous physiographic units for a small drainage basin*

Slope unit, vegetation and geology maps are superimposed to produce a map of physiographic units, as a basis for hydrological instrumentation and modelling (from D. Huson, I. Simmers and E. Seyhan).

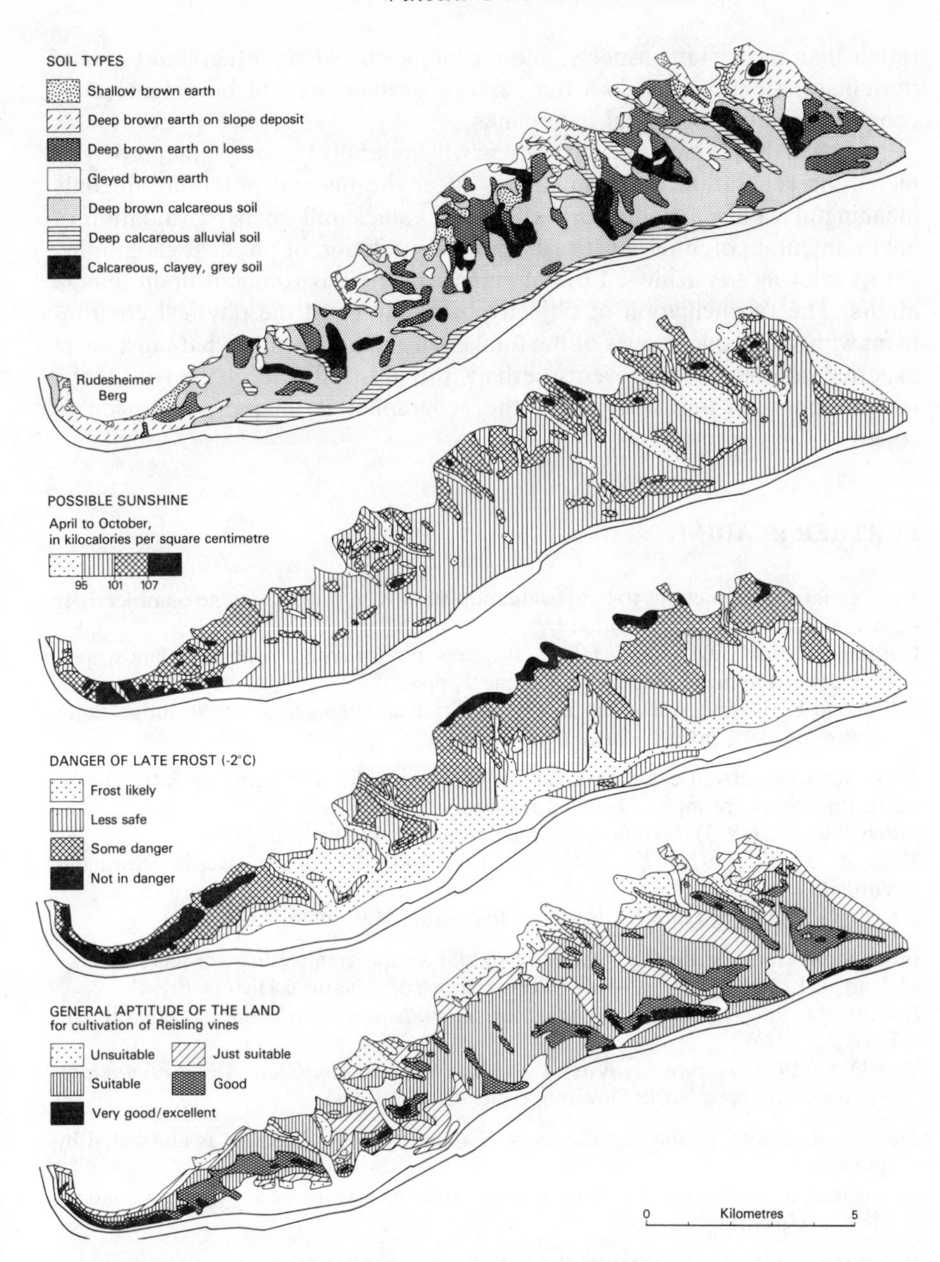

FIGURE 4.2.9 *Evaluation of terrain for a specific purpose – the cultivation of Riesling grapevines*
This is an application of topoclimatic scale investigation, and illustrates how synthesis between individual distributions can produce an effective evaluation (after H. Zakosek, W. Kreutzt, W. Bauer, H. Becker, E. Schroder (1967): Die Standortkartierung der hessischen Weinbaugebiete. – Abh. hess. L.-Amt Bodenforsch., 50, Atlas kt. I, II, III u. VI; Weisbaden).

rather than more static aspects, and techniques by which integrated views of the environment are used as the basis of evaluations will be enhanced by computer analysis of digital data banks.

Technical developments such as these will hopefully be complemented by an increasing realisation that evaluations of environmental potential are only meaningful within a framework of human values and needs. Evaluation of environmental potential must rest upon a foundation of physical geography, yet its relevance is achieved by the impact of the environment upon human affairs. The reconciliation of objective evaluations of the physical environment with subjective images of the total environment, within what must be an experiential domain, represents perhaps the most difficult challenge for the future – but also one with which the geographer is uniquely equipped to cope.

FURTHER READING

Short general introductions to land evaluation and to the general debate on objectivity versus interpretation are provided by:

Gardiner V. and Gregory K. J. (1977) 'Progress in portraying the physical landscape', *Progress in Physical Geography*, volume 1, pp. 1–22.

Hammond K. R. and Adelmann L. (1976) 'Science, values and human judgement', *Science*, no. 194, pp. 389–96.

Three texts on terrain evaluation can be suggested, the first being an introduction whilst the others are more advanced reviews:

Mitchell C. W. (1973) *Terrain Evaluation* (London: Longman).

Vink A. P. A. (1975) *Land Use in Advancing Agriculture* (Berlin: Springer-Verlag).

Stewart G. A. (1968) *Land Evaluation* (Melbourne: Macmillan, Australia).

Illustrations from the fields of planning and landscape architecture are included in:

McHarg I. L. (1969) *Design With Nature* (New York: Natural History Press).

Lovejoy D. (1973) *Land Use and Landscape Planning* (Aylesbury: Leonard Hill Books).

Way D. S. (1978) *Terrain Analysis: A Guide to Site Selection Using Aerial Photograph Interpretation* (New York: Dowden, Hutchinson & Ross).

The use of remote sensing data sources in a variety of applications is illustrated by chapters in:

Townshend J. R. G. (ed.) (1981) *Terrain Analysis and Remote Sensing* (London: Allen & Unwin).

Books which include reference to the use of geology, vegetation and soil as bases for evaluations are:

Costa J. E. and Baker V. R. (1981) *Surficial Geology* (New York: John Wiley).

Kuchler A. W. (1967) *Vegetation Mapping* (New York: Ronald Press).

Davidson D. A. (1976) *Soils and Land Use Planning* (London: Longman).

4.3
Natural Hazards – Adjustment and Mitigation

John Whittow
University of Reading

Any study of hazards will necessarily involve an understanding of both human and physical systems and of their interaction with each other, since no hazard can exist unless it is perceived and in turn provokes a human response. Thus, while trying to avoid a deterministic stance, most geographers believe that human systems of resource utilisation, which have often been taken to demonstrate how mankind has 'controlled' its environment, have rarely been able to operate independently of geophysical and biophysical systems. Even slight changes in the operation of these physical processes can threaten mankind and be seen as hazards. A river, for example, during periods of 'normal ' flow offers a positive inducement to human activity; only during a period of flood (or even drought), when its régime fluctuates, will it become a menace. Natural hazards have been defined, therefore, as 'those elements of the physical environment harmful to Man and caused by forces extraneous to him'. This 1964 definition by Burton and Kates has recently been termed the 'dominant view', but this standpoint has also been challenged (see below).

BACKGROUND TO HAZARD RESEARCH

Hazards are usually divided into two groups – natural and human-induced. The former (often termed Acts of God) exhibit different magnitudes and contrasting spatial and temporal patterns. The latter range from domestic, industrial and travel hazards to global scale hazards of environmental pollution. A comparison between natural and human-induced hazards illustrates that the former kill greater numbers but at less frequent intervals. Nonetheless, it is important to realise that senility, disease and malnutrition account for the vast majority of deaths, although there is a significant dichotomy apparent between the developed and the developing world. In the USA a mere 0.05 per cent of the population die from natural hazards, while in a poor country like Bangladesh as many are killed from natural hazards as die from cardiovascular disease in the United States. The greater vulnerability of poorer nations to the threat of natural hazards is a recurrent theme in this overview.

Serious hazard research began during the mid twentieth century, with geographers playing significant roles. Initial concern with causal agents was intended to assist in the formulation of policies to mitigate the environmental impact by means of engineering technology. After two decades, however, some geographers abandoned the positivist approach which dominated early process studies and turned increasingly to the phenomenological outlook which underpins perceptual and behavioural studies. Investigations of physical processes, of course, continued to be of great importance if remedial measures were to be implemented, but in order to explain the different patterns of human behaviour response it was also essential to discover how people perceived risk. Only when the physical and human variables could be interrelated would it be possible to make credible recommendations for management of the resource base.

It was questions such as these which led Professor G. F. White to investigate the American flood problem and to his discovery that despite an enormous government expenditure, flood damage in the United States continued to escalate. He was ultimately to pose a fundamental question which encapsulates the entire study of natural hazards: how does man adjust to risk and uncertainty in natural systems, and what does an understanding of that process imply for public policy? The following year he outlined a research paradigm whose five objectives have inspired many geographical investigations:

1. Estimate the extent of human occupation in areas subject to extreme events in nature;
2. determine the range of possible human adjustments by social groups to these extreme events;
3. examine how people perceive extreme events and the resultant hazards;
4. examine the process of choosing damage-reducing adjustments;
5. estimate what would be the effects of varying public policy upon that set of responses.

Physical geographers may wish to add a sixth objective, although it is probably implicit in the other five:

6. evaluate the hazard's dimensions in order to predict the degree of impact and the spatial dimension of the risk zone.

Indeed, a recent plea has been made for British geographers to assess such impacts in order to change management practices and through them influence policy formulation. Geographers are in a favourable position to pursue these objectives because of their holistic approach, and their training enables them to make relatively simple measurements of the seven hazard variables that have been recognised: *magnitude* and *speed of onset* can be measured instrumentally and critical thresholds identified; determination of *frequency* and *duration* is a statistical exercise, as is a recognition of *temporal spacing* (a distinction between sequential and random events); the hazard's *areal extent* and *spatial dispersion* (i.e. linear, diffuse or nucleated pattern) is a mapping exercise well within a geographer's capability, particularly if aerial photographs are available. In this respect there is a paucity of risk-mapping in Britain, in contrast to the considerable progress made in the United States, even if much of this has been undertaken by geologists.

It is one thing to have a knowledge of hazard parameters, it is quite another to be able to implement damage-reducing measures, simply because adjustment and mitigation policies differ between world societies. While developed nations have, in many instances, been able to cushion themselves against most natural hazards, the developing countries appear to have become more disaster-prone. This is not to say that natural hazards are proliferating, but that many Third World populations appear to be becoming less capable of withstanding the vicissitudes of the physical systems without enormous external aid. From this recognition has sprung the so-called theory of marginalisation, in which disaster-proneness is seen as a result of relief aid merely reinforcing the *status quo* of underdevelopment that produced such initial vulnerability in a society already experiencing a deteriorating physical environment.

THE DEVELOPMENT OF HAZARD STUDY

Although people have sought damage-reducing measures for centuries, the documented history of hazard research is relatively brief. Most of the early work stemmed from North America – notably from G. F. White and R. W. Kates in Chicago during the 1950s and 1960s, and also from I. Burton and K. Hewitt in Toronto during the 1960s and 1970s. These geographers, cooperating with psychologists, sociologists and applied scientists, can be said to have triggered the diffusion of hazard studies throughout the world. Today there is much greater awareness of risk, despite the continuing reluctance of the general public to undertake remedial measures. Nevertheless, many decision-making bodies are showing a greater willingness to incorporate hazard modelling into public policy planning programmes.

Initial research in America was concerned largely with flooding, following the US Flood Control Act of 1936, but it was the technological advances of the Second World War that gave a boost to earthquake, volcanic and meteorological hazard studies. Furthermore, the post-war reconstruction period gave numerous opportunities for countries to incorporate hazard mitigation regulations in their new planning policies. Significantly, such research units as the Disaster Research Group of the United States (1961), the National Centre for Disaster Prevention in Japan (1962) and the Centre for

Disaster Studies set up in the 1970s in Queensland, Australia, were concerned mainly with their own national problems. It is true that the United Nations Disaster Relief Office (1972) attempted to co-ordinate disaster relief funds and reconstruction policies at a global scale, but their early work was often marked by incompetence and muddle which unfortunately detracted from their three-volume *Guidelines for Disaster Prevention*. Much more influential and relevant were the series of *Natural Hazard Research Working Papers* published by the University of Toronto from 1968 onwards, and the *International Journal of Disaster Studies and Practice* launched in 1977.

By the 1970s several developed nations had a government department responsible for hazard policy planning, although this implementation was usually in such countries as USA and Japan which experience hazards of high magnitude and frequency. The remaining countries reacted more slowly, either because they were unable to afford the research and development (most Third World countries) or because their losses from natural hazards were relatively low (most European countries).

The British contribution

Because Britain is visited fairly infrequently by high magnitude hazards it is hardly surprising that risk-assessment studies have progressed at a more leisurely pace than in North America, with most research being confined to academic institutions. Apart from floodplain development restrictions contained in the 1947 Town and Country Planning Act, there has been very little central government legislation on hazard mitigation. The Health and Safety Commission's Advisory Committee on Major Hazards (1974) is concerned essentially with industrial hazards, reflecting the major source of environmental risk in Britain. Significantly, the most important and costly flood-prevention measures ever undertaken in Britain, namely the Thames barrier, were the work of a local planning authority. It is true that much research into the physical and social dimensions of natural hazards was originally funded by Britain's research councils, and an important five-volume NERC report on flood studies appeared in 1975. So far as drought is concerned, however, the inadequacies of British water resource planning were revealed when the so-called 100-year event of 1976 was followed by another severe drought in 1984. Geographers have made important contributions in this field, not least of which is a drought atlas of Britain published by the Institute of British Geographers. Clearly an atlas of all British hazards is long overdue. It has recently been claimed that Western Europe as a whole lags a long way behind North America in hazard research, despite the government-funded engineering works of the Rhine Delta Planning Area (following the 1953 Low Countries flood disaster) and the Lower Elbe region (after the north German floods of 1962).

Some engineering works for hazard mitigation have also been carried out in Third World countries, but they are generally insignificant when compared with those of the developed nations. More characteristically, the world's poorer countries are recipients of disaster relief funds following catastrophes which their own governments seem powerless to alleviate. It is noteworthy, however, that during the last decade several hazard studies have been carried out in Third World countries by local rather than overseas researchers. Thus, they too are beginning to become more risk conscious and to take steps to reduce those risks.

HUMAN PERCEPTION AND RESPONSE

Natural hazards only exist in relation to human activity, aspirations and needs and one of the major geographical contributions to hazard research is in the field of risk perception. In fact the first significant symposium on perception studies was organised by the Association of American Geographers and stemmed directly from flood perception research. Today a great deal is known about the ways in which people appraise the hazard, analyse the alternatives available and decide to *accept* the risk, *avoid* the risk or *modify* the impact – such a decision is termed an adjustment. The interdisciplinary perception research between geographers and psychologists was seen by Bruce Mitchell, a Canadian geographer, as 'some of the earliest empirical work in what has become recognized as the behavioural approach in geography'.

Numerous perception studies have shown that although people who have experienced damage from a hazard exhibit a greater propensity to adopt adjustments, other factors, such as age, income and education, are of considerable importance in any attempt to explain the type of adjustment. Nevertheless, to date, no satisfactory model has emerged that gives a significant correlation between personality traits and adjustment behaviour. An interesting case study involving floodplain dwellers in Shrewsbury, England, has demonstrated that perception of the flood hazard came far below such social hazards as traffic noise and vandalism, and that a mere handful of respondents had made any positive adjustments, such as preparing a means of flood defence or taking out insurance cover. This degree of fatalism or 'risk minimisation' was reflected in the large majority of respondents who believed that it was someone else's responsibility to carry out flood control measures, organise warning systems or set up programmes for evacuation and relief. This expectation of a centralised or collective adjustment appears to be commonplace in most world societies but, paradoxically, there appears to be a marked reluctance by individuals to heed hazard warnings or move away from a risk area. A distinction can be made, however, between those unwilling to adjust and those unable to adjust due to lack of wealth, technology and mobility. Many studies have commented on the fatalistic attitudes of many Third World respondents whose inability to migrate meant that they have had no alternative but to bear the loss.

One of the most influential models of behavioural responses in the face of natural hazards is that of Burton, Kates and White who outlined, in 1978, the alternative ways in which society copes either by *adaptation* or *adjustment*. Because biological adaptation is seen to be an extremely slow process it is suggested that cultural adaptation is of more significance. Geographical studies have shown, for example, that a society may adopt a nomadic life-style in order to cope with recurrent drought. More recent work has suggested that if the nomad's economy is altered by relief aid or innovatory techonology it could lead to a sedentary life-style, an enlargement of herds, overgrazing and increased desertification – a classic case of marginalisation. The model in question also distinguishes between *purposeful* and *incidental adjustments*. The former can be defined as an adjustment designed primarily to cope with hazards, for example, building on a river terrace above the flood level, taking out insurance or planting drought-resistant crops. The latter is not primarily hazard-oriented but one which can help reduce potential losses, for instance, the improvement of communication systems can also increase early-warning time; improvement of transport networks can also increase escape possibilities; introduction of better quality housing can also improve hazard resistance. The model recognises four modes of coping separated by thresholds (Fig. 4.3.1).

The first mode (loss absorption) depicts the developed society as being so dependent on its technological 'cushion' that it may initially be less prepared to cope with a disaster than a less developed society. Nonetheless, its absorptive capacity is greater in the long term because of its well-organised infrastructure and widespread expertise. The mode of loss absorption is separated from loss acceptance by the threshold of awareness. Once a society recognises the hazard's characteristics it can accept the loss by either bearing or sharing it. Bearing the loss is the most common response in most societies, i.e. do nothing. If the hazard's return period is short-lived, however, the affected people often attempt to share the burden by borrowing or moving temporarily to share accommodation with friends or relatives. Loss acceptance yields to loss reduction when the action threshold is crossed and positive measures are taken to reduce the loss by corrective and/or preventive actions.

If any or all of these prove to be unsuccessful the next threshold, that of intolerance, will be crossed and drastic action is called for. Relocation of settlement (e.g. the capital of Belize was moved inland following a hurricane disaster) or radical changes in resource use (e.g. cessation of water abstraction from beneath Tokyo and Venice to alleviate ground subsidence and related flooding) are two of the more common adjustments. Finally, the model distinguishes different degrees of adjustment according to the development stage achieved by the particular society. A folk society appears to be more willing to adopt modifications of behaviour and land-use practices, namely, to co-operate with nature rather than attempting to manage or control it. Thus, costs are low and technological requirements negligible. But this type

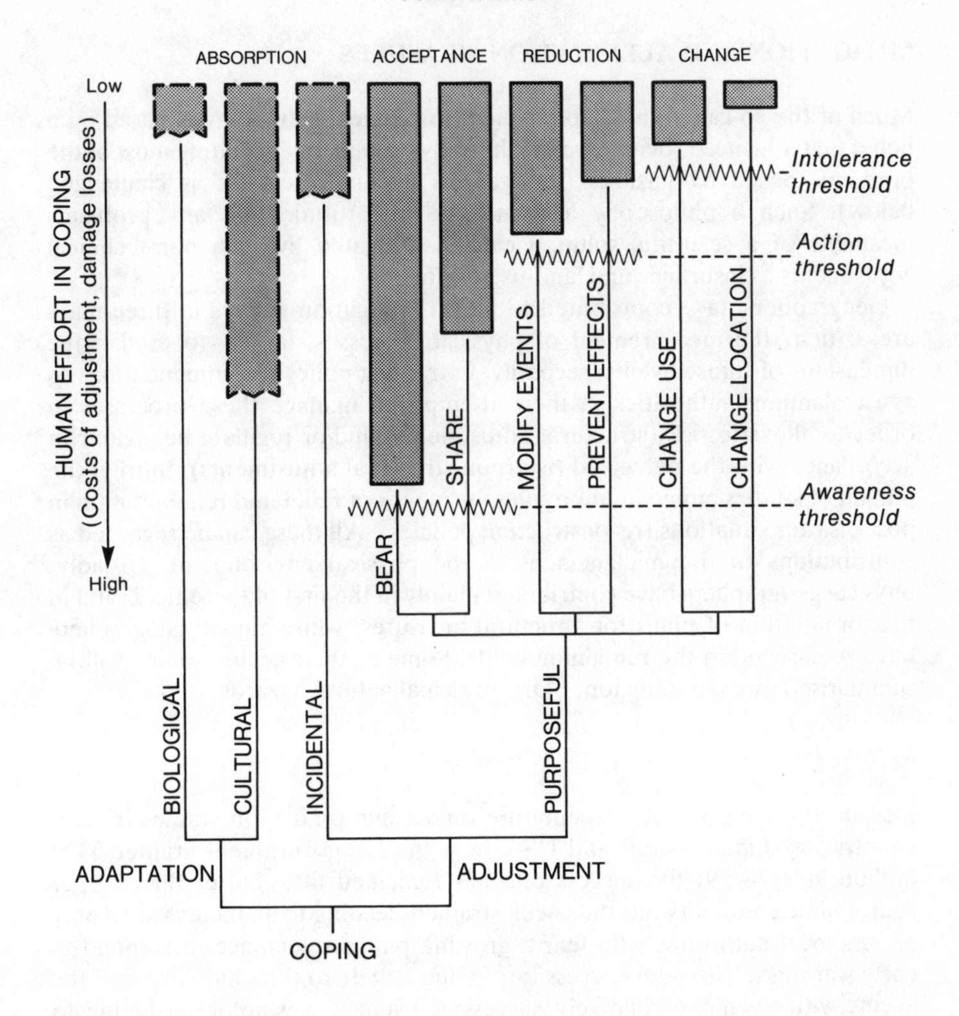

FIGURE 4.3.1 *Modes of coping with natural hazards (after Burton et al., 1978)*

of adjustment is only effective in the face of low-magnitude hazards; a high-magnitude event will probably require massive overseas assistance. By contrast, the response of an industrial society is seen as shifting the method of coping from bearing the loss personally (or even sharing with relatives) to that of sharing the burden nationally by means of insurance and/or financial loans. Because of its greater technical expertise, the industrial society is more concerned with managing or controlling rather than simply co-operating with nature. As this society moves into a post-industrial age it is finally suggested that there will be a greater propensity to reduce loss by considering a much wider range of mitigation and alleviation measures.

MITIGATION AND ALLEVIATION MEASURES

Much of the so-called 'applied' element in physical geography is based on a belief that advanced societies are technically capable of alleviating most of the problems posed by hazards (though this 'technological fix' is challenged below). Such a philosophy also includes an attitude that any problems incapable of a scientific solution can be mitigated by such non-technical adjustments as insurance and land-use zoning.

Geographers have contributed to hazard mitigation studies in three main areas: first, the measurement of physical processes, largely to predict the dimensions of future events; secondly, in framing policy recommendations to assist planning authorities in their attempts to manage these processes in order to alleviate risk (structural adjustments) and/or regulate behaviour in accordance with the perceived risk (non-structural adjustments); thirdly, the pursuit of studies aimed at improving the means of relief and rehabilitation in post-disaster situations (reconstruction policies). All these can be regarded as contributions to the management of the physical environment. Broadly, physical geographers have contributed mainly in the first of these fields and in the formulation of plans for structural measures, while human geographers have researched in the remaining fields. Some of their achievements will be summarised by examining four of the principal natural hazards.

Earthquakes

Despite the considerable expenditure on seismic prediction studies in such countries as Japan, USSR and USA (e.g. the US government granted $120 million in 1978–79) the success rate has remained little better than 50 per cent. Public credibility has thus been strained, leading to an increased anxiety among local authorites who fear a growing public reluctance to respond to early warnings. The poor success rate is due largely to the randomness of the event, with even the relatively successful Chinese seismologists failing to predict a catastrophic earthquake in Tangshan in 1976. Sophisticated geophysical instruments have pinpointed crustal movements but the timing and magnitude of the tremor remain difficult to forecast. An important breakthrough has been the 'Dilation Theory' expounded by Russian scientists, in which the arrival time-pattern of seismic waves at a recording station gives some warning of both magnitude and imminence. Although verified by other world scientists this theory has proved incapable of predicting all seismic events.

A warning period of at least two weeks is probably essential to evacuate a large city successfully, although it is claimed that the longer the warning the greater will be economic blight caused by enforced closure of industry and commerce. Compensation for loss of output could be extremely high and lead to complex litigation if the prediction failed. Consequently, some have argued that seismic-prediction funding should be directed to improved building design, construction of risk maps for land-use zoning, and in planning disaster-relief programmes. Most countries have now made it mandatory for such high-risk artifacts as nuclear reactors and dams to be built to withstand earthquakes. By the 1980s it had been recognised that most earthquake casualties relate more to construction standards than to seismic magnitudes, a conclusion particularly relevant to Third World nations where shocks as low as 4.9 on the Richter scale have caused high death tolls because of the poor building standards. The UNESCO guidelines for new building codes in seismic zones (e.g. replacing top-heavy stone or tile roofs with wood or corrugated iron) appear to have gone largely unheeded. Unhappily, UNESCO's much publicised earthquake reconnaissance missions, aimed at co-ordinating Third World programmes of earthquake engineering and disaster relief, have also achieved little. Not surprisingly, the handful of expensive shock-proof buildings already constructed are in California and Japan.

Severe weather

In contrast to the uncertainty of seismic prediction, progress in weather forecasting is marked by several spectacular successes. This has resulted from the regular seasonality of events, partly from the worldwide instrumental networks and finally from the rapid development of satellite photography. Geographers, by virtue of their interests in remote sensing, have played an important role in the latter field.

Radar-tracking of tropical storms has saved millions of lives, especially when people could take advantage of early warnings as in Galveston, USA (1969) and Darwin, Australia (1974). By comparison, the lack of monitoring, inadequate early warning systems and the virtual impossibility of evacuation due to underdeveloped transport facilities, has meant that recurrent tropical storms in southern and eastern Asia, for example, have wrought fearful havoc year after year. But property losses from severe weather continue to rise even in the developed world, particularly where people continue to inhabit coastal risk zones. Paradoxically, the affluent residents of California's Malibu Beach continue to rebuild their storm-battered homes despite endless opportunities to relocate. National and local zoning ordinances have failed to halt this type of coastal development in many countries thus making it necessary to build expensive sea defences.

To improve weather forecasting, the World Weather Watch (co-ordinated by WMO) has linked 8500 land stations and numerous weather ships with 140 national weather centres. Additionally, by using satellite photography of Atlantic cloud formations the Global Atmospheric Research Programme (1974–78) has co-ordinated hurricane research with studies of West African drought. Geographers have also contributed to studies of tornadoes and thunderstorms both in the United States and in Britain, showing that although 40 million Americans are at risk from tornadoes there has been minimal investment in protective measures because of their relative ineffectiveness and the sporadic occurrence of the hazard.

Much remains to be done on the effects and costs of snow, fog and ice on British transport systems, although Manchester University climatologists have developed an interesting Fog Potential Index, and a team of meteorologists and geographers at Birmingham University has modelled the relationships between motorway surface temperatures, driving hazards and salting requirements. Previously, most frost hazard research by geographers has been concerned with agriculture. Finally, it is important to note the contributions of geographers to the study of former climates in order to understand atmospheric change and future trends. The US National Center for Atmospheric Research in Boulder, Colorado, is one of the leaders, while in Britain the Climatic Research Unit was founded at the University of East Anglia in 1972.

Floods

The flood hazard causes almost two-thirds of the world's deaths from natural hazards and an even greater proportion of its material losses. This is the price mankind pays when competing with rivers for the use of their floodplains or when building injudiciously on lowland coasts. Floodplain management policies are slowly moving away from the traditional reliance on structural adjustments, for four main reasons: first, increased expenditure on engineering works has failed to decrease flood losses; secondly, over-optimistic belief in the infallibility of engineering technology has actually increased the flood risk by encouraging people to encroach into hazard areas believing them to be safe; thirdly, in contrast to the relative cheapness of non-structural adjustments, the costs of many engineering structures are becoming prohibitive; finally, the improvement of flood forecasting, more rigorous land zoning and the greater availability of insurance have together given greater credibility to non-structural measures.

In a threefold classification of structural adjustments a distinction can be made between those (corrective) engineering schemes which modify the river course to fit the flow (embankments, channel enlargement, flood-relief channels and intercepting channels) and those which regulate the flow to fit the course (reservoirs and washland storage). A second group, flood abatement schemes, depends on such upstream land-use changes as afforestation to delay runoff in order to reduce flood peaks downstream, thus obviating dependence on engineering works. The third group of structural adjustments, flood-proofing measures, can be either temporary or permanent. The former is entirely dependent on a realistic flood-warning time, while the latter includes such modifications as raising buildings, flood-proofing cellars and ground floors, together with the installation of pumping equipment in order to modify the impact.

In summarising non-structural (preventive) adjustments, four main responses can be recognised. The most common is flood forecasting and warning, but this is effective only in large river catchments (e.g. the Mississippi) where warnings can be given several weeks in advance. Such technological advances as satellite monitoring and radar scanning of precipitation are already improving the flood prediction success rate, but it has recently been concluded that the previously held assumption, that floodplain residents will respond rationally to a warning, is quite unjustified. Response to coastal flood warnings, however, appear to be rather more positive, especially if storm surges are forecast. Paradoxically, the unusual phenomenon of a tsunami (a seismic-generated ocean wave) has sometimes proved to be a tourist attraction, often with tragic consequences. The predicted arrival of the 1964 Alaskan earthquake tsunami on the Californian coast, for example, drew no less than 10 000 sightseers to San Francisco's shoreline.

Improved flood forecasting has made the second of the non-structural adjustments, that of insurance, a more efficient means of alleviation, although a more reliable correlation between degree of flood risk and scale of premium is still being sought. Some have claimed that compulsory insurance in risk areas would ensure that only economically viable development would occur. Such a claim would rely heavily on careful zoning of land use.

Ideally, floodplain zoning would clear the vulnerable riparian zone of all permanent structures, but in most cases this is impractical both socially and economically. Instead, a three-zone division has long been advocated: a prohibited zone; a restricted zone (essential buildings only plus recreational amenities); a warning zone (where residential development is permitted) – with all three zones being linked to flood recurrence times and levels. The warning zone, for instance, would stand above the expected twenty-year flood level. Some drawbacks have been recognised, including the false sense of security engendered by the plan, the creation of economic blight by lowering property values in the high risk zones, and the considerable difficulty of precise definition, despite the use of aerial photographs.

Like insurance policies, the final measure, that of public relief funding, is simply a means of reducing the loss after the event and is, therefore, the least useful means of alleviating the impact. Nevertheless, like all disaster relief funds, flood relief is one of the most common adjustments, since humanitarian responses are often the sole means of alleviation for many poorer nations.

In conclusion, it is important to emphasise that the most successful flood mitigation programmes are those which have replaced piecemeal strategies with multipurpose river basin planning along the lines of the United Nations 1969 report on *Integrated River Basin Development*. Geographers have long been collecting the hydrological, meteorological and pedological data which assist in the formulation of flood adjustment policies (Fig. 4.3.2). In Britain much of this work has been initiated and/or co-ordinated by workers at the Flood Hazard Research Centre of Middlesex Polytechnic.

Modify the flood	Modify the damage susceptibility	Modify the loss burden	Do nothing
Flood protection	Land-use regulation and changes	Flood insurance	Bear the loss
(channel phase)	Statutes	Tax write-offs	
Dikes	Zoning ordinances	Disaster relief	
Floodwalls	Building codes	volunteer	
Channel-	Urban renewal	private activities	
improvement	Subdivision regulations	government aid	
Reservoirs	Government purchase of	Emergency measures	
River diversions	lands and property	Removal of persons	
Watershed treatment	Subsidised relocation	and property	
(land phase)	Floodproofing	Flood fighting	
Modification of	Permanent closure of low-level	Re-scheduling of	
cropping practices	windows and other openings	operations	
Terracing	Waterproofing interiors		
Gully control	Mounting store counters on wheels		
Bank stabilisation	Installation of removable covers		
Forest-fire control	Closing of sewer valves		
Re-vegetation	Covering machinery with plastic		
Weather modification	Structural change		
	Use of impervious material for		
	basements and walls		
	Seepage control		
	Sewer adjustment		
	Anchoring machinery		
	Underpinning buildings		
	Land elevation and fill		

FIGURE 4.3.2 *Adjustments to the flood hazard (after Sewell, 1964)*

Drought

Unlike the three previous hazards, drought is a pervasive threat, lacking a sudden dramatic impact. Nonetheless, it is the most far-reaching of all hazards, undermining the economy and threatening the very existence of certain nations. Because drought destroys food supplies and brings starvation and disease in its wake, it is one of the most serious hazards faced by modern society. Despite the construction of expensive reservoirs, desalination plants and irrigation schemes, together with the introduction of national water resource programmes, even such major nations as the USA and USSR have sometimes been powerless to save their harvests. But while they can regard drought as a temporary setback, for the countries of sub-Saharan Africa it is a matter of life or death, particularly since the drought which began in 1972 has coincided with a world recession. There have been numerous international conferences on desertification in addition to considerable research and investigation by such organisations as FAO, UNESCO and UNICEF. The resulting volumes of analysis, explanation and advice, many of them stemming from geographers, have undoubtedly shed much light on the drought problem but, because of its gigantic scale and because it is a hazard not easily amenable to the 'technological fix', very little practical progress has been made.

In the summer of 1984 the European Space Agency (ESA) announced a major breakthrough in monitoring the African drought by means of remote sensing from Landsat. Plant growth, water balance and soil characteristics can be assessed from the satellite photographs and experts believe that impending African droughts can now be more easily forecast. However, it has been pointed out that because of its prohibitively high costs, the Landsat space technology is inappropriate to poor African countries and serves merely to give the developed nations a better inventory of Africa's mineral resources. A more useful and cheaper source of information appears to be the satellite photographs from the US National Oceanic and Atmospheric Administration (NOAA).

The relief aid organisations now realise that introducing high-cost technology to drought-stricken Africa is a waste of precious funds and should be replaced with rudimentary equipment which can be manufactured and maintained locally. But technology, by itself, can offer only the mirage of a solution. Thus, instead of simply adopting the so-called scientific approach (weather modification, etc.) or the technological approach, the drought stricken African countries will have to look, initially at least, to a behavioural solution because their problems arise not from poor technology but rather from a malfunctioning of their current socio-economic systems. This notion is developed in Chapter 2.5, and its broader implications are explored in Chapter 4.4.

It is not only Africa that suffers from aridity, for 1983 was a disastrous year for both Australia and the mid-west farmers of the United States. Many American farmers, faced with a failed harvest, high interest rates and low prices, went bankrupt, their 'cushion' of insurance and bank loans no longer proving effective. Because droughts are most frequent and more widespread in Australia, however, its federal government has prepared a ten-point plan which incorporates both technological and behavioural adjustments: increased scientific research into cause and effect; better education in drought mitigation; better water storage; better fodder storage; increased shelter for livestock to lower dehydration; better financial backing; improved distillation plants along stock-routes; improved groundwater monitoring; increased irrigation to diversify agriculture; improved transport facilities to evacuate stock from hazard zones. Following the disastrous 1983 drought most of these recommendations were introduced by the Australian authorities while at the same time the United Nations finally produced a Plan of Action to combat drought at a global scale.

A REASSESSMENT

By the 1980s hazard studies had largely moved away from the initial stance of environmental determinism, although some geographers have clung to the belief that natural disasters can be explained simply in terms of fluctuations in the operation of physical processes. Contemporaneously, there has grown up a substantial body of researchers in human perception and risk assessment. Kenneth Hewitt has recently challenged the 'dominant view' that disaster can be attributed to nature and that society is incapable of doing anything about it without adopting a technocratic approach. He believes that the 'technological fix' has blinded all nations and even the international agencies (especially UNDRO) to any alternative assessment of the hazard problem, because technology is a creature of the most powerful, wealthy and centralised institutions. It is suggested that the hazards debate has recognised socio-economic and political factors as being paramount in understanding why disasters occur. Hewitt believes, therefore, that there are 'decisive human ingredients' in today's hazard studies. He concludes that perception of and response to hazards are not simply functions of hazard mechanisms or past experience, but depend more significantly on socio-economic factors and other societal pressures and goals; and that natural disasters are characteristic rather than accidental features of the places and societies where they occur – natural hazards will cause disasters essentially in countries already undergoing rapid sociocultural and political change and ongoing environmental impact. This is the basic premise of the theory of marginalisation referred to above. Finally, however, Hewitt recognises that totally abandoning the 'dominant view' leaves behind an unbalanced view of natural hazards, and a danger of replacing environmental determinism with social determinism. Thus, he finally admits:

> It would be wrong to suggest that events associated with flood or earthquake in no way reflect the nature of these geophysical processes. It would be indefensible to argue that disruptions occasioned by disaster produce no distinctive, even unique, crisis phenomena. There are particular aspects of hazard that can be helped by improved geophysical forecasting. Nor are any foreseeable actions going to remove the need to bring emergency assistance to ill-equipped victims of natural calamities.

Thus, it is probable that for the foreseeable future geographers will continue to research the problems of natural hazards by following both the 'dominant' and the 'alternative' approaches.

FURTHER READING

The first three suggestions for reading represent the development of the topic through the work of perhaps its most influential team of North American researchers. The first is a classic paper which not only forms the basis of the mainstream of geographical thinking in the field, but is the first co-operative work between two leading figures. There follows the first major book on the subject, and the most influential. It contains several classic case studies, including one of Shrewsbury. Third is an important collaborative work which summarises the state-of-the-art in hazard research and also offers a model of 'coping' with hazard:

Burton I. and Kates R. W. (1964) 'The perception of hazards in resource management', *Natural Resources Journal*, vol. 3, pp. 412–41.

White G. F. (ed.) (1974) *Natural Hazards: Local, National, Global*. (New York: Oxford University Press).

Burton I., Kates R. W. and White G. F. (1978) *The Environment as Hazard* (New York: Oxford University Press).

Three more recent books provide invaluable summaries moving from the broad perspective to more specialised dimensions. The first provides an overview of worldwide hazards which emphasises the dichotomy between developing and developed nations in their responses to threat. The second is a modern review of British hazards, including many that are human-induced, and the third concentrates on just one hazard:

Whittow J. B. (1980) *Disasters: The Anatomy of Environmental Hazards* (London: Allen Lane).

Perry A. H. (1981) *Environmental hazards in the British Isles* (London: Allen & Unwin).

Ward R. (1978) *Floods: A Geographical Perspective* (London: Macmillan).

Finally, two more advanced studies develop the argument, first, by looking at a perspective which describes the socio-economic and political effects of a European earthquake in contrast with the North American literature, and second, by reassessing the directions followed by previous hazard studies (chapter 1, by Hewitt himself, is fundamental reading since it challenges the 'dominant' view):

Geipel R. (1982) *Disaster and Reconstruction* (London: Allen & Unwin).

Hewitt K. (ed.) (1983) *Interpretation of Calamity* (Boston: Allen & Unwin).

4.4

Geography and Conservation: The Application of Ideas About People and Environment

Andrew Warren
University College London

'Nature to be commanded must be obeyed.'
Francis Bacon

The connection between people and the environment, built and natural, is the crux of both geography and conservation. Geographers study and conservationists worry about the environment over the same breadth of space and time. Geographers distinguish themselves from those who, like psychologists, look at the connections in finer detail. Conservationists distance themselves from other environmental managers like farmers, foresters and engineers, whom they accuse of having more limited goals. Both geographers and conservationists believe that they have to integrate many disciplines.

There are differences. Geography is an academic discipline that has many other interests. Conservation, whose main aim is the practical management of the environment in the long term, has links with other academic disciplines. A threat of doom would scatter geographers to bibliographies and conservationists to smallholdings. Nonetheless, their mutual present interests cry out for co-operation. For geographers, who have been returning to the theme of the relation between people and the environment in the last decade, the conservation movement is a hand-specimen of a lively, cosmopolitan attempt to come to terms with nature. Its history shows how people constantly structure the network of their environmental goals, understanding and management.

Conservationists, for their part, have seen a burgeoning of their literature, professional competence and political influence in the last decade, and are set fair for even faster growth. They are at a point where they must examine their philosophy and objectives very carefully if they are not to lose the ground they have gained. In this they can learn from the geographer's critical thinking about people and the environment, and from physical geographers scientific approach to the systems they wish to conserve. Only geographers, among the academics who interface with conservation, claim to offer the tools from both natural science and social enquiry that conservationists are constantly calling for.

This chapter is an extended commentary on the dictum at its head. In the end its implications will be rejected: people do not so much 'command' or 'obey' nature as accommodate it. The chapter covers only environmental conservation, whose aims and methods are very different from those of conservation of the cultural heritage. Nor does it discuss the conservation of non-renewable resources, a field which involves much more economics and technology, and much less natural science than does the conservation of renewable resources, which is the main focus here.

SCIENCE, IDEOLOGIES AND MOTIVES

The first lesson that a study of the history of conservation teaches geographers is that long-term environmental management must be based on both scientific understanding and a system of morals and values. The driving force in conservation is anxiety about the future, that is, a mismatch between an ideal and a prediction. A programme without a clear objective and a reliable method of prediction would be without the means of conservation. I will look first at the necessity for science, and then at the unavoidability of a system of values and morals.

Science and conservation

Our traditions of scientific prediction and of environmental anxiety appeared together in the scientific revolution of the seventeenth century, and have been closely related ever since. A good example of the continuing dialogue between environmental anxiety and scientific research is soil conservation which, in practical terms, began with the monumental work of Hugh Hammond Bennett, the founder of the US Soil Conservation Service. One of his campaigns used what would today be thought of as a limited knowledge of fluvial geomorphology to predict that soil erosion would result in the serious silting of navigable waterways. The prediction launched a programme of research into sediment delivery ratios which now shows that he overrated the danger. Nevertheless, this, and many of the other traditions of research that he so powerfully helped on their way, have vastly improved predictions about the environment, and so led to much better conservation.

In nature conservation, the sources of Rachel Carson's deep anxiety about pesticides in her book, *Silent Spring*, were the results of scientific research. The interpretations she and others made about the concentration of toxins along food chains have not all stood up to scrutiny, especially as regards aquatic systems, but testing of the hypothesis accelerated in response to her outrage. Even in Britain, the whole edifice of research into the environmental effects of pesticides is said to have stemmed from that one book. The story again shows the strong links between the conservation ethic and scientific research.

This dialogue between science and anxiety has been very productive, and is anyway quite unavoidable, but it does have its dangers. A theme that dominates the history of conservation is the repeated distortion of scientific judgement by anxiety. Such distortions stem in part from misunderstanding of the complex relations between science, ideology and conservation, and the exploitation of this difficulty for ulterior motives. Since these seem to be fundamental components in any process of adjustment of people to their environment, it is worth looking at them in more detail.

Science and ideology

Many modern environmental writers have condemned the scientific revolution of Francis Bacon's time as the first wedge in a widening rift between people and nature. Some have even rejected the scientific view of nature: they see both 'commanding' and 'obeying' as alienating people from the environment. The discussion will return later to objections to 'command'. It must confront the thornier problem of 'obey' here.

Bacon sketched a utopia ruled by scientists who were to discover values and even morals. Only they would know how to 'obey' nature. Most scientists today would condemn this prospectus as grossly misleading, for they see that science has its limits, and that the important process of discovering values and morals is beyond these. But they also see the relations between science and ideology as two-way. In one direction, systems of morals and values are constantly adjusting to scientific discovery. In the other, historians of science point out that scientific hypotheses must come from the culture milieu (including as we have seen conservation). They cannot come from the facts alone. The flow from ideology to science also includes the influence of values and morals on the choice of research and the interpretation of results. Thus the dangers of alienation between people and nature, and of the moral hegemony of science, lie not so much with science itself as with those who overrate its powers or its threats.

Ideology and conservation

There is more of a feeling of inherent values and morals in conservation than in science, but there are many false trails here too. It is puzzling for a start that environmental conservation has been urged for a bewildering variety of reasons, many of them contradictory. Resources have been conserved for the successors of very exclusive classes (such as South African Whites or Iranian Royalty); for the betterment of the whole of mankind (as in the World

Conservation Strategy); for the sake of other organisms; or for God. Most active conservationists in Britain today see themselves as radical, but the Conservative Bow Group sees conservationalists as 'natural Conservatives'. They could, of course, be either. The urge to conserve neither follows directly out of, nor wholly validates, any one ideological position.

Nevertheless, environmental conservation is not completely ideologically neutral. The conservationist stand finds more echoes in some ideologies than in others, and it also has some easily identifiable enemies. Romanticism is one notion that resonates with it. Communalism is another ally; the necessarily broad scale of environmental conservation demands, almost by definition, concern for the community, and communal action. Extreme *laissez-faire* economics or anarchism could not encompass conservation. Exploitative capitalism is rightly one of the prime targets for the conservationists' fury, whether it be for whales, shale oil or uranium. Nevertheless, the ideology behind each of the allied systems is independent of the environmental roots of conservation. This is simply to reiterate that conservationists, like scientists, cannot find codes of behaviour and systems of value in nature alone.

It is hard to find this final admission in the writings of conservationists, or in those of some of their geographical counterparts. Most of their ideological stances blur the distinctions between science and ideology, in a whole series of monist heresies. One of these is the assertion that conservation can dispense with ideology altogether, being simply a matter for science and technology. Another is the Baconian variant that scientific predictions dictate an ideology. This particular view becomes mischievous when it produces the claim that conservation or ecology can be a political ideal or even a religion.

Ulterior motives

The persistence of the claim that environmental conservation contains distinct moral imperatives begs an explanation. What motivates the deliberate or unconscious need to confuse ideology and environmental prediction? Freud disparaged the 'oceanic feeling' for the environment as simply puerile, but Marx's condemnation of Malthus is more interesting. Malthus's hypothesis was that population, if not controlled by war, pestilence or moral restraint would soon outstrip the ability of the environment to supply food. Marx attacked the theory as a crude defence of capitalists against the proletariat.

Some Marxists now go further and suggest that there are no environmental limits, but they in their turn can be accused of distortion, this time in order to recommend revolution on the grounds that there will be no limits to resources under their kind of socialism. Nevertheless, the germ of truth in Marx's original accusation cannot be denied, and it is not only Marxists who now denounce conservationists as defenders of the privileged classes and nations. That other polemical economist, Kenneth Galbraith, put it this way: 'the conservationist is a man who concerns himself with the beauties of nature in roughly inverse proportion to the number of people who can enjoy them'.

The charge is really that science and ideology are confused in order that certain policies chosen for ulterior motives are seemingly sanctioned by a scientific imperative. This is a common enough tactic. The added dimension in conservation is to play on anxiety about the future. The charge is serious if the aim is short-term sectional gain, but it is still inexcusable in a case for long-term conservation.

An exotic example illustrates the role of the ulterior motive. When earth scientists first encountered the pastoral bedouin in Palestine, they brought with them a history of prejudice that dated to the Greeks and the Babylonians. They also entered an environment familiar to them as 'the promised land', but scientifically unknown. Walter C. Lowdermilk, a soil conservationists, and one of Bennett's protégés, was one of the most influential of these early scientists. His condemnation of the land-use practices of the indigenous inhabitants of Palestine had more to do with preconception than with science. What he was doing, consciously or unconsciously, was using his reputation as a scientific conservationist to play upon anxiety for the future of the Palestinian environment in order to condemn one particular segment of society. He made very little attempt to understand the practices he condemned.

The charge of confusing ideology and science for an ulterior motive can be taken into the conservationist camp. In looking through the evidence of two public inquiries, Richard Grove claimed to have found evidence that ecological or scientific 'values' had been used as a disguise to defend the symbolic value of Amberley in Sussex and Wastwater in the Lake District. But there are no independent 'scientific values'. If the meaning of 'scientific values' had been taken as 'value of science to the community', then this could only have been measured by non-scientific means. Science is too important to leave to scientists. If the meaning were 'value to scientific research', the open question is how to value the research? It was the wider values that the inquiries should have been addressing. Science should have been no more than the predictive part (essential as that was) of an overall argument that defended clearly articulated values of other kinds.

The tactic of some of the objectors was deliberately to confuse these issues. What they quite correctly perceived was that the inspectors and other decision-makers not only held science in high esteem, but above all, found its certainties easier to cope with than confused arguments about other values.

The disguise of the real issues at these inquiries was perpetrated as much by the scientists as by the other objectors, either as a way to jobs for the boys, or because they were as bound up as anyone in the delusions about science. A charge of venality has sometimes been thrown at ecologists, and could now also be levelled at those who raised the old spectre of soil erosion in an attempt to rescue the Soil Surveys of Britain in the 1980s, perhaps in a cynical jostle for government funding. But it is both more charitable and more probable that the scientists were themselves ambivalent about the purpose of science. Some ecologists still join in the general belief in the independence and absoluteness of scientific values, but many soil scientists have unquestioningly been party to the widely held belief that losing soil is an absolute evil.

The truth, unpalatable though it may be to conservationists, is that predictions and evaluations of the future usually lead to action only when they have been useful for some short-term purpose. Their enlistment to ulterior ends does not, however, automatically invalidate them. Scientific predictions, and environmental evaluations obviously need to be allowed for in planning the future. It is sorting the real from the sham that is the problem. This can only be done by analysing both nature and the culture that relates to it. This requires three things: an appreciation of the scientific acceptability of claims about the future; an understanding of the way allocative decisions are made; and a knowledge of the cultural framework within which the predictions and allocations are made. This is a combination that geographers believe they are able to deliver.

ENVIRONMENTAL CONSTRAINTS

Are there environmental imperatives?

Even if the environment does not hand out rules for behaviour or value, some have argued that it does prescribe some very specific advice on how the land should be used: nature here at least demands to be 'obeyed'. At base, this position is no more than environmental determinism, a quagmire from which geographers have slowly dragged themselves. They know that the idea leads along many false trails. Only two can be followed here: the unique 'balance of nature' and the idea that land use should be sustainable.

The inadequacy of the notion of natural balance has been explored theoretically in Chapter 1.2. Its shallowness can be further demonstrated with a very simple example. Soil profiles are usually fed from beneath by weathered material and drawn upon from above by erosion (Figure 4.4.1). These two processes are usually in dynamic equilibrium. When enough soil has accumulated, weathering slows down, because deeper soil provides more 'insulation' for the rock beneath. The soil eventually reaches a depth that is maintained as the surface is slowly lowered: if an erosional episode removes a slice of soil, weathering replaces it from beneath. But the equilibrium is not unique. A sustained increase in the rate of erosion would usually quicken the rate of weathering and produce a shallower but still steady depth of soil. Each erosion rate has its associated weathering rate and depth of soil. A farming régime which encouraged fast, but not too fast, erosion would produce a new, artificial equilibrium. It might even be more acceptable: shallower soils are sometimes less leached of nutrients than deeper ones. Of course, there are limits to this model, for accelerated soil erosion is often a serious problem.

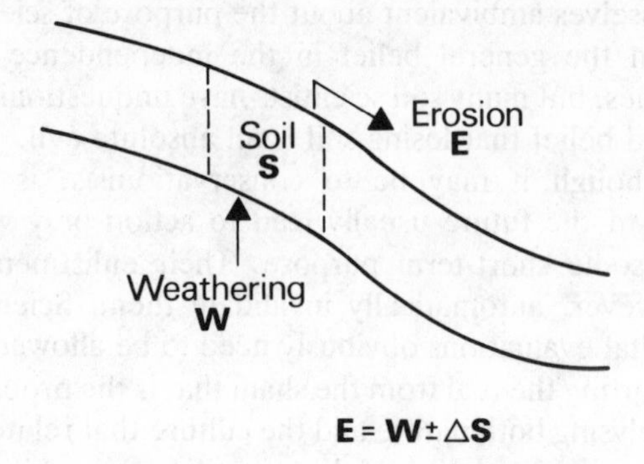

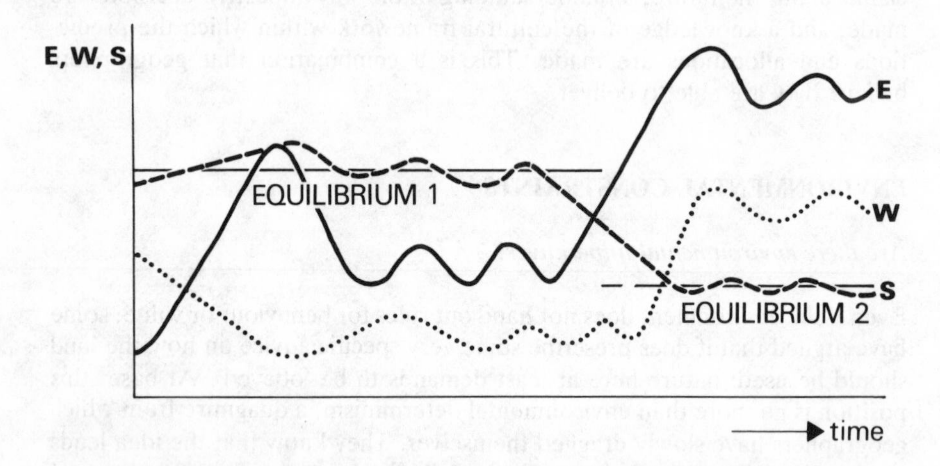

FIGURE 4.4.1 *A soil 'system' showing the balance of inputs of weathering and outputs of erosion*
The graps shows how these can combine to give different equilibrium depths of soil.

The model has been used here simply to show that there are no unique balances in nature.

Theoretical ecologists have developed this point, for they too find that there can be any number of equilibrium points in ecological systems (which are orders of magnitude more complex than the soil). Each particular mix of species, arrived at with a considerable measure of randomness, will have its own unique kind of equilibrium. Even in 'natural' ecosystems, chance disturbances are always causing lurches from one equilibrium point to another. Any economic system that used an ecological one would in turn have to discover its own distinct balance.

The insidious danger in the idea of unique balances is shown by the case of the Sinai/Negev border. There is a strong contrast between the tall shrubs on the Israeli side and the apparent nakedness of the land on the Egyptian side of the northern end of this frontier. Because there is no difference in the climate or soil, the contrast in vegetation could be seen as an accusation that the nomadic bedouin in Egypt are guilty of 'desertification'. The implication is that the bedouin should change their land-use system, and further that the land-use system of the Israelis is to be preferred.

The science of grazing lends little support to the argument. Theory and controlled experiments in southern Israel both demonstrate that several different kinds of equilibrium are possible. First, under grazing, the optimal, and evidently sustainable, strategy for sheep production in these semi-arid areas is to keep the shrubs grazed low. This is what the bedouin do, and though their pastures may well be producing suboptimally, there is no evidence, on the ground, that the system is not in equilibrium. The persistence of the bedouin way of life in Sinai for millenia implies that an equilibrium could have been reached. In fact, they may choose a suboptimal equilibrium for other reasons (see below).

Second is the very different kind of equilibrium in Israel, where there was very rapid regrowth of shrubs to a stable cover after the bedouin were excluded in 1949. This happened because the Israelis made little use of the sands, hoping to set them aside for defence or nature conservation. Incidentally, the rapid recovery of the vegetation in Israel demonstrates that the bedouin system had not damaged the environment irreversibly.

So there is a fair measure of theory and evidence behind the assertion that there are two quite different kinds of balance in the same physical environment, each adjusted to a distinct culture. Both are producing forage at suboptimal levels, each for apparently good reason. Many other kinds of environmental demands could be envisaged in the same area and each would be matched by its own unique environmental equilibrium. There is no environmental argument to use against either the bedouin or the Israeli settlers.

The search for unique balance is the search for an environmental philosopher's stone. The environment gives few clear messages about its long-term use. Nevertheless, although the choice of equilibrium points may be nearly infinite, there are some that are pretty unmistakably unacceptable. This notion will be reconsidered shortly.

A related popular notion about the environmental imperative has more value, but also the seeds of some dangerous ambiguities. It is that land use should extract crops at a rate that it can sustain. An example is the proposal that British farmers who farm in this way should receive some kind of tax incentive. But just what is sustainability?

Consider even the very simple land-use system of the Sinai bedouin. They sow, reap, and tend flocks, but the sum of these activities is never anything like enough to support most families. Their balancing act has had also to call on a large set of other activities since the time of Abram. These include salt-digging, raiding, caravaneering, trading, smuggling, wage labour, guiding and so on, all bringing in income from several south-west Asian countries. Profits from these activities have always been reinvested into the flock. Should some of them be included in a calculation of sustainability, and if so, where is the line to be drawn?

The measurement of the sustainability in an Israeli kibbutz or a British farm meets the same difficulty, but now in an order of greater magnitude. Are the inputs of fossil energy, minerals, and capital to be included or not? Should this year's inputs alone be added in or those or the last decade? The first and most basic problem in all discussions of sustainability is seen to be the drawing of boundaries round land-use systems. These are now seldom smaller than a nation state, and commonly much bigger. Most also depend on capital and other inputs accumulated over many years. No one can deny that sustainability is a vital objective: but equally there are very few systems for predicting or judging it.

Less ambiguous constraints

One way out of this problem is to examine all the components of a land-use system to see if there are some that are more demonstrably unsustainable than others. In farming, the soil is often the critical link, for many managed soils are near a limit where they are about to be lost by erosion, and this loss would certainly bring down the whole superstructure of agriculture.

The soil is in fact one of a number of critical environmental systems that reach unambiguous environmental limits, for in many soils regeneration is almost impossible. The real stony limit is the loss of a shallow soil over a bedrock that weathers very slowly. Another example of a real environmental limit is the loss of a species, for it is hard to envisage its replacement (see Chapter 2.5). Soils and species combine in ecosystems, which can also reach unacceptable limits when they become too simplified to maintain any kind of equilibrium. This happens, for example, where pesticides have been used to excess. A third limit that is often discussed is energy, but this is much more problematical.

The notion that conservationists must draw the line at the loss of soils, species and stable ecosystems, is the same as saying that the primary end in conservation is to retain options for the future. This is a popular definition of the aims of conservation, and not a bad one either. Of course, one could imagine a concerted effort to regenerate a soil, even in the most inhospitable conditions. There have even been claims that species can be recreated, and, it follows, so could ecosystems. The fact that the costs would be prohibitive, and usually far greater than those involved in conservation, merely demonstrates that there is a spectrum of resources that begins with cheaply replaceable ones and ends with soils over hard rock, and ultimately with species.

Conservation always demands efforts that might be used elsewhere. A decision to conserve will always be part of a wider system of allocation. The discipline which concerns itself almost exclusively with allocation is economics, though the idea of replaceability has been considered to be too long term for economic analysis until recently. Conservationists should welcome this change of heart. But there is more to allocation than economics: it also depends on systems of value, belief and morality.

VALUES, MORALS AND REASONS

The environment holds very powerful meanings for people. An attack on it can release howls of rage, and there is a real, practical need to stop shouting and to try to be more articulate about the reasons for conservation.

Accepted reasons

There are two kinds of resource that most governments now agree to conserve, and these conservation programmes have been drawn into the framework of day-to-day decision-making, including economics.

The first widely accepted goal is the conservation of the supply of some material goods. Most states now attempt to conserve soil, and many agree to tax energy specifically to conserve it. But though accepted on paper, these policies come up against numerous conflicts in practice. Strife enters when, for example, a programme to conserve the soil for the good of the nation or of future generations jeopardises the immediate livelihood of its present tillers. Energy conservation too comes up against the fact that to tax fuel is to discriminate against the poor. To control the burning of fossil fuels in order to prevent the growth of carbon-dioxide levels (and so perhaps to halt the rise in air temperatures), or the acidity of the rain, is far from being a simple policy. The trade-offs of the present against the future and of one group against another are never clear-cut and can easily entail prevarication.

The second resource that is very widely acknowledged to need conservation is wilderness, perhaps in a British context better called 'peace and quiet'. Few states do not have some form of national park or reserve system, even if only in name. Two reasons are widely given. The first is that they are baselines for the scientific study of nature. The second is that they act as sacred areas in which the modern world can be escaped. For most people, it is not science, but technology that alienates them from nature, and they crave the alternative of being able to encounter nature through technique (by using their knowledge of survival in the wild, for instance), rather than through technology (the whole panoply of modern housing, heating and transport). Reservation as a policy is also not without controversy, for the declaration of a wilderness deprives people of other resources.

Conflicting values

The questions of distribution in conserving economic resources and wilderness are, nevertheless, of a different kind from the many battles that embroil conservationists. Allocation is the bread and butter of economics and politics, but the conservationists' lot is to engage in endless tangles over seemingly irreconcilable systems of value. Even agreements to conserve material resources and wilderness are themselves based ultimately on values that are not usually dreamt of in the philosophy of most politicians and economists.

Controversies over the value of the environment may often be very messy, but they are far from trivial. The high costs of modifying German power stations, rerouting the Alaska pipeline, and replacing whale products; the violence of the fight to keep power-boats out of the Grand Canyon; the alarm of the comfortably retired community in Amberley, mentioned earlier; and the unprecedentedly long debate about the Wildlife and Countryside Act in Parliament in 1981, all show the power that these issues hold. The immediate and strong reactions that they provoke demonstrate that, far from the readjustment of values that some pundits claim to be necessary, the reasons for conservation are readily appreciated by very many people. Conservation rests on a resurgence of deep common values, rather than on new ones.

Deeper reasons

Many reasons have been advanced for conserving the environment. Their diversity explains the difficulties that conservationists have had and will continue to have in evaluating the world about them. Environmental evaluation will never reach the precision of economics, just because the first value in the environment is quite simply that it has many values. Only five of these underlying reasons will be examined here. They fall on two axes: reductionist to holistic: and teleological (purposive or value-centred) to deontological (duty-centred). In the end they all mesh into distinct cultural views of the world. If these really are the dimensions of the ways in which people relate to the environment, then they should hold interest for geographers.

The first two reasons are firmly reductionist and teleological. The very first is to reduce the problem to material values. This can be done in two rather contrasting ways. Market economists reduce everything to exchange value: pollution can be seen as a factor that reduces the economic value of property, or as an external cost to industry. Recreational resources can be costed by the money that visitors are willing to pay to go to see them. Some Marxists, by contrast, see nature valued simply as part of the naked self-interest of powerful classes. Both explanations require conservationists to pay attention to the distribution of wealth and power in society.

The second explanation is still teleological and reductionist: it is sociobiological or psychological. Our bodies and minds are seen to be adjusted by evolution to the environment. We have inbuilt needs that are programmed to respond to nature. Diverse environments, for example, attract the efficient searchers, who have survived because searching conveys evolutionary advantages. We like diversity too because it offers more security. 'Obey' to the psychologist has a very determinist ring. These explanations require conservationists to discover the principles of environmental psychology.

The third explanation is also teleological, but more holistic and more humane. It stems from semiology (the study of signs). Semiology sees people as incorrigible searchers for meaning, who use the environment as an endless source of symbols. People neither obey nor command the environment but bend it to their needs. Poetry is the most obvious expression of this kind of manipulation, but it has many others. Soil is a symbol of permanence; air and water of cleanliness; eagles of freedom; seals of helplessness; the oak of England. No part of the perceived environment escapes without a meaning.

It makes no difference to the power of the symbols that the soil also symbolises dirt and that the oak is also a symbol of Germany. The vigorous, successful and very popular campaign against spruce in England has lasted for over a generation, simply because to the English spruce is a symbol of the alien, and of the Kafkaesque power of the Forestry Commission. In Norway and British Columbia, on the contrary, spruce symbolises home, and untouched nature.

Conservationists are embarrassed to admit to the power of symbols. They hide it with smokescreens of bogus argument, as they did at Amberley and Wastwater. One of these scientific red herrings is that fewer invertebrates depend on alien trees, such as sitka, than on natives, such as oak. In fact *Nothofagus*, the beech from the southern hemisphere, probably supports more invertebrates than some native species, but is just as alien a symbol. In another diversionary tactic, the now scientifically rejected argument that diversity begets stability, is used to hide the real reason for its appeal as a symbol of the bounty of nature, and as a source of endless surprise.

If symbolism is an inescapable part of environmental evaluation, then conservationists have to understand it, and aim to conserve species and environments that have great symbolic value. But they must also have their eyes open to changes in symbolic fashion, and to the need to keep options open for these.

The fourth set of explanations is holistic but this time deontological. It depends on the belief that we have duties to the environment, or that other things (even stones) have rights. It is those who hold these views who are aghast at Baconian ideas of 'commanding' nature. Ideas of environmental duty have been rather neglected by conservationists and geographers, but they are being wakened to them by vociferous animal rights activists. The notion of obligations is, it can be argued, widely and readily understood. If this is so, conservationists would have to pay much more attention to moral philosophy, and to the environmental conscience of very many people.

The fifth and last explanation is unambiguously holistic. It incorporates elements of all the others, and is the particularly geographical view that environments come to be integral parts of whole cultures: their material wealth, their symbolism, their systems of morality and their functioning. The way in which English culture has assimilated its countryside is an excellent example.

ENVIRONMENTS AND CULTURES: TWO WAYS OF VALUING THE SAME ENVIRONMENT

The culture of the north Sinai bedouin is built around their herds. The needs of the animals dictate the yearly cycle of movement. Families move widely but cautiously during the winter season of migration, avoiding spring and summer pastures and places with pests, but making for where the rain has fallen and for salt-licks. This brings them into close contact with the environment. They understand where salt is most likely to be found. They attempt to understand the weather. Of the 186 plant species known to the modern taxonomists who have visited the region, the bedouin distinguish 180, and have a use for each one. More important, the bedouin social structure depends critically upon the flocks. Because the flocks only provide a very small part of their sustenance, they can be allowed to increase for a more important, and not strictly material reason, namely, exchange in the system of marriage alliances, which in turn control the redistribution of wealth. It is this vital social need rather than strict food production that controls the size of the herd. Thus the pastures are grazed low as an almost calculated outcome of the social structure.

To the Israelis, the Negev just across the border from Sinai means something entirely different. Their occupation of this harsh land is the most potent symbol of the regeneration of their culture, and it demonstrates the superiority of their technology and social organisation, notably the hyper-socialist Kibbutzim. Ben Gurion's argument was that settlement of the Negev showed a pioneering spirit in an environment in which it was not only the climate that was hostile. The Israelis today see the Negev as a recreational lung which, as in most western countries, can also help to conserve their natural heritage.

The almost identical environment of the Negev and Sinai has therefore been integrated deep into two quite different cultures – one essentially Western; the other very Eastern. Rather than 'command' or 'obey' the environment, each has assimilated it in its own way. It would be philistine to call for change in either land-use practice without trying to understand its relation to the environment and culture. And yet that is what some so-called conservationists have done. Environmental conservationists of a newer breed should involve themselves with the people who use the resources they seek to conserve. They must aim to conserve what people value and feel duties towards, always remembering to keep the options open for changing needs.

CONCLUSIONS

It seems that conservationists can be served by the broad approach to the environment for which geographers pride themselves. They need the very best natural science that geography can focus on the environment, and they also need the geographer's critical eye on ideology, material interests, feelings of obligation and cultures.

Geography and conservation are nevertheless fundamentally different. Geographers are academics while conservationists have practical aims and must work in a political environment. This means that they often have to cut academic corners to produce programmes that satisfy their deep convictions. They have often to play sectional interests against each other. They may even feel that they have to indulge in prevarications in which scientific or emotive smokescreens are blown into the path of the opposition. But conservationists have to heed academic analyses of their objectives and reasons. This shows them to be unavoidably complicated. The conservation movement will always be an alliance of many interests, and will have to accommodate them to be effective.

To the geographer, environmental conservation may be an outlet by which their skills can be applied to the work-a-day world. This may be welcome to those who want to bow to the increasing pressure to be relevant. But conservation means much more to academic geography. This chapter has tried to show that the conservation movement is a hand specimen of people adapting to their environment: the first blind, fearful rush that throws out talk of 'command': the reaction with meek suggestions of 'obeying': and finally the discovery that people must learn to live with the findings of more careful analysis (both of environment and of their own society). Perhaps the maxim should be 'nature to be conserved must be assimilated'.

FURTHER READING

Three books can be suggested as a basis for a general introduction. The first is an up-to-date, well-written and thorough expansion of some of the ideas in this chapter. The second is a very useful guide to the ecological basis of conservation, and to some of its current concerns in the countryside. The third is a passionately argued, easily read and comprehensive account of nature conservation:

Pepper D. (1984) *The Roots of Modern Environmentalism* (London: Croom Helm).
Green B. (1981) *Countryside Conservation* (London: Allen & Unwin).
Maybey R. (1980) *The Common Ground. A Place for Nature in Britain's Future* (London: Hutchinson).

Two more specifically ecological studies follow. The first is a collection of essays covering the ecological basis of conversation, the application of ecology to conservation and the organisation of nature conservation in Britain. The second contains further discussion about the Sinai–Negev border area, the question of carrying capacity and the effects of acid rain:

Warren A. and Goldsmith F. B. (eds) (1983) *Conservation in Perspective* (Chichester: John Wiley).

Warren A. and Harrison C. M. (1984) 'People and the ecosystem: biogeography as a study of ecology and culture', *Geoforum*, vol. 15, pp. 365–81.

Looking towards the development of ideas, Williams can be recommended as a seminal work on the place of the country in British culture, whilst the IUCN statement is rather bureaucratically worded, but nevertheless an important support for the doctrine that natural resources must be conserved in the process of development:

Williams R. (1973) *The Country and the City* (London: Palladin).

IUCN (International Union for the Conservation of Nature and Natural Resources) (1980) *World Conservation Strategy* (Switzerland: Gland).

Finally, the journal of the British Association of Nature Conservationists provides a quarterly source of short readable articles on contemporary issues in nature conservation under the title:

Ecos: A Review of Nature Conservation (Funtington, nr. Chichester: Packard Publishing).

Environment, Management and the Future of Physical Geography

4.5 Planning and Management: Physical Geography and Political Processes

Edmund C. Penning-Rowsell
Middlesex Polytechnic

THE CONTEXT OF APPLICATIONS IN PHYSICAL GEOGRAPHY

When it comes to environmental planning and management, physical geographers are not in charge of making the key decisions, nor should they be. The skills of physical geographers in environmental data collection and analysis makes them experts, but experts operating within a wider political system. Within this system the key decisions are made in response to a range of pressures, not just to those exerted by geographers as scientists. This may well mean that our aspirations of applying the findings of physical geography to meet environmental needs may be deflected by a whole host of factors. These could be the inadequacy of legislation, insufficient finance, an inappropriate structure of government and political power, or an ambivalent and apathetic public. A clear scientific analysis does not necessarily lead to a clear planning or management 'solution'.

The corollary of this is that to exploit the findings of physical geography to the best advantage, in planning environmental modification or managing resources, scientists must be aware of the economic, social and political context in which they work. Only with such a keen awareness in this area can the geographer attempt to be a pilot rather than a mere pawn in the evolution of events. Thus, for example, the management applications from Ron Cooke's investigations of the causes of landslides in Los Angeles must link in with the political structures and economic interests within the city's organisation. Without these links the fundamental scientific research may well be ignored and its full value will thus not be realised.

The physical resource base and social context

To help us focus on the key issues, any sphere of environmental modification, planning and management can be seen as operating at the interface between society and physical resources (see Fig. 4.5.1). Seen like this the planning system is a mechanism which seeks to make adjustments to both society and the physical environment, so as to provide for peoples' needs. In turn, the physical environment affects the way decisions are made, which are also affected by their social context. The components of Figure 4.5.1 are:

The social context The myriad goals of society are influenced by the powerful forces and elements in the structure and working of society which direct and guide social activities, including all forms of planning. These guides include the country's legal framework, the structure of government and political power, the national economic performance and policy, and – perhaps less important – our social attitudes and public preferences.

The legal framework consists of prescribed legal rights, restrictions and enabling legislation, together with basic common law. Legislation such as the Control of Pollution Act 1974 and the Wildlife and Countryside Act 1981 affect the use we make of our environment. In turn these enactments reflect political power which is articulated through a structure of government and government departments where many crucial decisions are made concerning environmental use. These decisions in turn are affected by national economic priorities and performance. Thus, for example, the decisions for a slow implementation of the Control of Pollution Act 1974, such that Part III concerning polluting discharges to rivers and canals was still not fully operational more than ten years later, reflects the turbulent economic circumstances in the later 1970s and early 1980s. Ultimately, however, political processes perhaps have to acknowledge – or mould – public opinion and preferences in so far as these determine social priorities and needs.

The resource base Our natural resource base endowment includes land, minerals, soils, flora, fauna, water and atmospheric resources. The characteristics of each affect its potential for planned exploitation or management. For society the most important characteristics of resources are their quality and quantity, and these two are also interlinked. Resource quantity and quality vary geographically. Land for economic agricultural exploitation is not infinite, for example, and to increase this resource requires large energy inputs or major capital investment for land reclamation or improvement. Air pollution has been a serious problem in the past, but only locally, and mineral availability for modern industry – for example, gravel for extraction – is highly localised. The planned exploitation of all these resources is thus intimately linked to these geographical distributions and their quality.

Other key resource characteristics affecting planning and management are the physical interdependencies in natural ecosystems, such that alterations in one part of our environment can have important adverse consequences elsewhere. Thus burning fossil fuels for electrical power generation may result in acid deposition downwind, which can adversely affect vegetation and stream water quality, and thus influence the life-supporting role of soils, water and their associated vegetational complexes.

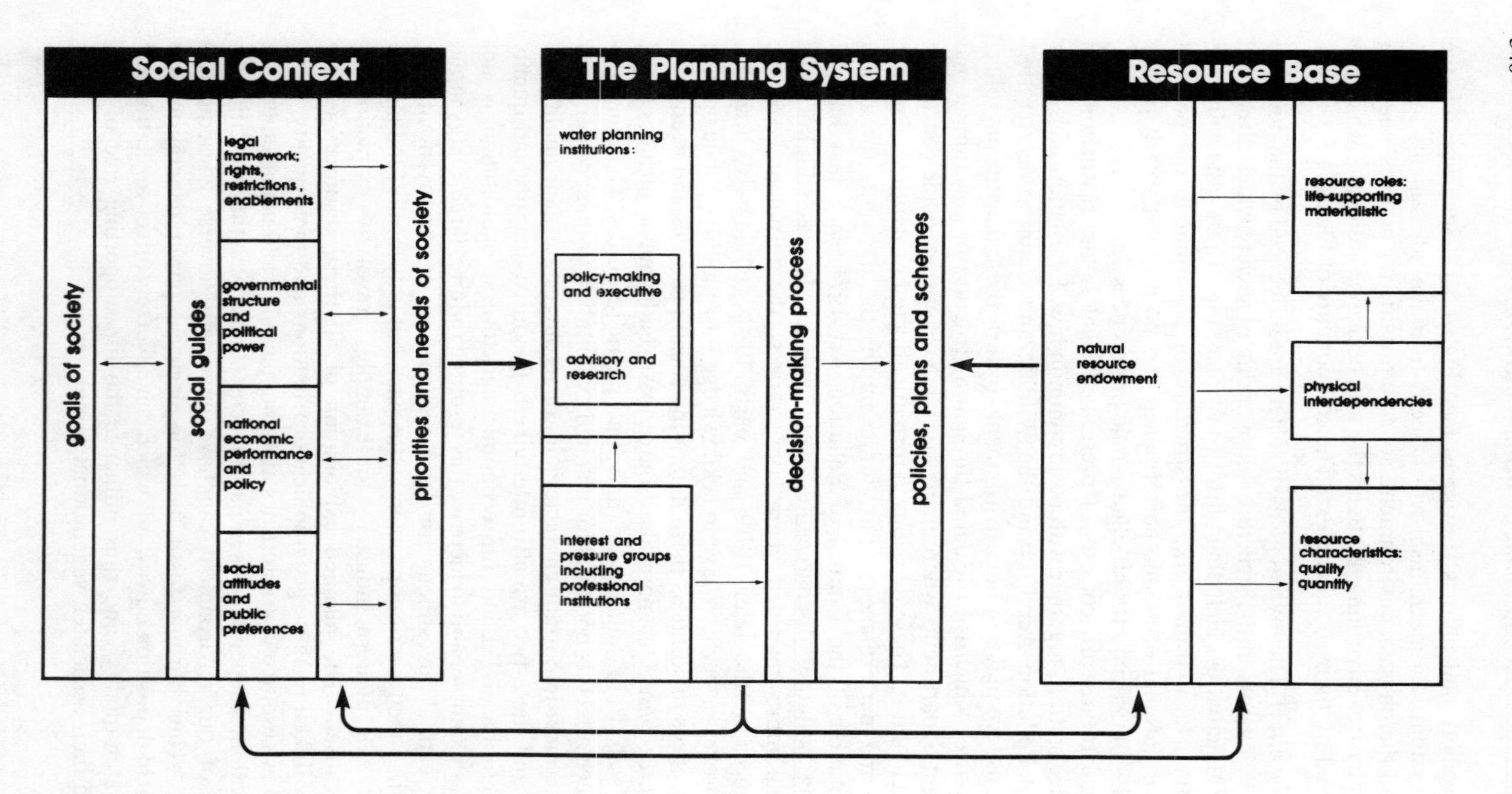

FIGURE 4.5.1 *The resource base and social context to environmental planning and management systems*

The role of resources such as air, land and water to support life is complemented by a less basic and more 'materialistic' role as inputs to wealth generation. Thus minerals are inputs to the chemical industry, water is widely used for cooling, and coastal areas are favourite sites for recreation. The planning of these resources must reflect their essential characteristics, such as their geographical distribution and the ability of the ecosystem of which they form a part to withstand interference without incurring long-term decline in resource productivity. It is undeniable that many other factors affect resource use but also that the physical characteristics of the resource base cannot be ignored.

The planning and management system Decisions to influence the environment in which we live, either intentionally or unintentionally, are generally made within institutions, and largely determined by government through legislation but also in the private sector within large corporations. These executive institutions draw upon research and advisory agencies, many of which contain geographers, and are also affected by interest and pressure groups. These groups attempt to influence the decisions of planning and environmental management institutions by lobbying local and central government and the specialist environmental agencies or private corporations, in an attempt to further their members' ideals and interests. These groups also include geographers who, for example, have influenced decisions in Britain concerning road construction programmes, nature conservation policy and water resource planning.

Decision-making, in theory, leads from a definition of goals and objectives through to plan implementation and monitoring. In practice, decisions may be confused, contradictory and rushed. They may be overwhelmingly influenced by crises rather than rational analysis. The result, nevertheless, will be some form of policy, plan or scheme for environmental modification or to change the social context perhaps by altering patterns of investment such as for air pollution control, coastal protection or wildlife habitat conservation. Alternatively, policies or plans may be designed to change public attitudes through education, for example, by increasing awareness of the complexity of physical processes and the dangers of positive feedback creating continuous and uncheckable adverse environmental modifications. All these plans are affected, of course, by the legacy of history which determines our pattern of resource definition and use, our attitudes and institutional arrangements.

INDIVIDUALS AND SOCIAL STRUCTURES: THE CASE OF FLOOD HAZARD ALLEVIATION

Analysing decisions and making policy recommendations remains far from easy, even given the overall framework and the questions that arise from the conceptualisation in Figure 4.5.1. To understand the detailed evolution of policy resulting from the many decisions on environmental modification, and the roles for geographers as scientists and professionals, requires detailed and rigorous analysis of the local and broader forces creating environmental modification, plans and individual schemes. Most of the examples quoted in this chapter come from Britain but similar situations could be found elsewhere, with detailed differences there reflecting contrasting cultures and political systems. The roles of geographers may also differ in different countries. Thus in Australia, physical geographers have been involved in basic data collection on a wide range of natural hazards, of which floods are a small but important part. In the USA, the emphasis recently has been on the way the relationship between government agencies affects flood hazard mitigation policies. The work of geographers there is heavily overlain by the statutory obligation for Environmental Impact Analysis as an aid to decision-making.

Two traditions which have emerged in this type of analysis are fully discussed in *Horizons in Human Geography*. First, there is the emphasis on the individual decision-maker who is seen as crucial to all decisions. This behaviouralist tradition has been popular with geographers, and sees the individual or at least the local community or institution as the main agent affecting policies. Their perception of problems and solutions is seen as crucial to environmental modification. The approach uses the techniques of interview and questionnaires to elucidate the factors influencing these individuals or agencies and thereby to understand the evolution of events.

The second tradition stresses the overwhelming significance of overall societal forces, and tends to reduce the role of individuals, agencies and even government to being directed solely at the support of those who own the resources to produce wealth within our capitalist society. More subtle arguments tend to put stress on the role of government and major corporations in reinforcing and supporting the capitalist system as a whole rather than supporting individual components such as institutions or single firms and entrepreneurs. This 'structuralist' tradition (wherein the structure of society is seen as the determining factor in all decisions) tends to involve the collection of economic data on the various interests involved in each case and leads to an analysis of the way that particular decisions affect different social classes. Nevertheless, it is difficult in this form of analysis to link local empirical research of the kind germane to physical geography to these large-scale theories of how society as a whole evolves.

Applying these ideas to the flood alleviation field we can see a number of characteristics that support both these approaches. In terms of institutional or 'structural' forces, decision-making is dominated in Britain by central government institutions including the Ministry of Agriculture, Fisheries and Food. Policies are 'driven' by legislation designed to subsidise agricultural production and promote profitable farming. The whole system is geared to the control of agricultural land drainage and urban flood alleviation is only involved because of the close hydrological links between the two. Water Authority policy is thus dominated by the government's grant aid system and by the power of the agricultural 'lobby' which is well represented on Authority Committees and Boards. The interests of all involved are served by the continuation of this channelling of state resources into subsidised local flood schemes and the central domination has been devised to control this system of central government expenditure. The institutional control mechanisms are not a Machiavellian conspiracy, but merely the natural process whereby the flow of financial resources in this field is controlled by government and parliament which are strongly influenced by vested interests. In this particular case these interests are dominantly those of the agricultural and landowning classes.

In the alternative perspective the behaviouralist can point to the importance of government civil servants in drafting the initial flood alleviation legislation, and promoting these powers over the last half-century, and their continued significance in supporting schemes to alleviate flooding. Individuals as flood victims can play an important role in pressurising their local County and District Councils, Water Authorities and Members of Parliament for investment in flood alleviation schemes. Pressure groups promoting or opposing such schemes are very powerful and can crucially affect events. The individual land drainage engineer in a Water Authority is often decisive in recognising flood problems, designing acceptable protection schemes and steering the proposals through his committees and to the Ministry of Agriculture for approval.

Whichever approach to policy analysis is taken can depend upon the political leanings of the analyst: this type of investigation is not value-free or neutral. Nevertheless, the 'flow chart' in Figure 4.5.2 may be helpful. This identifies some of the main forces influencing decisions, non-decisions or decisions not to make decisions (which can be more important than anything else). It recognises that there are important 'local' influences such as the personal characteristics of decision-makers, the power of local groups and the importance of technical data and other considerations. It also itemises the major 'structural' forces influencing decisions and recognises that a major occupation of those in powerful decision-making positions is with legitimating what they do: post-rationalising decisions taken perhaps incorrectly, 'faking' decisions to coincide with a shift in events, and changing aspirations and ideologies to promote the outcomes they desired.

Geographers may find themselves fitting into this 'flow chart' in a number of places. As members of pressure groups they may attempt to influence the decisions made in government departments or large corporations. They may occasionally act as catalysts by highlighting issues which need research or new policies. Geographers as scientists provide technical data and the interpretation of research results to feed into decision-making. These roles may be scientific or managerial, professional or amateur, and each is affected by the political context and value systems in which the individual works.

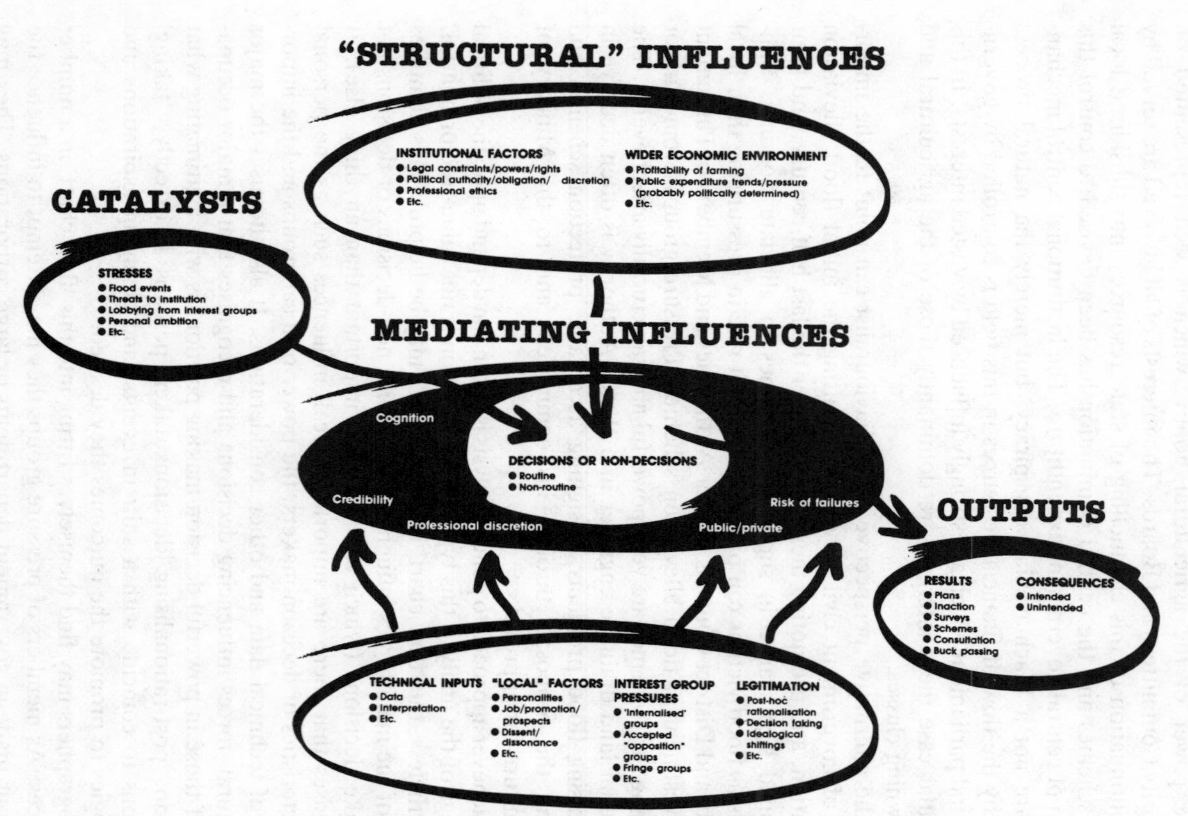
"STRUCTURAL" INFLUENCES
INSTITUTIONAL FACTORS
● Legal constraints/powers/rights
● Political authority/obligation/ discretion
● Professional ethics
● Etc.
WIDER ECONOMIC ENVIRONMENT
● Profitability of farming
● Public expenditure trends/pressure (probably politically determined)
● Etc.
CATALYSTS
STRESSES
● Flood events
● Threats to institution
● Lobbying from interest groups
● Personal ambition
● Etc.
MEDIATING INFLUENCES
Cognition
Credibility
Professional discretion
Public/private
Risk of failures
DECISIONS OR NON-DECISIONS
● Routine
● Non-routine
OUTPUTS
RESULTS
● Plans
● Inaction
● Surveys
● Schemes
● Consultation
● Buck passing
CONSEQUENCES
● Intended
● Unintended
TECHNICAL INPUTS
● Data
● Interpretation
● Etc.
"LOCAL" FACTORS
● Personalities
● Job/promotion/ prospects
● Dissent/ dissonance
● Etc.
INTEREST GROUP PRESSURES
● 'Internalised' groups
● Accepted "opposition" groups
● Fringe groups
● Etc.
LEGITIMATION
● Post-hoc rationalisation
● Decision taking
● Ideological shiftings
● Etc.
OTHER INFLUENCES

FIGURE 4.5.2 *Structural, mediating and other influences on decisions and non-decisions*

PHYSICAL 'PROBLEMS' AND POLITICAL DECISIONS: TWO EXAMPLES

Two examples will illustrate the interaction of physical geographical factors and political–economic forces. Both concern flood alleviation schemes, but the ideas could easily be applied to other circumstances.

Lincoln flood alleviation: the power of vested interests

The city of Lincoln last experienced major flooding in 1947 (Fig. 4.5.3). However, upstream channel improvement since the mid-1960s, designed primarily to protect agricultural land there from flooding, created an increased flood risk in the town's industrial and working-class residential areas. Given this problem, and the paucity of flood records on the River Witham, a mathematical model was 'built' in 1982 by consulting engineers Halcrow Water. This model showed that some flooding was likely every ten years and that the flood occurring once every 100 years would affect over 3380 properties. Table 4.5.1 shows that the likely flood damage at 1982 prices from this major event would be some £43 million including major disruption to the city's enormous engineering firm Ruston-Bucyrus which manufactures cranes and excavators (Fig. 4.5.3).

The full economic benefit from protecting the area from floods of severity up to and including (but not exceeding) this once in 100 years event was calculated to be £18.6 million for a flood alleviation scheme designed to last at least fifty years. This figure of £18.6 million is the accumulated total of flood damages over these fifty years, assuming damage from the full range of floods including that causing the £43 million damage but recognising that the most severe floods are infrequent and that the real value of money declines with time (as reflected in discounting). The most economically rational use of this £18.6 million would be to invest it to protect just the industrial concerns from flooding, perhaps with a ring embankment, while allowing the residential properties affected to continue to suffer a flood risk. This perhaps somewhat callous 'solution' was rejected by Anglian Water Authority as being politically unacceptable given that ratepayers such as those in Lincoln in fact pay the majority of the Water Authority's operating costs.

A system of two upstream reservoirs was thus proposed (Fig. 4.5.3) which would fill with flood water on average once very ten years to 'syphon' off the peak of the flood hydrograph and thus protect Lincoln from the more serious events. However, the farmers of the land affected, including the Lord Lieutenant of Lincolnshire, opposed the whole scheme. Their grounds were that the compensation they were offered by the Water Authority for the legal right to flood this land was inadequate. A public inquiry was therefore called to settle the dispute. In the meantime, the industrial base of Lincoln was severely hit by national economic recession which forced a major closure of part of the Ruston-Bucyrus site. This site alone had accounted for over half

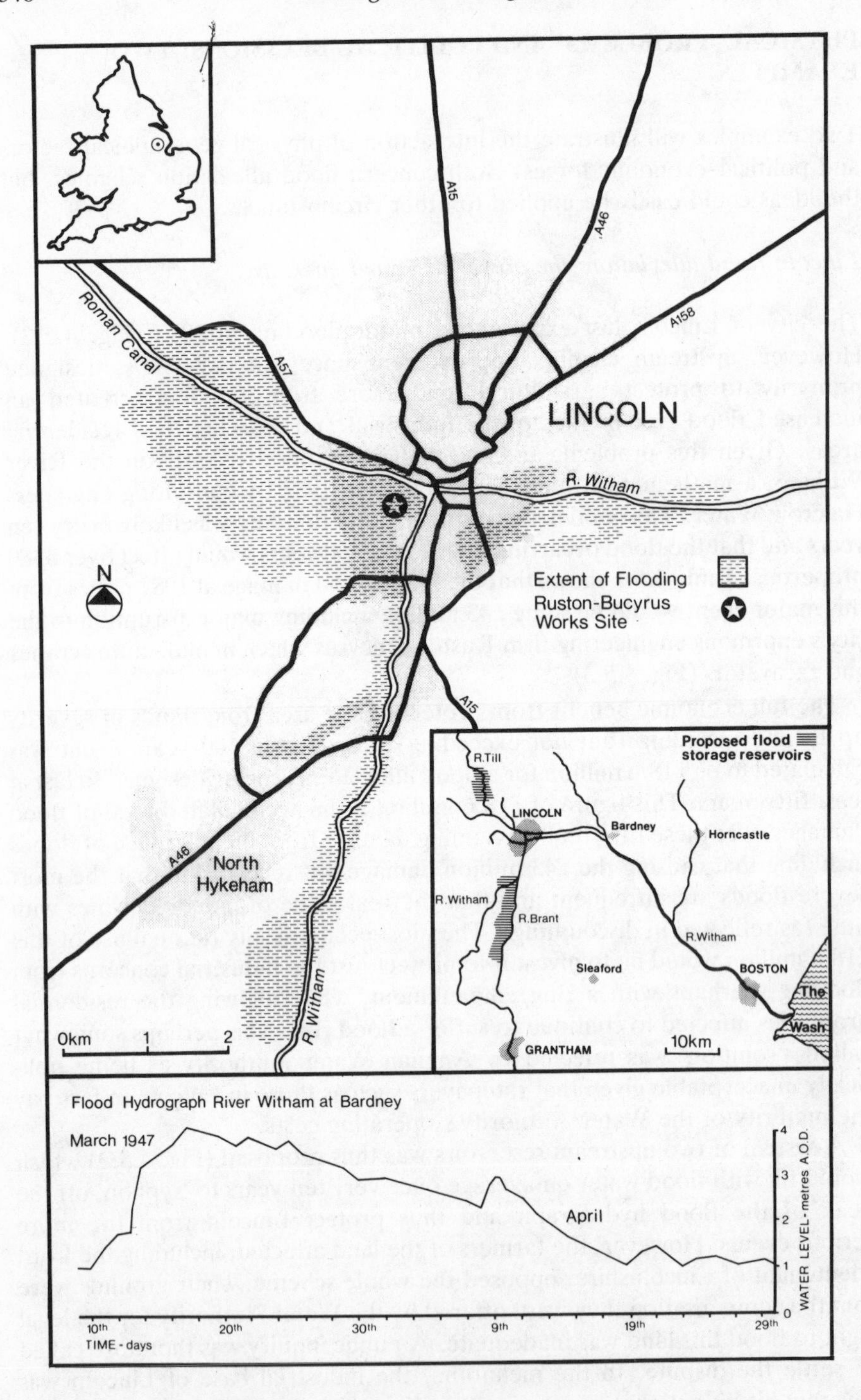

FIGURE 4.5.3 *Flooding and flood alleviation for the city of Lincoln*

TABLE 4.5.1 *Flood damages predicted for a range of flood severities from the 10-year to the 100-year event*

Lincoln Flood Alleviation Scheme: benefit assessment results (event damages)

	Return period of flood (years)							
	5	10	20	30	40	70	100	
Direct Damages								
Sector 1	0	44	119	1111	1773	2884	3152	Residential
	0	24581	67606	554317	151203	2767268	2952506	damage (£)
Row percent	0.0	0.8	2.3	18.8	53.2	93.7	100.0	
Col percent	0.0	4.9	11.3	7.0	4.4	7.0	7.4	
Sector 4	0	1	1	1	1	2	3	Hotels, etc.
	0	9921	11720	12847	13411	24966	27410	damage (3)
Row percent	0.0	36.2	42.8	46.9	48.9	91.1	100.0	
Col percent	0.0	2.0	2.0	0.2	0.0	0.1	0.1	
Sector 5	0	3	4	78	111	145	151	Retail and related
	0	465052	517121	802125	2054200	2757357	2872690	damage (£)
Row percent	0.0	16.2	18.0	27.9	71.5	96.0	100.0	
Col percent	0.0	93.1	86.5	10.1	5.7	7.0	7.2	
Sector 6	0	0	0	4	4	8	9	Prof. offices
	0	0	0	25890	30943	76297	78774	damage (£)
Row percent	0.0	0.0	0.0	32.9	39.3	96.9	100.0	
Col percent	0.0	0.0	0.0	0.3	0.1	0.2	0.2	
Sector 7	0	0	1	7	13	23	24	Public buildings
	0	0	1540	12110	26710	43971	45707	damage (£)
Row percent	0.0	0.0	3.4	26.5	58.4	96.2	100.0	
Col percent	0.0	0.0	0.3	0.2	0.1	0.1	0.1	
Sector 8	0	0	0	14	31	44	44	Industrial
	0	0	0	6537866	32376642	33709681	33821639	damage (£)
Row percent	0.0	0.0	0.0	19.3	95.7	99.7	100.0	
Col percent	0.0	0.0	0.0	82.3	89.8	85.6	85.0	
Total directs	0	48	125	1215	1933	3106	3383	Total properties
	0	499554	597987	7945155	36073109	39379540	39798726	damage (£)
Row percent	0.0	1.3	1.5	20.0	90.6	98.9	100.0	
Indirect losses								
Trade loss	0	6070	6070	33710	101004	137308	142114	Losses (£)
Traffic disr'n	0	0	552	2323	71297	52335	73270	Losses (£)
Indust. prod'n	0	0	0	0	1923077	3307591	3307591	Losses (£)
Emerg. services	0	5000	25000	45000	60000	75000	75000	Costs (£)
TOTAL	0	510624	629609	8026188	38168487	42951774	43396701	Damages/losses (£)
	0.0	1.2	1.4	18.5	87.9	98.9	100.0	Row percent

the benefit of the whole flood alleviation scheme and it thus appeared in 1984 as if the whole carefully analysed scheme to protect Lincoln might be jeopardised by the joint forces of farmers' opposition and national economic recession.

However, this was not to be the case. A full reappraisal of the consequences of flooding the industrial premises in Lincoln by researchers at Middlesex Polytechnic showed that the economic recession had had two effects. First, flood damage potential was indeed reduced, owing to the factory closure. Secondly, however, the costs of flood protection were also reduced by civil engineering building contractors reducing their tender prices to seek contracts in the difficult economic circumstances which had radically reduced the demand for such construction work.

The situation still hung in the balance owing to the farmers' opposition. Nevertheless, it transpired that this opposition was more apparent than real. Immediately before the public inquiry, at which the farmers would probably have received considerable adverse publicity for opposing the flood protection of the 3152 houses, they negotiated a new deal. With the tactic of forcing a public inquiry, against the background of their considerable political power *vis-à-vis* the Water Authority, the farmers effectively doubled the compensation on offer and neatly 'won' the case.

The Soar Valley drainage proposal: a Water Authority victory?

The 1983 House of Lords judgement on the Soar Valley drainage scheme proposed by the Severn–Trent Water Authority was an important event in the long-standing saga concerning the environmental consequences of agricultural land drainage schemes. The case for the scheme to drain the Soar Valley between Leicester and the Trent confluence near Nottingham (Fig. 4.5.4) was based primarily on the anticipated increased agricultural production arising from lowered water tables, which would allow more intensive dairy farming and some conversion to arable cropping. The valley has a history of flooding affecting agricultural land, the scattered villages in the valley, and also severely disrupting the cross-valley road communications.

The Severn–Trent proposal was legally complex, however, because the Soar in places takes in a statutory navigation, designed in the early nineteenth century, which alternates between the river channel and the canal which was constructed to bypass river meanders. This complication meant that the drainage proposal could not be carried out by the Water Authority alone without additional powers being vested through a special Parliamentary Bill.

This Bill was promoted unopposed until the Council for the Protection of Rural England (CPRE) was alerted to this and promoted opposition through Lord Beaumont. In parliamentary terms, however, this opposition came too late. Nevertheless, by special dispensation a Select Committee was established to determine whether the Authority had paid sufficient attention to economic and environmental considerations in its proposals.

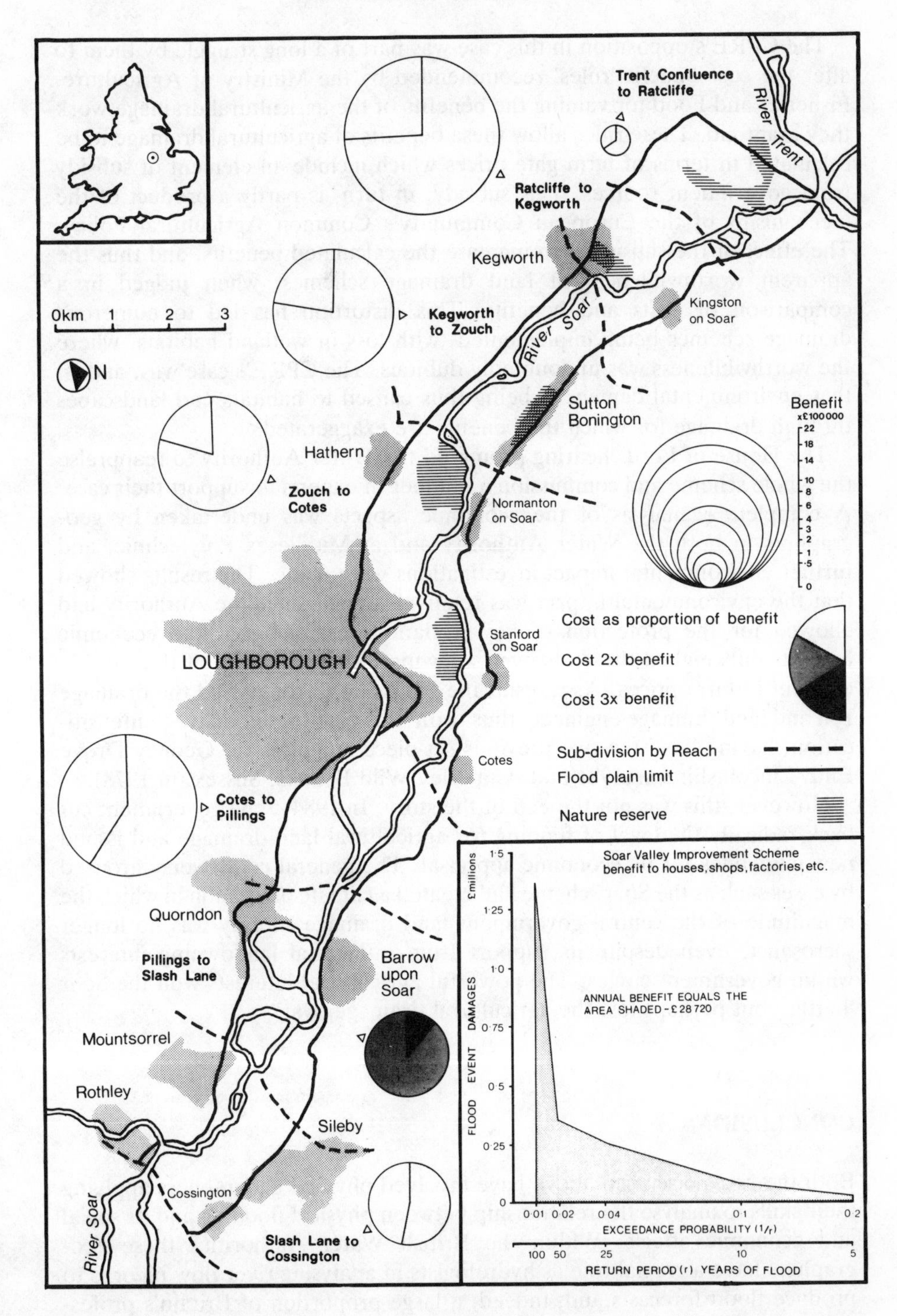

FIGURE 4.5.4 *The benefits and possible impacts of drainage of the River Soar Valley, Leicestershire*

The CPRE's opposition in this case was part of a long struggle by them to alter the cost–benefit 'rules' recommended by the Ministry of Agriculture, Fisheries and Food for valuing the benefits of the agricultural drainage work they grant-aid. These rules allow these benefits of agricultural drainage to be calculated in terms of farm gate prices which include an element of subsidy from government sources. This subsidy, in turn, is partly a product of the mechanisms of the European Community's Common Agricultural Policy. The effect of the subsidy is to increase the calculated benefits, and thus the apparent worthwhileness of land drainage schemes, when judged in a comparison of costs and benefits. This distortion has led to numerous drainage schemes being implemented, with loss of wetland habitats, where the worthwhileness was undoubtedly dubious. The CPRE's case was, and is, that environmental damage is being thus caused to habitats and landscapes through drainage for which the benefits are exaggerated.

The House of Lords hearing prompted the Water Authority to reappraise the whole scheme and commission a number of experts to support their case. A complete reanalysis of the economic aspects was undertaken by geographers in both the Water Authority and at Middlesex Polytechnic, and further environmental impact investigations were made. The results showed that the environmental impact was relatively small, since the Authority had allowed for the protection of key wetland sites, and that the economic benefits outweighed the costs even allowing for the removal of the subsidy element in farm prices. As a result the House of Lords passed the drainage Bill and land drainage engineers thus 'won' this case for agricultural intensification, having 'lost' the two previous set-piece enquiries (at Gedney Drove End, Lincolnshire, in 1981 and Amberley Wild Brooks, Sussex, in 1978).

However, this was not the end of the story. In 1984–5 the government cut back radically the level of funding for agricultural land drainage and issued new guidelines for its economic appraisal. The general controversy aroused by cases such as the Soar scheme had created a climate of opinion in which the magnitude of the central government land drainage subsidy was no longer sacrosanct, even despite its support from influential landowning interests within government circles. The powerful agricultural interests won the Soar 'battle', but perhaps lost the agricultural drainage 'war'.

CONCLUSIONS

Both the cases discussed above have involved physical geographers applying their skills to analyse the relationship between physical flooding and its social and economic effects. Within the British Water Authorities these geographers have a crucial role as hydrologists in analysing river flow records to produce flood forecasts and, indeed, a large proportion of Britain's professional hydrologists have geographical backgrounds. Physical geographers at

Middlesex Polytechnic have analysed and used these flood forecasts, and their associated flood extents, to quantify the economic impact on the local, regional and national community. In both cases these contributions have been made within teams including engineers, hydrologic and hydraulic scientists, agricultural economists, landscape architects and ecologists. The issues involved have been scientifically, technically and politically complex. The ultimate decisions and results took into consideration a careful appraisal of the physical factors, but were also crucially affected by powerful economic and political forces.

If physical geographers want to continue to maximise their effectiveness in environmental planning and management, they must understand the social, political and – above all – the economic context in which they operate. Attempting to change the political power structures and economic environment in which decisions are made – pushing 'against the tide' – is a difficult, and probably a fruitless task. This is because these forces in our society are so powerful.

However, it may be possible to work 'with the tide', by analysing the various interests involved in a particular environmental issue and advocating research and policies which both have the support of those interests and a wider scientific credibility and value. Working 'within' organisations may also provide opportunities for changing policies or advocating particular research agendas.

This approach applies equally to the flooding problems discussed in this chapter as to the many other areas of concern to physical geographers. For example, remote sensing is highly significant in military terms as well as for resource management, and the skillful scientist may be able to exploit the subject's strategic significance to obtain support for other remote sensing research and application into topics such as crop failures in the Third World or patterns of gully erosion, to name just two areas. At a more local level, the glacial geomorphologist may work best in co-operation with the companies seeking gravel extraction, rather than wholely against them. The soil scientist can contribute to wise agricultural intensification, rather than ignoring such tendencies completely. Fundamental research into polar geomorphic processes may be a valuable 'spin off' from providing essential data for exploiting the region's strategic significance. Physical geographers analysing hazards can help the insurance or building industries, but only if they provide the right analysis at the right time.

This policy of working 'with the tide' may involve attempts to compromise the scientist's objectivity. We can recognise that scientists are not neutral but they should approach their work with objectivity, rather than merely providing data solely to support particular interests or viewpoints. Pressures will be exerted, however, to provide decision-makers with what they consider to be the 'right' answers, and the alternative may simply be being ignored. Physical geographers operate within a complex and overwhelmingly important social context and if they ignore this they will misread their true environment.

FURTHER READING

For further material on the politics of flood alleviation and agricultural drainage, see:

Penning-Rowsell E. C., Parker D. J. and Harding D. M. (1986) *Floods and Drainage: British Policies for Hazard Reduction, Agricultural Improvement and Wetland Conservation* (London: Allen & Unwin).

Handmer J. (ed.) (1987) *Flood Hazard Management: British and International Perspectives* (Norwich: Geobooks).

An analysis of the overall water resource planning and management field is given by:

Parker D. J. and Penning-Rowsell E. C. (1980) *Water planning Britain* (London: Allen & Unwin).

For a broader analysis of resources and physical geography, with some stress on institutional–political factors, see the first of the following references. The second provides a case study which is an excellent analysis of the political processes behind a major environmental issue:

Mitchell B. (1979) *Geography and resource analysis* (London: Longman).

Blowers, A. (1984) *Something in the air: corporate power and the environment* (London: Harper & Row).

The final suggestion offers the opportunity to pursue in depth the wider philosophical background to environmental policy, planning and management:

O'Riordan T. (1981) *Environmentalism*, 2nd edn (London: Pion).

4.6
From Description to Prescription: Measurements for Management

Malcolm Newson
University of Newcastle upon Tyne

In consolidating the management orientation of many aspects of physical geography noted in previous chapters, we can hardly do better than to note that in the first edition of *Progress in Physical Geography* in 1977, R. P. Moss offered a methodological structure for the subject as a means of distinguishing 'between good and bad geography'. Possibly the biggest change in our subject since then is that we feel less need to judge our work by the strictures of epistemology, the theory of knowledge and its acquisition. 'Never mind the quality, feel the weight' might be the new slogan for some, especially those who, like the author, have had to leave the world of induction and deduction for the world of production. Even if we do not actually apply the weight criterion to our work, the age of declining national wealth and financial cutbacks has promoted others who use it to sit in judgement on our efforts.

TOWARDS PROFESSIONALISM

Moss's article has proved a useful model for what has happened in physical geography because he concluded by suggesting what we could regard as the second great interface for the subject – that between prediction and prescription. If we consider the move from qualitative description to statistical prediction (commonly called the Quantitative Revolution in the Cambridge and Bristol coffee bars in the 1960s) as the first hurdle in gaining the respect of the scientific community, real maturity has been attained at the second stage, that of putting prediction to the test in management. Moss warned that whilst the strong tests inherent in deductive strategies were not out of place at the interface with management or planning, those two fields apply their own constraints to science by setting goals (see Chapter 4.5). Nevertheless, he hinted, systems analysis would see us through because it stresses explanation as a fundamental platform from which to launch into either one of the newfound arenas for the subject – prediction (because it is numerate) and prescription (because it is applied).

In the coffee bars of Bristol (I cannot speak for Cambridge) two decades ago, it was notable that physical geographers were worried most by the time they were forced to 'waste' by gathering their own data. Not for them the Census returns or the warmth of the Public Records Office but the daily (nightly!) task of installing weirs, making or borrowing samplers and, if time was left, designing experiments. It is no accident that the first data compilation available to many who wanted to gain analytical skills, without the initial Wellington-boot phase, was the Surface Water Yearbook. Not surprisingly, therefore, it is through hydrology and fluvial geomorphology that physical geography was first given the luxury of what are now called data bases (though climatologists had long used theirs for mapping). At last there was time and energy to spare for experimental design, learning statistics, calculus or modelling and for putting those heady surfaces through hyperspace. Regrettably, the gifted modellers grew slightly away from those who remained committed to fieldwork, and to this day there remain very few of us who can do both with equal ability or even link up with a complementary soul. Most importantly, however, both types of approach have attained rewarding levels of influence both inside and outside academic life. What has sustained them is competence; what has maintained physical geography as a cognate field is relevance. This is an era in which society needs generalists and communicators, but they must also be highly competent. Arguably, society gets no closer to the modern version of the eighteenth-century complete man or woman than in an enthusiastic geographer!

However, the purpose of this chapter is far from that of historical review; neither is it to indulge in self-congratulation on the maturity rapidly being attained by our subject. It is to suggest a structure for the way in which physical geography applies itself to the urgent tasks of environmental management, thereby answering a powerful contemporary poser: 'Have we become professional?' There is a subsidiary question: 'Have we moved from an ethos to an ethic of physical geography?' To find out, we must particularise. The reader is offered a somewhat biased review, or rather a partial one – restricted to those areas of environmental management in which the author has had some experience of both research and application. The review begins with flood prediction, which allows human communities to protect themselves against a major natural hazard. Later on, it deals with the conservation of plants and animals in wetlands which are often the victims of land drainage and flood protection. Clearly, in the application of his new skills, the physical geographer runs with both the hounds and the hare, often on alternate days; yet he can retain objectivity. In practice, the realisation that the reverse of the resource coin is usually hazard, and vice versa, allows him to deal with both.

ASPECTS OF MANAGEMENT

Flood prediction in Britain: Geographical inputs

The success of engineers in the field of river management can be judged from the fact that rivers have been successfully managed for 3000 years, whilst scientific hydrology is but 300 years old. In Britain, engineering experience has served the nation well, though it is evident that notable failures, such as dam disasters, have prompted the quantum leaps in flood prediction. The dam collapse at Dolgarrog (1925) led to the Reservoir Safety Provisions Act (1930) and to the compilation of an 'experience curve' linking peak recorded (estimated) flood discharges in Britain to catchment area. This curve formed the basis of adding further data for forty years, to include such an 'act of God' as the Lynmouth disaster. There were few river gauging stations in Britain at this time and so little opportunity existed to investigate flood parameters other than the peak flow (which can be estimated by field survey at ungauged sites using debris marks). The sample was too small to split and so no attempt was made to look at the spatial variability in flood response.

For the first time, in 1960, a map appeared of regional variability in the flood intensity/catchment area relationship; it showed what physical geographers might take as axiomatic – greater likelihood of high floods in the wet uplands of the west. But this was still mapping by engineers, who remained largely ignorant of the ability to quantify landscape morphometry which formed the very basis of the early period of quantification in geography. The United States led the way in the incorporation of morphometry with a series of massive regional analyses to predict the mean annual flood.

Applying morphometry forced a major and disappointing realisation; namely, that only the simple, repeatable indices, measurable from the most available maps, would be acceptable as input into a national study of flood predictions. Nevertheless, the fieldwork wing of the subject had, by the time of the major British flood study of the early 1970s established that the stream network and the saturated areas served by the network were the twin bases of flood response. It was therefore possible to concentrate mapwork effort (for 1100 catchments!) upon variables such as stream frequency and soil–climate factors leading to saturation. These variables attained a high degree of success in predicting a wide range of flood parameters, and the eventual guidebook to flood prediction throughout the British Isles has one complete volume (of five) devoted entirely to maps.

Water resources and upland use: from measurements to maps

Engineering hydrology perceived a national need for the Flood Studies of the 1970s; there was something of a carrot, therefore, in persuading the professionals to use new and unfamiliar facts and figures from another discipline. However, fieldwork in physical geography has also produced sticks with which to beat other disciplines who were hitherto unaware of the relevance of a spatial dimension to their topic. During the 1970s there was a rapid expansion of research catchment studies in Britain, mostly in the uplands in order to escape artificial influences (there seldom being sufficient resources to set up conclusive studies of the effect of these influences). Geographers, driven largely by academic curiosity, participated in a revolution in the way hydrologists consider slopes to provide runoff to channels. Those 'partial areas' of the drainage basin which provide most runoff during floods are still the focus of research and application in hydrological modelling. However, it was to the phase of the hydrological cycle preceding slope runoff – i.e. evaporation – that physical geography, rather surprisingly, became drawn.

The effects of the progressive afforestation of the British uplands upon runoff volume and timing are of crucial interest to reservoir operators (largely engineers). Evaporation is a process understood best by physicists; so where does physical geography come in? First, the strength and versatility of fieldwork in the subject has often put geographers in the responsible role of gathering catchment data, even of designing instrumentation to do so. Secondly, our talent to see an underlying spatial variable behind what appear to be conflicting results from different environments can assist in providing general conceptual models, even if other disciplines are needed to make them work. Thirdly, a familiarity with maps makes it easy both to know them as data sources and to use them as vehicles to turn results into guidelines.

In the case of afforestation and water resources, the efforts of the multidisciplinary Institute of Hydrology have illustrated all three aspects of the geographical contribution. First, the major research catchment efforts have been staffed largely by physical geographers and environmental scientists. After ten years of careful data collection in the Plynlimon catchments, mid-Wales, it was possible to conclude that conifers more than doubled the rate of evaporation, largely through the action of the interception process. The water industry did not take up the implications of this confirmation of the earlier work by Frank Law until the Forestry Commission published expansion plans for plantations in the uplands. In a short period it was necessary to provide a simple basis for extrapolating from Plynlimon to the rest of Britain and to explain why results from lowland forests showed there to be little or no influence on evaporation.

Within two years the water industry had produced a map showing those areas of Britain with a 'water constraint' to further afforestation, effectively a 'hands off' map, at least until further research is completed. Physicists provided the basis for a general model explaining regional discrepancies in terms of microclimate but it cannot be used for specific applications without sophisticated weather stations. Consequently a simple method was required and a single equation linking rainfall and evaporation figures from maps to the interception effect, and thus to the effects of afforestation, became the basis of the industry's response. Of key importance to the method is the relatively extreme gradient of climatological variables from west to east, upland to lowland, in Britain, especially rainfall duration and windspeed.

Flood prediction on ungauged catchments derives hydrology from mapped variables. Applying the results of research catchment studies (whether of forest evaporation or slope runoff) puts hydrology on a map for simple utility. It is possible that geographers had considered that they would grow away from maps in their new scientific role. Acting to counter that notion is the power of a regional or national prediction or prescription in map form (since it crosses disciplinary boundaries) and the realisation that, although Britain is a developed country, it is badly mapped for some key variables and is mapped hardly at all in its remoter regions. Thus, for the full objective planning of land use around reservoirs, maps of sufficient detail for existing vegetation, measured rainfall, windspeed, etc. are not available.

Soil and water management in Britain

Possibly the long history of human occupancy in Britain makes us rather unwilling to accept that artificial influences on the natural environment are real. In more extreme climatic zones and more recently settled continents the effects are all too obvious in flash flooding, soil erosion and river instability. In Britain, there was for many years a virtual moratorium on official research into soil erosion on agricultural land. Papers suggesting man-made causes of increased flood peaks have been largely ignored. However, physical geographers have quite a 'nose' for the resource/hazard paradox, and have slowly begun to establish the magnitude of the effects. It is always tempting to monitor and publish results for spectacular examples in order to shock a reactionary audience. There is thus a danger that the small scale, at which most field research is practical, will yield more spectacular results than it is wise to extrapolate to management scales. However, this is no longer the case in lowland soil erosion; 'improved' agriculture has extended the area at risk quite considerably since the war, and despite the prevailing low intensity of rainfall in Britain several 'hot spots' have been identified in terms of rainfall energy. One of these, the Peak District, has been the topic for urgent studies of late. Here the erosion risk is to land of high amenity and conservation value. There is considerable visitor pressure, and another influential area of research in physical geography has been peat erosion and the influence on footpaths of trampling by sheep and humans.

Reservoir sedimentation rates higher than previously recorded were among the more startling conclusions of the Peak District study, and strangely one of the remedial suggestions was the afforestation of the area. This idea enshrines observations abroad that soil erosion follows tree clearance; indeed the same was true of Britain in the Neolithic and Bronze Age. However, modern plantation forestry requires more ground disturbance and recent studies of sediment yields from coniferous plantations reveal order of magnitude increases in erosion, mainly along drainage and plough lines. Two major cases of reservoir sediment pollution have been reported downstream of recently ploughed upland slopes. In this case the geographer, having grown up with pictures of contour ploughing abroad, wonders why soil and water managers here see no problem in draining or ploughing normal to the contour. A possible answer is that machinery is unstable when travelling sideways across a slope. The oft-quoted need to 'get the water away' by draining ignores the fact that one man's loss is another's gain in terms of flood peaks and sediment load!

The problem of scale in applying our results in physical geography is nowhere more pressing than in assessing the influence of land management changes on flood flows. Whilst it is relatively simple to prove the increased intensity of floods from ill-drained peat when ditched, or the reduced intensity of floods from permeable soils when tile drained, how does this very local result relate to the scale at which it is politically relevant and therefore managerially acceptable? In the case of flooding there are other complications, including non-stationary climate and the fact that locally accelerated runoff may in fact ease flood hazard if the natural response time is much longer (and vice-versa). The scaling-up process is a demanding research challenge for both fieldworkers and modellers in geography; we should aim to become capable in it simply because, as a working guideline, scale is eschewed by the other disciplines involved.

Land drainage and river channel stability

Perhaps the most surprising example of the immense relevance of physical geography, but of its general exclusion from application, is that of fluvial geomorphology. There is no doubt that British river environments are not as spectacular as those abroad. We generally lack the climatic and physiographic proclivity to river disasters caused by regular extremely high discharges. Therefore, our rivers may almost be considered steady state systems in the engineering time-scale; a thousand year flood in 1947 altered the Great Ouse very little. Even in the uplands our slopes are generally blanketed by soil. There is a tendency for this cover to be disrupted during periodic phases of upland development, and there has been a general post-glacial removal of sediments from the uplands to the adjacent piedmont river zone of gravel beds and composite banks. These are our active rivers and the most likely to show complex response rather than the predictable 'regime' behaviour so essential to the river engineer.

The main job of river engineering is flood protection and land drainage – designing channels or maintaining them to a capacity and grade whereby they can accept the excess water from farmland and conduct it seawards without hazard to other farmers and settlements along the way. It tends to be a compromise, aided in Britain by our short, relatively steep rivers. Thus piecemeal, largely structural, approaches to creating channels have had a good measure of success. It would possibly be the geographer's prejudice that such efficient channels would need comprehensive design from source to mouth, but in practice most schemes are local.

Erosion and deposition in rivers are not considered as problems by our legislation, except where they threaten the capacity of the floodway or undermine flood banks, etc. Thus erosion protection is less often provided in 'natural' headwater areas than in those reaches where the river, as an efficient flood carrier, cannot be stable without erosion protection. Our most reactive and changeable channels, therefore, have neither the economic relevance nor are managed sufficiently for geomorphological approaches to predicting river erosion to be easily accepted as relevant.

Nevertheless, physical geography is perhaps poised to make a contribution because there is an increasing demand for 'natural' rivers, at first in terms of bankside habitats and amenity value but, if the lesson of 'channelisation' in the USA is a guide, eventually in terms of equilibrium channel thalwegs and planforms – using, rather than combating, natural processes. Is our contribution adequately prepared? The published evidence is that two decades of river research have yielded an impressive body of empirical knowledge about British rivers. It is no longer necessary to teach fluvial geomorphology from North American examples. That the material is largely site-specific and highly empirical should not detract from its relevance. Engineers realise the value of local information and, yet again, the strength of our subject in the design and deployment of field techniques is of advantage. Only some of the empirical lines of enquiry are weak. For example, it is unlikely that a continuing emphasis on hydraulic geometry will be of benefit to prediction; neither will continuous regional survey of channel change achieve more than declining scientific yields. For a bigger input into channel management, fluvial geomorphology has either to perform field experiments, to structure a process-based model of channel morphology (measuring sediment transport, flow resistance, etc.), or to improve the predictability of channel reaction to external or internal stimuli. The message of fieldwork in Britain is that reservoirs, towns, mines, etc. all produce channel changes in capacity or planform or both. The answer needed by the engineer is where, how, by how much and, most difficult of all, when? This field is perhaps a case of a large volume of material of relevance waiting for a bridging simplification, or a legislative change. To take the argument a stage further requires a change of example.

Conservation of wetlands

Fluvial geomorphology has the potential to guide civil engineers towards channel designs which are acceptable for conserving or creating wildlife habitats. Physical geographers have also become directly involved in the fight to save Britain's dwindling wetland habitats. This was once a very boggy island, but though the successful settlement of lowland Britain's clay vales is an impressive tribute to land drainage engineering, the loss of further wetlands now would seriously diminish the European stock of wet wilderness.

In fact only the demands for intensive post-war agriculture have brought us to a state in which rivers no longer spill on to their floodplains in winter. If they do return to their natural flood safety valve, the nation calls it a disaster. Having lost many floodplain washlands through flood protection and fens to pumped drainage, the remaining smaller wetlands are most vulnerable. The spread of field drainage, impressively mapped by physical geographers in recent years, makes most landowners willing to 'reclaim' these 'wastes' for food production. The smaller the wet area, the larger its relative perimeter open to drainage influence.

The obvious route to a conservation role for physical geography is through biogeography. However, after a mainly classificatory phase, that division of the subject is often now involved with systematic studies of the soil, plant and animal relationship at the detailed level of nutrient or energy cycling. Such systems calibrations are most relevant to autecologies or crop production. Conservation of complex habitats and communities of plants or animals have remained mainly the province of the biologist – though geographers have a contribution to make. For example, in dealing with wetlands, biologists cannot achieve conservation by purchasing land and fencing it off against external disturbance. Their resource can ebb away beneath the fence to the drains of their neighbours. More importantly, the crucial areas of supply (e.g. springs in a fen basin) and removal of water (e.g. an open ditch) often remain in unsympathetic hands outside their management control, even after extensive land purchase.

To the geographical conservationist, the hydrology of wetlands is foremost a problem of definition. The myriad of terms used by biologists often scares engineers and agriculturalists into adversarial stances; simpler structures involving basic hydrological systems and quantities are preferable prior to management. The water balance of a wetland is difficult to measure because of uncertainties about boundaries. Nevertheless, very simple field observations of flow routes, water temperature/conductivity and so on will frequently provide a first approximation. It should be remembered that much of this work has an urgency imposed by the threat to the site. It is often the case in applied study that the bigger the problem, the less time there is to work upon it, a situation which plumbs the depths of any ad hoc definition of 'professional'.

If there is time for further hydrological study it normally attends more to the storage term in the water balance than is usual; unfortunately the peat substrates often encountered are among the least researched in the key terms of conductivity, saturation, etc. Seasonal fluctuations and extremes are of more than normal interest because of the vulnerability of the site during dry spells and the potential resource to the site during very wet spells. The effect of plant cover on water content is also of interest, particularly since it is impossible to ignore the fact that hydroseres tend to evolve rapidly and it is improper to assert external causes for an internally motivated mechanism of drying out. A further basic technique of great value to wetland management is topographic survey.

Although construction is not a geographical skill, positive wetland management does not end at establishing the water balance or rate of drainage; excavation, sluicing, bunding or damming are generally required to harness and store the essential wetness, often by taking it from neighbouring land where it is not required.

Stream acidification and acid rain

This review of topics cannot end without reference to the environmental management problem which has taken a firm grip on public imagination in the 1980s. It demonstrates many aspects of applied studies, not the least the way in which external steerage of science is now a fact of life – journalists have almost begun to act in judgement of research. 'Acid rain' is hardly a topic for which one would expect a close involvement of physical geography, unless directly through karst solution studies which have relied on acid inputs for centuries! However, once again, the pure science aspects of the topic have required both spatial and historical treatments before they can be carried forward into the management arena for application.

Most chemists understand the recent history of acid rain. Improvements to urban air pollution in the 1960s and 1970s came about by controlling domestic smoke and by moving factories out of town. Heat generation was moved from the sitting room to the power station. Pollution 'took a ride in the country', and into other countries too following the introduction of high stacks. Once in the atmosphere, the downwind recipients of dry or washed-out acidity from sulphur and nitrogen compounds can, theoretically, be mapped by climatologists – thus the problem becomes geographical. By the same token, the effects of acidity on the aqueous environment will be intimately connected with the land phase of the hydrological cycle and its spatial–temporal aspects.

The most frequently quoted examples of acid pollution of the natural environment are those of fishless Scandinavian lakes and dying German forests. Herein is a major geographical difference: whilst Scandinavia is clearly downwind of much European industry, an essential element of continental climate is the anticyclone; thus still-air pollution, with photochemical reactions a complication, is far more likely away from Europe's Atlantic coasts. Clearly, Britain is currently officially unwilling to manage the environments of other nations, but the problem has now come home with the discovery that upland waters in Britain have acid pollution. Largely because of the strong fieldwork element to their work and the specialism of spatial hydrological studies, physical geographers have become involved in a major way. The historical element is also crucial: acidification dates back to vegetation and soil changes in prehistory and a crucial biological and land-use record awaits the investigator of lake sediments. Work so far indicates an intensification of acidification with the coming of the Industrial Revolution. Whilst air pollution is clearly a factor, the Atlantic origin of most upland precipitation tends to rule out home-grown pollutants as a direct cause, unless dry deposition dominates acid inputs. If so, winds from an easterly quarter, generally rare in our western uplands, bring the pollutants and Atlantic rain washes them off. Such a model explains why acid episodes are the main form of upland stream pollution, why snowmelt is often acid and why conifer plantations, with their large surface areas for receipt of dry material, are known to produce the most acid streams in such areas.

One thing is clear to the geographer, beside the seemingly unending complications to the subject: this is a regional science, with inputs and environmental effects varying with location. Acidity is the first pollutant which requires routing through the complete land-phase of the hydrological cycle and the topic has given a fillip to students of hydrological pathways and processes. As a pollutant, the hydrogen ion has been largely missed by the water industry (outside the fisheries interest) because it occurs in upland tributaries and during floods. Monitoring is inadequate in both time and space and there is thus an ideal opening here for the geographical field-worker.

Given the topic's complexity it is obvious that official agencies will react unpredictably with regard to management; there is a widespread assumption that Britain has taken a political stance against emission control. A more valid reason for caution would be that a single option at source might have many regional variations of effect. So far, the major direct intervention has been to lime streams and lakes, a local cure for a local ailment. A decade ago most upland farmers limed the land! A final local input from geographers is seen in Wales, where the exacerbating influence of forests on stream acidity is well proven, a map has been produced by geographers at the Water Authority, isolating those areas which would be susceptible to further stream damage if more plantation forestry were permitted.

COMMON THEMES IN APPLIED STUDIES

The above accounts of ways in which physical geography has been applied to problems of environmental management have identified certain common factors. From them one would gather that society has already identified certain of the problems, and steers the research towards the eventual management action. In other cases, the geographer is a radical, working from outside the decision-making framework to identify present or future problems – that is, working strategically, in the manner formerly thought by government to be the true utility of science. Whilst the accounts identify hydrological topics like flooding and acidification as the most obvious cases of direct applicability and geomorphology as the most obviously strategic field, the partiality of my selection is partly at fault. One group of geomorphologists, dealing with slope stability and the spatial predictability of slope failures, has been applying geographical skills to engineering problems for some years now. It is a measure of success in this field that the closest professional identity of the British Geomorphological Research Group is with engineering geologists. The same skills have been applied to coastal problems too, not only to the slope processes of cliffs but the 'channel' processes of longshore sediment sources and sinks. Another major omission above is the long-standing interface between physicists and mathematicians who dominate meteorology and the physical geographers who dominate climatology. It is a measure of the continuing role of geographical climatology that it has made significant inputs to at least two of the fields described above (regional variability of evaporation and acid rainfall).

If an albeit distant memory of undergraduate climatology can be a professional strength in the applied arena, what other geographical strengths can be identified along with the common threads of the topics reviewed? First, one is forced to conclude that fieldwork is crucial. Admittedly the bias of the author helps explain this but, even so, the last decade has seen Geography Departments in schools, colleges and universities become acknowledged as consumers of laboratory and field equipment, data recorders and computers. In some cases biologists have been overtaken as the scientists most concerned with environmental monitoring.

Secondly, perhaps because we are relatively recent newcomers to the scientific stage we are still acknowledged integrators, linking disciplines such as physics, biology and engineering. The way in which the field of multidisciplinary research has grown in Britain in the last two decades is due in no small way to the arrival of competent geographical scientists able to hold their own – and indeed that of others! A third related skill is the ability to communicate well. By keeping fundamental aspects of our work simple, geographers tend to hold multidisciplinary efforts together and also relate material to the community. One of the only elements of mistrust which has arisen with other disciplines has been with those which have no traditional skill in communication. It should be remembered that stoicism and conservatism have been a feature of professional life in Britain! One feature of our late arrival on the management scene is that we have normally investigated the approaches of the other disciplines involved as well as our own; the hardship of the generalist is that he must know everyone else's job in some way before he begins his own.

The core area of identity in the topics reviewed is, however, our skill at making spatial, temporal, or joint predictions, often backed by explanations. One should note above how often regionalisations were of use, how often mapping was the start or outcome of the methodology, and how even the traditional skill of field survey can become essential. Our human geography colleagues have long realised that planning and politics are two major users of simply-presented mapping. It might have appeared during the 'quantitative revolution' that the historical dimension was dying. Instead, we have recovered from an obsession with contemporary process to near a point of balance from which present and past can be linked to great explanatory benefit. Society respects guidance on 'memory' in natural systems, perhaps because it gives a broader sense of discrimination between crisis and trend.

Finally, it is a source of continuing surprise to the author just how simple are the problem-solving exercises offered by society as customer to the physical geographer as contractor. 'Simple', however, must not be confused with 'easy', and this fundamental dichotomy is highly unsettling to the new graduate.

PREDICTION, PRESCRIPTION AND PROFESSIONALISM

If our graduates are unsettled when they enter 'the real world', how can we help them, particularly bearing in mind the fact that they may well work as part of a team with professionals or will feed their results into a world of professionals? The skills identified in the previous section may help to guide the curriculum without dragging it off an internally guided path to scientific excellence. However, to return to the theme of the introduction, the judgements made of our work in contemporary terms may follow just those values which Moss deliberately avoided: elegance, simplicity, plausibility and utility.

The dichotomy between internal and external drives to science is a preoccupation for many in an era of declining resources for research. Science as a distinctive and pre-eminent form of knowledge has been undermined, and the stages at which sciences are susceptible to external steerage by society are both pre-paradigm and post-paradigm. During consolidation phases internal scientific methods rule. Structuralist moves, too, have undermined the independence of science from value judgements made on the basis of societal ethics, Galileo thought he had attained this independence for eternity!

Compromise is almost bound to arise for two reasons. Without a continuing pure and strategic science effort, internally driven, what stock of knowledge can the applied scientist use? Secondly, most of us are afraid of flying too close to the sun in terms of prescription. We may get disappointed by the existence of politico-economic filters which come between our work and its application, even when the work is bespoke. The decision-making framework affects environmental management, and its results are opposed by society if they are perceived to threaten the preferred order. This may relegate our results to a minor stage of decision-making by 'gradually narrowing choices'. We may well, therefore, begin to side with the view of Kogan and Henkel when they state: 'At the more baffling levels of policy formulation the scientists may fall away because the issues become predominantly those of values and allocation rather than of the discovery of fact, reanalysis of concept and formulation of scientific conclusions.'

If these were not hard times we might well decide to get back to fieldwork. Indeed, patient environmental monitoring has its applied value – but it is not experimental science. Neither do monitoring and monetarism go well together – both equipment and labour is expensive and the preference is for 'fire brigade' approaches. Is this, then, the time to seek a professional structure for our activities? Educators will hopefully forgive their temporary exclusion as professionals when I define professionalism as a unified, standardised set of occupational responses to external stimuli. The advantages of this to physical geography would be, for example, to identify and promote the potential of the subject. It would allow a formal approach to error, thus building confidence in making the rapid response often required by environmental management demands. Another reason why occupations seek to become professions is to incorporate an ethical stance; we would, it seems, be well served by a conservation ethic, though Warren's earlier chapter has demonstrated the pitfalls of a simplistic conservation approach.

However, professions have other sides, much less well suited to physical geography. They have the danger of becoming self-interest groups, and this is incompatible with our preoccupation with the natural environment. Secondly, they tend not to be radical; if not overtly conservative, they are slow to react. Innovation must be incorporated by experience before it becomes established, as must new recruits. Johnston has discussed the problems faced by applied human geography in the political arena; physical geography is less likely to clash with politicians on aspects of values, but the warning should be heeded.

There seems little option except another compromise, between pedagogy and practice. Physical geography must be two-faced to serve its unique value in society. It must continue to be known by its enthusiasm; if lack of structure is a corollary it may not detract as much as we might think. The main task at present is to maintain confidence and produce competence; to act professionally by offering quality *and* weight. The combination of effervescence and maturity is seldom a virtue in a wine, but readers should taste old ale.

FURTHER READING

A starting point is a paper which summarises very succinctly the various routes to the acquisition and use of knowledge. It is a good guide to discipline even if, as with a parent's strictures, you go off and ignore it later!

Moss R. P. (1977) 'Deductive strategies in geographical generalization', *Progress in Physical Geography*, vol. 1, pp. 23–9.

The argument is taken further by the next two references. The first is an analysis of the way in which science is becoming steered by society, and is therefore losing sight of some of Moss's articles of faith. Johnston then offers another glimpse of the wicked ways of the world, and how to hold your head up as a simple scientist in a manipulative system (against which you just might feel rebellious):

Kogan M. and Henkel M. (1983) *Government and Research* (London: Heinemann Educational Books).

Johnston R. J. (1981) 'Applied geography, quantitative analysis and ideology', *Applied Geography*, vol. 1, pp. 213–9.

Two views of management follow, one summarising the relationship between scientific and managerial perspectives in the natural environment within a coastal context, and the other describing governmental and private structures surrounding the practice of hydrology (and some fluvial geomorphology) in Britain:

Clark M. J. (1977) 'The relationship between coastal zone management and offshore economic development', *Maritime Policy and Management*, vol. 4, pp. 431–49.

Parker D. J. and Penning-Rowsell E. C. (1980) *Water planning in Britain* (London: Allen & Unwin).

Finally, something to aim for! The situation in the United States is one where professional barriers are lowered to allow the wise counsel of the environmental scientist a more direct say in management. Try the exercises: would you make a consultant?

Dunne T. and Leopold L. B. (1978) *Water in Environmental Planning* (San Francisco: Freeman).

PART III
PERSPECTIVES ON PHYSICAL GEOGRAPHY

5.1
Perspectives on the Atmosphere

Roger Barry
University of Colorado, Boulder, USA

The most striking aspect of the development of atmospheric research over the last decade has been the widening interest in problems relating to climate on all time- and space-scales. This is reflected in the increased number of scientific journals publishing climate research studies and of specialised textbooks. The topics of interest range from questions of mesometeorological and topoclimatic scale (urban climates and mountain/valley circulations) to global problems concerning increases in atmospheric aerosols, trace gases and CO_2; from historical impacts of climate on society, to possible human-induced desertification effects on climate and 'nuclear winter' scenarios; and from observational studies of climate variability to paleoclimatic reconstructions using numerical climate models. The roots of this revitalisation of climatology can be traced to the Global Atmospheric Research Programme (GARP) of the World Meteorological Organisation (see their Publication No. 16 of 1975). A report prepared in conjunction with the GARP focused on the physical basis of climate and climate change. It documented the nature and scope of 'the climate system' which encompasses not only the atmosphere but also the oceans, biosphere, cryosphere and lithosphere. Factors internal to this system produce weather and climate variability on an interannual time-scale, while other factors external to it may account for longer term changes in climate.

The components of the climate system interact with one another over a range of time- and space-scales. For example, the atmosphere and biosphere are strongly affected by the diurnal and annual cycles. The oceans are principally affected by seasonal changes in their upper layers, but in the deep water respond only over 10^2–10^3 years to changes in surface temperature. Snow cover and sea-ice exhibit strong seasonal variability whereas ice-sheets only change substantially over millenia. Lithospheric variations, apart from volcanic events, are even slower. Interactions between these various components of the climate system are accordingly complex, but understanding them better is of high scientific priority. The framework for improved understanding is provided via the three streams of the World Climate Research

Programme (WCRP) which is a direct outgrowth of the GARP. The first is concerned with atmospheric responses to surface boundary forcings (ocean heat anomalies, snow and ice cover) on a seasonal time-scale, and the second with interannual variability and its relationships with surface forcing. The third deals with long-term variations of climate and responses to perturbations such as those caused by changes in atmospheric gases or aerosol loading.

Specific subprogrammes that have been developed for the WCRP include study of the interactions between the Tropical Oceans and Global Atmosphere (TOGA), global cloudiness and its variability through the International Satellite Cloud Climatology Project (ISCCP), and the analysis of anomalies of surface heating and moisture transfer through the International Satellite Land Surface Climatology Project (ISLSCP). Global mean cloudiness and its regional distribution are still poorly known, yet cloud cover is the major factor determining the planetary albedo and thereby the earth's energy balance. Increased cloudiness increases the albedo, tending to lower surface temperatures but increases downward infrared radiation. The net effect of these opposite tendencies varies latitudinally and geographically, although on a global basis the albedo factor appears to be more than twice as effective as an increase in infrared absorption. Thus, increased/decreased cloud amount should cause global cooling/warming. The albedo effect is greatest in the tropics, but during winter in the polar regions only the IR factor is operating. Global cloudiness does not vary significantly on inter-annual, decadal, or longer time scales.

The study of atmospheric variability has been greatly facilitated in recent years by improved global data sets and the availability of general circulation models (GCMs) for simulation experiments. There has been renewed interest in atmospheric teleconnections and related persistent modes of circulation anomalies. The major ones are the North Atlantic Oscillation, the North Pacific Oscillation and the Southern Oscillation. Linked with the last of these is the El Niño phenomenon in the eastern South Pacific, and the two are sometimes jointly referred to as El Niño–Southern Oscillation (ENSO) events. A pronounced ENSO sequence in 1982–83 gave rise to major climatic anomalies in the Pacific sector and was associated with anomalous conditions even further afield.

Since each circulation mode gives rise to distinct regional climatic anomalies, there is interest in attempting to identify their precursors. It is not clear, however, whether they arise through stochastic variability of the atmospheric circulation or whether they are forced by surface heating anomalies. Theoretical studies of large-scale flows show the likelihood that multiple flow equilibria exist in nature, confirming earlier rotating dishpan studies. These multiple states give rise to the irregular alternation between zonal and blocking types of flow, for example. Nevertheless, the timing and location of such modes may well be influenced by specific surface forcings. An intrinsic problem hindering the isolation of such forcing effects is the infrequency of close analogues of particular circulation modes.

On long time-scales, the causation of ice age regimes continues to be a source of fascination. It is now generally accepted that the initial forcing and the overall timing of global cooling/warming events over the last million years is determined by the astronomical variations in the earth's orbit about the sun, as first detailed by Milankovitch in the 1920–30s. The main periodicities of *c.*100 000 years relating to the eccentricity of the earth orbit, 41 000 years relating to the tilt of the earth's axis, and 23 000 (and 19 000 years) relating to the precession of the vernal equinox, are all identifiable in the paleoclimatic records from ocean sediment cores and ice-cores from Greenland and Antarctica. However, the 100 000 year period shows up more strongly than is expected theoretically, indicating that other factors come into operation. Calculations of the changes in global climate attributable to these orbital effects made by Hansen *et al.* in 1984 show that the direct cooling amounts only to ~ 0.2°C. Additional global average cooling of ~ 4°C is caused by feedback effects involving decreased water vapor and clouds (1.4°C–2.2°C), more extensive land-ice (0.7°C–0.9°C) and sea-ice (0.6–0.7°C), reduced CO_2 concentrations (0.3–0.6°C) and vegetation changes causing increased surface albedo (0.3°C). Nevertheless, it is striking that the latitudinal zones where the maximum effects of each orbital periodicity are theoretically to be expected indeed show precisely such effects – the tilt cycle is detected in high latitudes, the precession cycle in low latitudes, and the eccentricity cycle globally (but being more apparent in middle latitudes where the other two are less effective).

Two major developments in the field of technology have had a profound and continuing impact on the atmospheric sciences. First, there is the growth of numerical modelling for all scales of atmospheric motion from the mesoscale (land/sea breezes, mountain/valley winds) to the global scale (detailed general circulation models and simplified energy balance climate models). The second development involves the evolution and application of remote sensing techniques across the entire electromagnetic spectrum. Given the variety of modelling studies only a few general remarks are appropriate here. There are four basic approaches to modelling the climate system.

1. General circulation models (GCMs) attempt to represent the complete three-dimensional behaviour of the atmosphere (radiative and dynamic processes and surface interactions) including synoptic processes on a grid scale of about 3–5° and time-steps of the order of a few minutes. Some atmospheric GCMs have been coupled to ocean GCMs which are integrated with much longer time steps.
2. Zonally averaged statistical dynamical models (SDMs), which eliminate longitudinal differences and therefore run much faster on a computer, can be integrated over longer time-intervals for climate experiments.
3. One-dimensional (zonally averaged, vertically integrated) energy balance models (EBMs) examine latitudinal changes in surface temperature to differences in energy fluxes. These have been widely used and analysed mathematically to assess the stability of global climate to changes in

external forcings. EBMs are also being extended to treat two horizontal dimensions and monthly time-steps.

4. One-dimensional radiative–convective models (RCMs) describe a vertical column throughout the atmosphere computing the vertical temperature structure from radiative processes and vertical heat fluxes.

Climate experiments are of two kinds. Most are 'sensitivity' studies which compare the model climate with and without some imposed change in boundary conditions (solar constant, atmospheric composition, surface characteristics) or in the level of parameterisation of various processes. A second kind introduces time-dependent forcings and examines how the model climate evolves in response to them.

Satellite studies provide climatic information over a wide range of spatial and temporal scales. Global assessments of planetary albedo and outgoing infrared radiation, as well as measurements of the solar constant and variations in solar output, provide one example. Surface variables such as snow cover and sea-ice extent can also be routinely mapped, with microwave techniques providing an all-weather capability. As yet, however, the records are of too short duration to determine long-term trends in these parameters. High resolution visible and IR data have contributed immensely to the understanding of synoptic and mesoscale systems and their interactions in tropical–extratropical latitudes. The frequency of tropical storms in the north-east Pacific Ocean was greatly underestimated prior to the development of satellite climatology and new categories of weather system, such as the tropical cloud cluster, have been identified largely through satellite analyses. Difficulties frequently remain, however, in meshing data obtained from satellites, aircraft programmes, and conventional synoptic surface and upper air stations. Cloud cover statistics illustrate this type of problem. Cloud amount measured on a satellite image at a point must be either zero or 100 per cent whereas the figure tends towards some intermediate value as the area examined increases. Ground observations represent an estimate for a circle of about 10km radius (300km^2 area), whereas statellite statistics are typically for a $2\frac{1}{2}°$ grid 'square' (an area of about 60 000 km^2).

The World Climate Research Programme has the major objectives of determining to what extent climate can be predicted and the extent of human influences on climate. In addition, there are the related World Climate Applications Programme and the World Climate Impacts Programme. In the light of these developments, areas where substantial interests and effort are likely to be focused over the next ten to twenty years include:

1. Expanded development of global climate models incorporating coupled ocean and sea-ice systems, with longer model simulations ($\geq$100 yr) becoming feasible on fifth-generation computers.
2. Intensified study of ocean–ice–atmosphere interactions utilising remote sensing data on ocean surface and ice characteristics.
3. Studies of atmospheric chemistry and global biogeochemical cycles

(involving closer co-operative research with biologists, chemists and oceanographers).

4. Study of the human impacts of climate anomalies and potential global climate perturbations due to volcanic (stratospheric) aerosols, tropospheric pollution (including acid precipitation effect), CO_2, and trace gases (ozone, methane, chlorofluorocarbons and nitrous oxide).
5. Studies of regional and global effects of mountain areas of the world, especially the Alps, Rocky Mountains and Tibetan Plateau.
6. Expanded and improved climate data archives and interactive access to at least some data sets via computer networks.

5.2
Perspectives on the Geosphere

Keith Clayton
University of East Anglia

The landforms of the earth are found in a very diverse range of environments. Indeed the combination of structure, uplift (the primary determinant of relief), rock type and the climatic and vegetational environment yields an almost endless variety of geomorphological systems. These can be investigated at many different scales of space and time. Few individuals even visit all six continents in a lifetime, let alone carry out comparative research in many different parts of the world. We rely on assembling our evidence through the published literature to bring sufficient diversity to tackle some of the basic problems of geomorphology.

Our attack on understanding geomorphology is based on the assumption that a scientific approach is feasible and will be profitable. So much of what we study is too large or complex to be taken into, or recreated in the laboratory, that experiments in the precise sense of that word are few and far between, whether in the field or in the laboratory. Yet we feel the need to try to isolate single variables so that we may examine them 'in a scientific way'.

The analytical approach to landforms which has become common over the last two or three decades is to proceed from process to form. It is argued that since landforms are created by the action of geomorphological processes over time, then careful observation, measurement and analysis of the operating processes will lead to the ability to relate processes to forms. Indeed, putting it in a more ambitious way, not simply to relate, but to predict the resulting landforms from the processes. In so far as understanding can become complete, and the process–landform relationships are found to be unique (i.e. one process = one landform) then it would also be possible to infer past processes from existing forms.

In those areas where most progress has been made (and this means areas where the processes are relatively frequent, and often also rapid) the simple process–form model is already replaced by a less certain relationship in which the twin concepts of equifinality and indeterminacy play a central part. In other words, the same forms may be produced by different processes, while one process may also produce more than one form. Thus in the understanding of alluvial channels, chance seems able to tilt development towards one solution rather than another to the balancing of channel size and form to the stresses placed on it and the sediment transport required for the equilibrium to be sustained. Concepts such as thresholds and relaxation times become essential to an understanding of the dynamic evolution of these geomorphological systems.

One of the classic pieces of work of the last twenty-five years was the paper by Wolman and Miller on 'Magnitude and frequency relationships in geomorphology'. They produced data to support the traditional view, which has so often been challenged, that it is the relatively frequent yet far from extreme processes which accomplish most geomorphological change. Indeed, in the context of channel geometry, they were able to suggest that the bankful discharge, occurring more frequently than once every other year, is the controlling discharge. The range of data they were able to quote in support of their view was limited, and there were some suggestions that in some areas (e.g. landsliding) occasional larger movement may have had the dominant role. It is a sad commentary on process studies of the last quarter century that they have failed to produce sets of data which would carry these observations forward. In other subjects, a paper as frequently quoted as magnitude and frequency would have stimulated data gathering and been subject to rigorous checks against new data before passing into the standard literature.

If we examine almost any approach to geomorphological analysis, we find that form plays a dominant role – yet most published work follows convention by placing it in the secondary position. No one studying rock creep would waste time by establishing field sites across the whole landscape – they would at once stratify the sample by selecting sites on talus slopes. They might then make sure that the sites were a random sample of talus slopes in the area, were representative of different climates (by scattering them around the world or more probably by varying altitude and aspect within a single region), or were related to some variable such as slope angle or rock type which would no doubt influence fragment size and angularity. In other words, the influence of landform on process would be assumed and be the basis of the field sample. We spend much of our time looking at the variations in process found on a particular slope-form or landform (e.g. stream beds) and relating these to subsidiary variables. Many designs are even more restricted – for example, the highly sophisticated examination of karst processes and forms, which is by definition restricted to one very narrow range or rock types.

The difficulty we meet through jumping to such selected samples of the landscape is that the assumptions we adopt in the initial stratification are never properly tested. To some extent errors in sample design, or unexpected results from comparison through the literature of apparently similar sites in different parts of the world, may throw light on the hitherto unquestioned assumptions about landform evolution that preface so many pieces of research. Some of these are so disturbing in their implications for a school of work that they may be ignored or explained away; a good example are the data on rates collected by Rapp from what would seem to be a periglacial environment in northern Sweden, yet which showed none of the pattern or overall rate of operation of different processes described (though usually without any measurement) from other contemporary and relict periglacial environments. Since another decade of fieldwork would be needed to replicate the data from another site, it was feasible to dismiss the data as 'unrepresentative'.

Gradually we do accumulate data which may be used to test the concepts dominating so much geomorphological research. Thus the rate data analysed from time to time by Young are beginning to be sufficient in number and of good enough quality to allow some preliminary findings on different climates, not simply different relief environments. Their message is salutary – variation related to relative relief is ten times that attributable to climate. The ratio of work on climate and relief as variables has probably been 10:1 the other way. It will be argued that subtle relationships require careful and voluminous research; conversely it may be argued that our understanding of the relationship between relief and rates or relief and forms is not so good that it does not deserve the bulk of our research effort.

Another problem of the process-dominated approach is the emphasis it places on work on relatively short time-scales, to the extent that even periods of 10^3 and 10^4 years can be defined as 'long'. Destruction of a mountain range has been shown to take 25–35 Myr, so it is necessary to be able to tackle the evolution of landforms over periods of 10^7 years for a full coverage of geomorphological time-scales. Work on periods as long as this is currently unfashionable and the geomorphological techniques by which it might be pursued have not been advanced much over the last few decades.

Classic work in geomorphology related the erosional history of land areas to the depositional history of adjacent basins – the 'correlated deposits' of continental work. Despite the ability to acquire more comprehensive records from marine basins, our ability to date these sequences more precisely, and the evidence of the significance of the marine record for modern work on the Quaternary, suprisingly little contemporary research is along these lines. Further, improved understanding of the rates at which landforms develop and our ability to date at least some of the features found in many upland areas (e.g. the use of speleothems in the English Pennines or the Canadian Rockies) promises that the erosional and sedimentary records may be linked far more closely than has so far seemed possible. It is high time that these methods were adopted more widely, for they promise to be productive on a different scale from so much contemporary work. It is a scale at which we need to get better understanding if we are to be able to handle landforms on a regional or world scale. And improved understanding of the relationships at that scale would allow a renewed and better designed attack on some of the detailed issues which remain.

One critical topic which such an approach would bring into prominence is the neotectonic history of the earth's surface. Study of neotectonics has been pursued in two areas, mountain ranges where recent uplift has obviously occurred, and glacioisostatic areas where the evidence of recent and often continuing uplift is again compelling. Literature on neotectonics has been more common in Eastern Europe than the West until recently, but that situation is rapidly changing, and it is time geomorphologists became more closely involved. There are strong suggestions that deformation has been recent enough to affect the overall form of the relief, if not individual landforms, over far more of the globe than most explanatory accounts assume. A combination of contemporary measurement of rates of both erosion and uplift, together with study of correlated sediments and the analysis by established methods of drainage-relief relationships, should see rapid progress in a most important field.

5.3
Perspectives on the Hydrosphere

Richard Chorley
University of Cambridge

Perhaps the single most important fact regarding hydrology in general, and of geographical interest in the hydrosphere in particular, is the rapidity of its growth and the variegation of its blossoming in recent years. In 1963, the *Journal of Hydrology* was 22mm thick, in 1983 it was 126mm. As we view the horizon of physical geography from the hydrological vantage point, it surrounds us. Increasing numbers of geographers are making journeys towards this horizon but they travel in groups along often diverging paths, and with the passage of time contact between them may become increasingly difficult. It is not a simple task to map out the easiest and most productive routes for such explorations, particularly as these two kinds of routes are not always identical.

If we wish to conjure up the prophetic genii to reveal the future to us we commonly polish the lamp of the present because, however great a hiatus of change may be, the work carried out by the main mass of scholars is essentially conservative and elaborative, rather than innovative. Therefore, we do not have to rub very hard to receive the prediction that much future work in hydrology will be similar to that being conducted at present – only more so and better. There will be more accurate observations of throughflow, more sophisticated diffusion models of pollution, more generalised statements of loose channel hydraulics and hydraulic geometry, more accurate predictions of storm rainfall, more satisfactory rationalisations of the storm hydrograph, more satisfactory understandings of the causes of drought, better models of climatic change, more sophisticated economic treatment of water resource planning and a more secure socio-economic basis for decision-making in hydrology. In short, there will be bigger and better Stanford Watershed Models!

However, notwithstanding the foregoing, radical changes do occur in scholarship and, in examining the future, it is helpful to look back on recent examples of such changes. No changes have been more significant in hydrology than those associated with Horton's infiltration theory of runoff of the 1940s and the dynamic basin concept of the 1960s. Both concentrated and directed future research and led to the integration of work in hydrology with that in meteorology and geomorphology. However, these advances took place mainly as the result of a broader view having been taken of allied advances in hydrology itself. In other words, to continue the Aladdin analogy, the lamp was rubbed sufficiently hard to reveal what lay beneath the individual superficial relationships. Thus, for example, it is clear that presently independent work on the recurrence intervals of meteorological, runoff and erosion events may well unite to produce a significant leap forward in future integrated research.

Nevertheless, there are some changes in all academic disciplines which are of such a drastic character that they are virtually impossible to predict, partly because they depend on the vision of some individual to set existing work in some completely novel framework. In other words, occasionally some Aladdin comes on the scene who feels it necessary to break the lamp altogether and to use the oil in a completely different way. How is it possible to predict what circumstances might lead to such an action and what might be the result? Here we come to what I consider to be the most important aspect of recent work in hydrology. This is the dynamically and conceptually central role played by the hydrosphere in linking and interpenetrating the atmosphere, the lithosphere and the biosphere (Bach, 1983; Flohn, 1984; More, 1967).

Figure 5.3.1 shows two facts very clearly. The first is this central and linking role of the operations of the hydrosphere between the atmosphere, lithosphere and the biosphere. It is no chance that the broadening horizons of physical geography over recent years have tended to throw the emphasis preferentially on hydrological studies. The second fact emerging from the figure is that the interactions between the hydrosphere and the other spheres occur over a very wide range of time-spans, from 10^{-2} to 10^{8} years. If we consider the times required for cycles to occur, for significant changes to arise, for turnover and equilibrium to occur, for significant transfers to take place, and for interactions to proceed, it is clear that the major prospective problems in geographical hydrology will be those involved with time. Geomorphology has taught us that short-term changes and long-term changes apparently require different fundamental methodological approaches – the functional and the historical. The real problems arise when time-scales of the hydrosphere interact with the generally longer timespans of the lithosphere and the shorter time-spans of the atmosphere. Modelling of this type requires the always difficult combination of functional and historical modes. It is perhaps encouraging to note that, in time-scale terms, the hydrosphere and the biosphere are very similar, which explains the many successes which are being achieved in integrative research between these fields, for example, in pollution modelling.

The earth–atmosphere system complex depicted in the figure draws attention to the time-scale differences to which I have previously referred. Turning first to the interactions with the atmosphere, it is clear that of these only two (volcanic and anthropogenic) need not have important hydrological

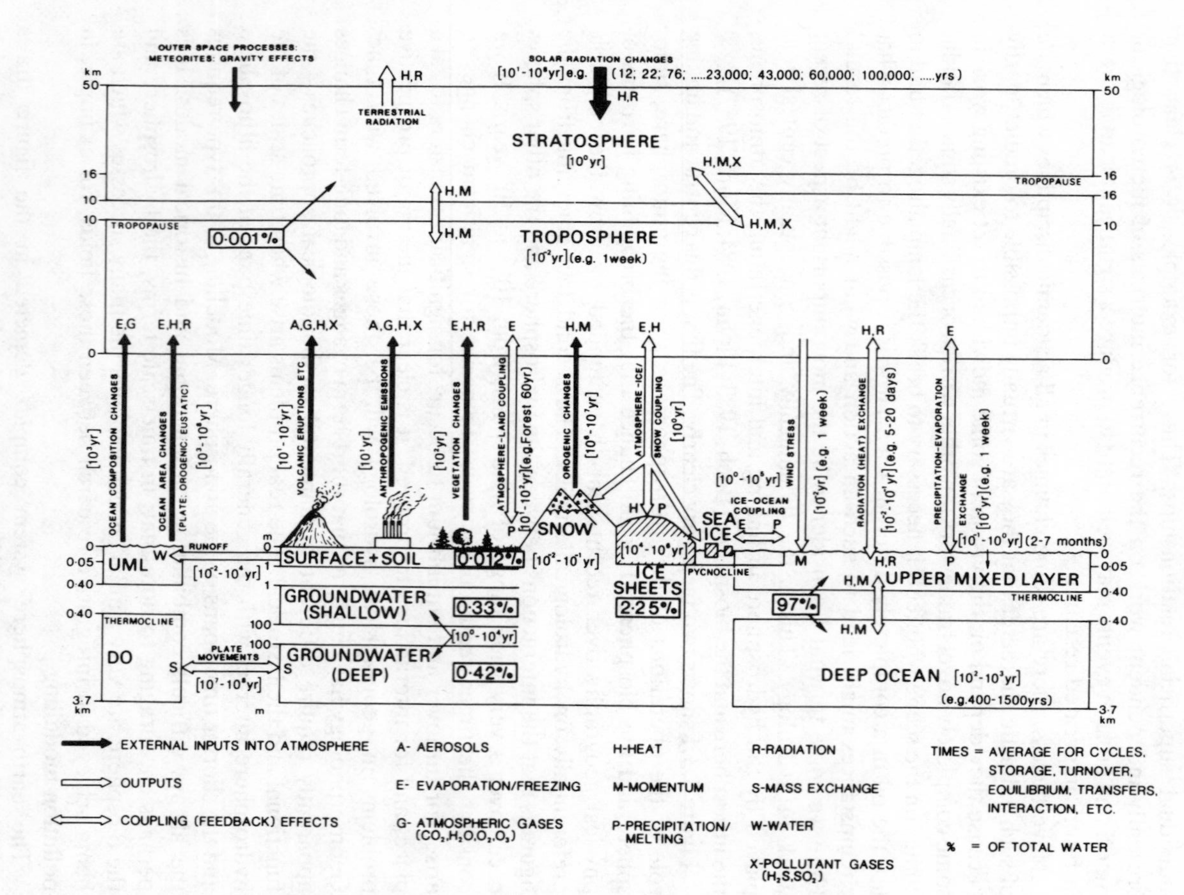

FIGURE 5.3.1 *The hydrosphere, atmosphere, lithosphere, biosphere system* (Adapted from Bach, 1983; Flohn, 1984; More, 1967)

aspects. Of the remainder, the biggest modelling problems are presented by those hydrological interactions operating on a time-scale averaging 10^3 years, namely deep oceanic circulations, sea-ice/ocean coupling and ocean composition changes. In contrast, coupling and exchanges on time-scales of 10^0 to 10^{-2} years are much more easy to deal with.

The interface between the hydrosphere and the lithosphere presents even greater difficulties because clearly disproportionate time-scales occur. The modelling of ground-water yield is more complex than that of surface flows. Water is discharged from the continents more rapidly than debris is removed and very much more rapidly than it is produced by weathering. For example, one of the biggest geomorphic problems resides in reconciling the short residence time of circulating water at the soil/rock interface with the generally very much longer time-span required for weathering to occur. Indeed, the identification of the nutrient cycle with the weathering cycle, of which it forms a part, is merely one aspect of this time differential.

The third interface, that between the hydrosphere and the biosphere, is, as has been pointed out, conceptually less intractable than the other two. The role of water in this interaction operates mainly via evapotranspiration and anthropogenic emission. Human activity as a source of vegetation and industrial changes operates at a rate which finds almost immediate response in the operation of the hydrosphere. If one believes, as I do, that the human biosphere in its broadest context forms the indispensible, unifying element of physical geography, then we must be happy that the general similarity of the relevant time-spans in the hydrosphere and the biosphere bodes so well for the future of hydrology as a key part of physical geography. However, we must not forget the other difficulties which arise when the negative-feedback processes of the hydrosphere encounter the increasingly prevalent positive-feedback surges in the human part of the biosphere.

I have suggested that the most difficult but productive future routes for geographical work on the hydrosphere will be temporal ones involving its differentially lagged interfaces. If this is true, then spatial attributes are also involved – indeed, this must be so if geographers are to make any worthwhile contribution to future work on the hydrosphere. Ocean–atmosphere lagged interaction is especially due to the geometry of the ocean basins and the hydrosphere–lithosphere interactions are dependent on the morphometry, not only of the earth's surface but also of the mass of the soil and mantle. It is thus in space–time modelling that physical geographers will make their most significant future contributions in the field of hydrology.

FURTHER READING

Bach W. (1983) *Our Threatened Climate* (Dordrecht: D. Reidel)

Flohn H. (1984) Introduction., In: Flohn, H. and Fantechi, R. (Eds.), *The Climate of Europe: Past, Present and Future*. (Dordrecht: D. Reidel)

More R. J. (1967) Hydrological models and geography, In: Chorley, R. J. and Haggett, P. (Eds). *Models in Geography*, (London: Methuen)

5.4 Physical Geography – Diversity and Unity: A Concluding Perspective

Michael J. Clark, Kenneth J. Gregory and Angela M. Gurnell
University of Southampton

Although both the *Horizons* books endeavour to avoid either implicitly or explicitly creating a new orthodoxy, it is important that they should offer a synthesis within which status can be assessed and priorities be identified. In the interests of brevity, this perspective can be exemplified by just two dimensions – first, relationships between human and physical geography, and secondly, the essence underlying the achievements of the multifarious branches of physical geography.

The period during which a system is actively changing from one steady state to another is known as 'relaxation time', but this is an oddly inappropriate label if applied to the present state of flux within the subject of physical geography. Whatever might be identified as the main characteristic of physical geography during the 1970s and 1980s, or of the twenty-four contributions to this book, it is clear that relaxation is not an appropriate descriptor. For whatever reason, complacent satisfaction in the status quo is rare indeed, and a new mood of academic and professional competition ensures that both productivity and innovation are on the increase.

Readers of both *Horizons* volumes could be forgiven for wondering if this confident prognostication is really justified, and could well argue that it is in human geography, with its radical rejection of the positivist straitjacket, that the more impressive innovation is to be found. Such a viewpoint highlights the great difficulty of drawing meaningful comparisons between substantially different entities. In terms of dynamism, both branches of the subject can look to the past two decades with pride, and challenge any other discipline to demonstrate a greater pace of change – most subjects undergo the occasional revolution, but few others have turned revolution into a way of life. However, if we try to assess success rather than either change or mere output, then we face the greater challenge of evaluating the aims of different parts of the subject. Immediately, comparison becomes meaningless, since there can be no doubt that the aims, and thus the achievements, of professional human and physical geography are very different. Furthermore, those aims are intensely debated within each of the subdisciplines, and the apparent consensus view changes frequently.

In such a situation, most commentators have either abandoned the attempt altogether, or have regressed to shallow and biased speculation which reveals more about the narrowness of the writer's awareness than about the real differences between human and physical geography. Perhaps the only other alternative is to risk diluting and diverting the issue in oversimplification, a risk that is justified only by the belief that a clear signpost may be more helpful than a detailed but blurred map. If we have the courage, we might suggest that the relationship between human and physical geography will not lie primarily at the level of either content (since the objects of study occupy a common world, but share relatively few attributes of behaviour) or technique (which, despite some real overlaps, is dominantly subject-specific), and must thus be sought in the realms of methodology and philosophy. Still greater academic valour might encourage us to submit that, for practical purposes, philosophy can be regarded as offering a framework through which to identify meaningful questions, whilst methodology provides a framework through which to seek meaningful answers. On such a basis, physical geography does appear to share with human geography a significant rejection of positivist philosophy, but differs in as far as it has retained and strengthened its reliance on scientific methodology.

Simplistic though this characterisation might appear, it does go some way towards illuminating the present relationship between the two branches of the subject, as well as pointing towards future enhancement of mutual respect and supportive activity. If human geographers denote the value-laden debate of current issues as a primary focus for their subject, then they will not be surprised to find their physical colleagues selecting either different targets or different paths towards the same target. Clarification of issues is a necessary precursor to the identification of practical problems to be solved, but the actual task of problem-solving then remains. Similarly, structuralist explanations of society or of its relationship to environment offer an exciting reconceptualisation through which greater understanding can be achieved, but the biophysical components of those structures remain to be understood both in their own right and in order to open up the possibility of prediction and management. These distinctions lie at the heart of the physical geographers' conviction that since the mid-1960s they have moved significantly towards an incorporation of socio-economic components into their thinking, and their fear that a substantial continuation of this drift would render them ineffective as natural scientists without achieving any marked increase in their service to social science.

It might seem, therefore, that the most profitable course for both branches of the subject could lie in some form of 'separate development' that avoided the less fortunate connotations of apartheid. Certainly, it is to be hoped that geographers have matured beyond the point where petty internal rivalries dominate the balance between their interests, or where resource allocation conflicts divert all concerned from the mutual advantages of each branch seeking to maximise the potential of its own strengths, rather than attempting to demonstrate either an aggressive superiority or a placatory obsession with merging its identity in some mythological centre ground. If this was to be the case, then what would be the attributes of the physical strand of this development? Readers of this book will wish to derive their own list of priorities, but many will include items from a shortlist of three recurrent themes.

First, physical geography has become, and will remain, an effective natural science with a strong reliance on the development and application of accurate monitoring, analytical and modelling techniques. Its focus on process will remain as strong as that on form, and the emphasis placed upon an understanding of material properties appears overdue for greater emphasis. Repeatedly in the foregoing contributions it has been apparent that advances in understanding and prediction, and thus in the ability to respond to management ideals, have been rooted in technical and technological developments. In part this reflects an explosion in the scale range that is accessible to geographical investigation, for the first time genuinely extending awareness and the prospect of mastery beyond the human scale to encompass the global and microscopic scales – the ultimate in extrapolation and reductionism. A related benefit of enhanced measurement devices, communications networks and (above all) the data-handling capacity and speed of microprocessors has been that real-time observation and management begin to become something more than a dream. It is impossible to imagine that physical geographers will forego the enormous potential of information technology as a sacrifice on the altar of idealism or humanism.

A second diagnostic characteristic of contemporary physical geography is the revival of interest in the concepts which motivate investigation and explanation. Spatial concepts are certainly of importance, but seem at the moment to be subsidiary to the elucidation of temporal behaviour. The new models for temporal change greatly increase the variety of behaviour that can be incorporated, and thus correspondingly increase the power of the model. At a pragmatic level, this should in due course reduce the number of occasions on which teachers, in the classroom or the field, have to acknowledge exceptions to the 'textbook' explanation. A similar leap in explanatory ability appears to be possible through the progressive integration of subsystem studies. The analytical approach, very necessary in the twenty years after the mid-1960s, is now giving way to a greater interest in synthesis (for example, linking the slope and the river, or integrating all the components within the mountain lands or arid lands) armed with the newly acquired details of the individual processes and features. Computational techniques are now beginning to allow the inception of such synthetic models, opening up an approach which feels inherently geographical, and may legitimately be regarded as one of the profession's unique contributions even in cases where socio-economic elements do not figure large.

Nevertheless, whilst the physical basis remains at the core of the subject and seems set to enjoy a revival in research and teaching emphasis, at least at higher education level, the third dominating theme of physical geography over the past decade has been its massive swing towards management-consciousness and the associated interest in socio-economic attributes. Not only is this seen as a direct contribution to socio-economic welfare and productivity (a notion supported by the greatly increased employment of physical geographers as consultants to commerce, industry and government), but also as a source of explanatory power for the 'pure' physical geographer. Few 'natural' systems are independent of socio-economic influence, and therefore the broadening of interest has rendered physical geography much better equipped to comment on real as opposed to ideal landscape or atmospheric systems.

In the 1970s the 'applied' trend was so powerful that it threatened the extinction of 'pure' fundamental studies. This problem had been faced somewhat earlier by the physical, engineering and medical sciences, and represents an extremely serious national dilemma, since research funding is often preferentially allocated to applied topics. Physical geographers are now beginning to awaken to the worries of fellow scientists and to the realisation that too great an emphasis on application to specific real-world problems can handicap the true innovation that is often rooted in pure studies. The real problems pose questions, but the answers may be more effectively sought through studies of more general applicability. Furthermore, whilst social conscience is a strong and justifiable motivator, it becomes philistine if it renders unacceptable the cultural value of understanding in its own right. Perhaps the purer physical investigations must proceed more positively towards familiarity with environmental design, whereas the applied studies remain more management oriented.

For these reasons, there are signs that the physical geography of the 1990s may be significantly different from that of the 1970s and 1980s. A stronger theoretical base is likely to be seen as a priority, providing a conceptual framework of greater rigour and greater flexibility than is currently available, with the macrocomputer model and data base becoming the successors to regression equations and maps. Socio-economic relevance will remain a central concern, but it will be geared to doing something relevant (often using a positivist-related methodology) rather than merely saying something relevant (a value-laden exercise for which speculative idealism is a sufficient foundation). As a countertrend, however, we are likely to see a renewed emphasis on specialist physical studies, albeit in an integrative mode, designed to avoid the suffocation and declining innovation that can be the product of an obsessive concentration on case-specific applications.

We opened the introduction to this book with a systems model of the study of physical geography, and it is thus appropriate to end with a related analogy. One of the most exciting implications of the conceptual advances signalled by John Thornes in Chapter 1.2 is the extent to which dynamical (as opposed to dynamic) systems allow us to envisage behaviour dominated by instability and by the possibility that a wide range of end-products might be triggered by a single group of initial conditions. If this is the case, then the strange attraction of physical geography to its many enthusiastic adherents might well be the joy of the unknown: the thrill and challenge of participating in a subject that is not just at a threshold, but at the brink of a change that could hurl us in any number of directions other than those prophesied above. In this way, science ceases to run the risk of acting as a deterministic straitjacket, and comes instead to offer physical geographers an exhilarating route to change and progress.

Whether beckoned by humanist ideals or spurred on by positivist working modes, the subject seems set to continue in a state of flux. We have tried to show in this book that the present diversity of physical geography signifies strength rather than weakness, yet to some geographers the lack of a clearly defined path towards a sharply focused target remains a source of frustration. Perhaps here the analogy of the horizon is helpful in two ways. First, since the horizon always recedes ahead of the traveller, it follows that enquiry in any subject is an ever-lengthening journey. Actually to reach our target would signify the end of geography, rather than the beginning. Second, when we view the world from ground level, the horizon seems complex and rapidly changing as we alter our viewpoint. But as we move to higher levels of observation – a tangible possibility with the civilian space programme – the globe comes to assume a more ordered shape and a more constant aspect regardless of the direction in which we look. Is the same true of physical geography? We have spent two decades re-examining the components of our world, but if we want simplicity, integration and order then we must move to higher levels – to theory and concept on the one hand, and to the scale-integrating power of computational technique on the other. Whether our quest is for understanding or management, such elevated viewpoints set the world of physical geography at our feet and open up a stunning prospect for the subject's future.

Index

Place names are not included and the index is confined to major items. References to figures and tables are shown in *italics*.

Also available from Macmillan
CRITICAL HUMAN GEOGRAPHY
A series edited by Mark Billinge, Derek Gregory and Ron Martin, University of Cambridge

PUBLISHED

Recollections of a Revolution: Geography as Spatial Science
Mark Billinge, Derek Gregory and Ron Martin (*editors*)

Capitalist Development: A Critique of Radical Development Geography
Stuart Corbridge

The European Past: Social Evolution and Spatial Order
Robert A. Dodgshon

The Arena of Capital
Michael Dunford and Diane Perrons

Regional Transformation and Industrial Revolution: A Geography of the Yorkshire Woollen Industry
Derek Gregory

Social Relations and Spatial Structures
Derek Gregory and John Urry (*editors*)

Geography and the State: An Essay in Political Geography
R. J. Johnston

Long Waves of Regional Development
Michael Marshall

The Geography of De-industrialisation
Ron Martin and Bob Rowthorn (*editors*)

Spatial Divisions of Labour: Social Structures and the Geography of Production
Doreen Massey

Conceptions of Space in Social Thought: A Geographic Perspective
Robert David Sack

The Urban Arena: Capital, State and Community in Contemporary Britain
John R. Short

FORTHCOMING

A Cultural Geography of Industrialisation in Britain
Mark Billinge

Regions and the Philosophy of the Human Sciences
Nicholas Entrikin

Strategies for Geographical Enquiry
Derek Gregory and Ron Martin